PRACTICAL EXERCISES IN PROBABILITY AND STATISTICS

FOR SIXTH FORMS · TECHNICAL COLLEGES
COLLEGES OF EDUCATION · UNIVERSITIES

With answers and hints on solutions

N. A. RAHMAN

M.A.(Alld), M.Sc.(Stat.) (Calc.)
Ph.D.(Stat.) (Camb.)
Senior Lecturer in Mathematical Statistics,
University of Leicester

GRIFFIN 18 20 LONDON

CHARLES GRIFFIN & COMPANY LIMITED
42 DRURY LANE, LONDON WC2B 5RX

Copyright © N. A. RAHMAN, 1972
All rights reserved

First published   1972

Royal Octavo, xiv + 338 pages
ISBN 0 85264 210 5 (Cased)
ISBN 0 85264 217 2 (Limp)

Set by E. W. C. Wilkins & Associates Ltd London N12 0EH
Printed in Great Britain by Compton Printing Limited
London and Aylesbury

To the memory

of

Sahibzada Saiduzzafar Khan

in gratitude

Preface

This collection of practical exercises in statistical analysis is the product of a personal philosophy of statistical education which rests on the principle that the currently fashionable dichotomy between "mathematical statistics" and "statistical methods" is artificial and detrimental to the education of professional statisticians. In my view a comprehensive understanding of the science of statistics emerges only when the beginner practises his skills on both theoretical and practical exercises. In an earlier volume (*Exercises in Probability and Statistics*: Griffin, London; 1967) I have provided a collection of theoretical exercises, and the present collection is a parallel one of practical exercises. However, this collection is more limited in scope, and the 200 exercises presented here follow closely the developmental sequence of my textbook, *A Course in Theoretical Statistics* (Griffin, London; 1968). In a strict sense the present volume is a practical companion to the *Course*, though the exercises can be used independently of the textbook. The exercises are specifically designed for the use of sixth-form pupils and first-year undergraduates who are studying statistics for the first time. The practicality of the exercises lies in the fact that computation, which is sometimes quite extensive, is an essential element in the solution of the exercises, apart from the inevitable application of statistical theory in meaningful practical problems.

In planning this collection, I have kept in view the three-part division of the *Course*, namely, Mathematical Introduction, Statistical Mathematics and Probability, and Statistical Inference. This division corresponds to the three chapters of the present book, namely —

Chapter 1: Elementary Computational Techniques and Numerical Methods (25 exercises);

Chapter 2: Statistical Mathematics and Probability (75 exercises);

Chapter 3: Statistical Inference (100 exercises).

The exercises included in Chapter 1 are based on an unconventional approach to the problem of practical work in statistical education. Arithmetic is often badly taught at school, and pupils are generally led to believe that the "higher" mathematics is largely, if not exclusively, non-numerate. This is a serious drawback from the statistical point of view, and makes it difficult for beginners in statistics to accept the computational discipline as an integral part of their education. It is therefore essential to break the student's implicit – and often explicit – opposition to arithmetical processes. In this context, the symmetries revealed by, for example, the simple differencing operation on polynomial and transcendental functions are particularly helpful, apart from the usefulness of elementary ideas of interpolation in practical statistical analysis. It is therefore to be hoped that users of this collection will devote some attention to the exercises in the first chapter.

Chapter 2 begins with a number of exercises on the more usual methods of summarising data and evaluating various constants of frequency distributions. Furthermore, an attempt is made to provide a series of exercises which illustrate the use of probability concepts and distribution models in a variety of practical situations. Although these exercises involve the calculation of meaningful expectations and probabilities, problems of statistical inference based on the ideas of estimation and tests of significance are deliberately excluded. It is my belief – supported by considerable classroom experience – that practical problems which involve the deeper concepts of inference should not be introduced until the ideas of probabilistic uncertainty, and of the probability model as an approximation to the behaviour of an observable random phenomenon, are clearly understood. The exercises in Chapter 2 should be particularly useful to teachers who wish to introduce interesting practical work in sixth forms. Of course, this presupposes that pupils will have access to individual desk calculators for carrying out the computations.

Chapter 3 is wholly devoted to exercises on statistical inference which are associated largely with the classical methods of estimation and tests of significance. However, some element of novelty is introduced by the use of three elegant but relatively neglected techniques of statistical analysis, namely, the generalised t test, Pitman's test for the equality of variances of two correlated variables, and the partitioning of the goodness-of-fit χ^2 into components with single degrees of freedom. Most of the exercises in this chapter are based on data gleaned from research papers in various applied sciences, although the statistical problems posed on the data are largely original. These exercises can also be regarded as a self-contained unit for use in a conventional first course in applied statistical inference for students of the natural and social sciences. In this context, it is perhaps useful to emphasise that a set of observations can often be analysed in many ways to answer different questions. The questions posed in the exercises are usually a few of many possibilities, and teachers using the book may find it helpful to consider other questions based on the data quoted.

In formulating those exercises which are based on published data, I have tried to give as much background information on the problems as I believe students can assimilate readily. Some of this information is, of course, directly relevant to the statistical analysis and its interpretation, but other details have a bearing on the motivational aspect of the educational process. The source of the data and other small points such as where, when, and why the experiments referred to were actually conducted do have an important psychological impact on the beginner; they suggest that he is, in fact, participating in the reconstruction of the analysis of data which not so long ago had a bearing on some problem of research interest. My experience over a decade confirms that this association, however tenuous, exerts a profound and helpful influence on the beginner. The encouraging results obtained by the use of these exercises at the University of Leicester, with groups of students having markedly different intrinsic mathematical ability, may be a useful indicator in other institutions interested in the development of statistical education.

The whole collection of exercises grew out of several years of practical work with first-year mathematics undergraduates at the University of Leicester who were studying statistics for the first time; practically all the exercises have been class-tested repeatedly, and some of them are examination questions set in the past at this university. In order to help users of the book, particularly those studying on their own, answers and hints on the solution of the exercises are given. Despite all reasonable care, it is unlikely that there are no numerical inaccuracies, and for any such the author alone is responsible. I should be grateful to readers for pointing out any errors or obscurities in the formulation of the exercises or in their answers and hints on solution.

I am obliged to the authorities of the University of Leicester for permission to reproduce past examination questions. I owe a major debt of gratitude to the authors of research papers, and to the learned journals in which they were published, for the data and ideas for many of the exercises. Acknowledgments of the sources of data and of copyright material are given in the text, and apologies are hereby tendered to author and publisher for any inadvertent omission.

I should like to thank Dr M.G. Kendall for initially encouraging me to undertake this project. I am also obliged to Professor H.E. Daniels and Professor D.V. Lindley who, as past external examiners at the University of Leicester, gave helpful advice for the improvement of some examination questions included in the collection. Many of my former students, who worked through these exercises, have also contributed significant improvements. They are too numerous to be mentioned individually, and I hope a general acknowledgment to their earnest endeavours will be acceptable as adequate. It is a pleasure to record my debt to Professor F. Downton for our earlier professional association which contributed in no small measure to the development of my philosophy of statistical education that underlies this book. Finally, my thanks are due

PREFACE

to the publishers and their consultant editor for their help and efficient co-operation in publishing this work.

N.A.R.

Leicester,

8th February, 1972.

Contents

Preliminary Notes

(1) To make the answers and hints on solutions more helpful to users of this book, in a few instances references are given to sections and solved examples in the same author's *A Course in Theoretical Statistics*. This book is referred to briefly as the *Course*.

(2) The use of symbols and abbreviations in the answers is generally in conformity with standard practice and, in particular, agrees with the usage adopted in the *Course*.

(3) The solution of many of the exercises requires the evaluation of square roots correct to several places of decimals. A square root can be calculated readily by using any one of the methods explained in the *Course* on pp. 8–12.

(4) In using recurrence relations successively to evaluate the point-probabilities of a discrete distribution, it is generally advisable to carry a couple of extra decimal places to guard against the accumulation of rounding-off errors. This precaution is also advisable in the tabulation of a polynomial for inverse interpolation.

(5) Of the 200 exercises, eighteen require monetary calculations which are based on the British decimal system. Readers unfamiliar with the British monetary system should note that, since 15th February, 1971,

$$\text{one pound} = 100 \text{ new pence (p)},$$

while in the earlier system,

$$\text{one pound} = 20 \text{ shillings} \quad \text{and} \quad \text{one shilling} = 12 \text{ (old) pence}.$$

(6) British (Imperial) units are adopted in such of the exercises as illustrate various applications of probability theory and statistics in industry and commerce, where, in practice, these units were still predominant in 1972. In a few other exercises, derived from experimental and research work, metric units are used, as is customary in scientific work.

The conversion table following these notes will enable the student to "metricate", as may be required, any of the exercises that are in British units, using (if preferred) convenient metric measures, e.g. substituting 50-kg sacks for 1-cwt sacks, 500-gram cartons for 20-oz, and so on.

Conversion Table

one inch (in.)	= 2·54 centimetres (cm)
one inch	= 25·4 millimetres (mm)
one foot (ft)	= 0·3048 metre (m)
one yard (yd)	= 0·9144 metre
one mile	= 1·609 344 kilometre (km)
one pint	= 0·568 245 litre (l)
one gallon	= 4·545 96 litres (l)
one ounce (oz)	= 28·3495 grams (g)
one pound (lb)	= 0·453 592 37 kilogram (kg)
one hundredweight (cwt)	= 50·8023 kilograms (kg)

1 Elementary computational techniques and numerical methods

1 Evaluate the following expressions correct to three decimal places:

    (i) $1685 \cdot 42 - (0 \cdot 543\ 901)^2 \times 835 \cdot 67$

    (ii) $2875 \cdot 39 - (1 \cdot 6749)^2 \times 496 \cdot 54$

    (iii) $48\ 394 \cdot 56 - 212 \times (-13 \cdot 58)^2$

    (iv) $\dfrac{996 \cdot 8201 - 3 \times (15 \cdot 9033)^2}{(15 \cdot 9033)^2}$

    (v) $3 \cdot 4592/(6 \cdot 0132)^{3/2}$

2 Evaluate the following numerical expressions correct to four decimal places:

    (i) $\dfrac{144}{19} + \dfrac{441}{32} + \dfrac{289}{26} + \dfrac{529}{24} + \dfrac{961}{53} - \dfrac{10\ 816}{154}$

    (ii) $\dfrac{82^2}{97 \cdot 0625} + \dfrac{39^2}{38 \cdot 6875} + \dfrac{50^2}{38 \cdot 6875} + \dfrac{22^2}{19 \cdot 5625} - 193$

    (iii) $\dfrac{366 \times (81 \times 90 - 72 \times 123)^2}{204 \times 153 \times 162 \times 213}$

    (iv) $\dfrac{6925 \times (897 \times 2238 - 2659 \times 1131)^2}{2028 \times 4897 \times 3135 \times 3790}$

3 Calculate the following square roots correct to four significant figures:

    (i) $(0 \cdot 8405 \times 0 \cdot 1595/71)^{\frac{1}{2}}$

    (ii) $(0 \cdot 002\ 768\ 9 \times 0 \cdot 005\ 392\ 8)^{\frac{1}{2}}$

    (iii) $(6 \cdot 857\ 42 \times 0 \cdot 097\ 64)^{\frac{1}{2}}$

    (iv) $\left[\dfrac{469 \times 7829 - (537)^2}{469 \times 495}\right]^{\frac{1}{2}}$

    (v) $(432 \cdot 7896/381)^{\frac{1}{2}}$

    (vi) $\left[\dfrac{48 \times 39}{48 + 39}\right]^{\frac{1}{2}}$

1

4 Evaluate the following expressions correct to three decimal places:

(i) $6\cdot54 \pm \dfrac{2\cdot074 \times 1\cdot876}{\sqrt{95}}$

(ii) $\left[\dfrac{19 \times 32}{19 + 32}\right]^{\frac{1}{2}} \times \dfrac{7\cdot845 - 4\cdot910}{6\cdot403}$

(iii) $\dfrac{1285\cdot43 - 32\cdot46 \times 12\cdot35}{(2978\cdot49 \times 645\cdot89)^{\frac{1}{2}}}$

(iv) $\dfrac{0\cdot6830 - 0\cdot5937}{\left[\dfrac{0\cdot6830 \times 0\cdot3170}{72} + \dfrac{0\cdot5937 \times 0\cdot4063}{138}\right]^{\frac{1}{2}}}$

(v) $\dfrac{3\cdot894 - 4\cdot201}{\sqrt{4\cdot201}}$

(vi) $\dfrac{(0\cdot5634 - 0\cdot5)\sqrt{84}}{0\cdot5}$

5 Determine, correct to the nearest integer containing n, the largest values of n satisfying the equations

(i) $1 - \left(1 - \dfrac{3}{n}\right)^{6} = \dfrac{5}{48}$;

(ii) $1 - 3\left(1 - \dfrac{4}{n}\right)^{2} - 4\left(1 - \dfrac{4}{n}\right)^{4} = \dfrac{13}{72}$; and

(iii) $1 - 6\left(1 - \dfrac{5}{n}\right)^{3} - 15\left(1 - \dfrac{5}{n}\right)^{6} = \dfrac{3}{4}$.

6 Calculate

(i) $1\cdot5384 \times 2\cdot716 + 9\cdot3271 \times 0\cdot319$

(ii) $0\cdot254\ 89 \times 0\cdot461 - 1\cdot0327 \times 0\cdot7130$

(iii) $\dfrac{7\cdot237 \times 0\cdot5125}{3\cdot001}$

(iv) $\dfrac{1\cdot5384}{2\cdot716} - \dfrac{0\cdot319}{9\cdot3271}$

stating the accuracy of the results.

7 Given the following 20 numbers (x_i, for $i = 1, 2, \ldots, 20$) in the table below, determine the exact value of $\sum\limits_{i=1}^{20} x_i^2$ by using a suitable simplifying transformation.

3·424	3·409	3·458	3·498
3·422	3·474	3·418	3·475
3·427	3·419	3·421	3·441
3·442	3·492	3·418	3·466
3·401	3·496	3·514	3·508

8 Evaluate, correct to five significant figures, the expressions for $P(r)$ defined by

$$P(r) = \binom{16}{r}\binom{36}{5-r} \Big/ \binom{52}{5}, \quad \text{for } 0 \leqslant r \leqslant 5.$$

9 Calculate, correct to six decimal places, the sums

$$S_1 = e^{-2}\sum_{r=0}^{3} 2^r/r!; \quad S_2 = e^{-2}\sum_{r=4}^{8} 2^r/r!; \quad S_3 = e^{-2}\sum_{r=9}^{\infty} 2^r/r!$$

Hence evaluate to the same accuracy

$$X = \sum_{k=1}^{3} (2k+1)\, S_k.$$

10 Determine the exact value of a_{12}, the coefficient of x^{12} in the expansion of

$$12!\left[\frac{x}{1!} + \frac{x^2}{2!} + \frac{x^3}{3!} + \cdots + \frac{x^{12}}{12!}\right]^8,$$

and hence evaluate, correct to six significant figures, the ratio $a_{12}/8^{12}$.

11 Given that x is so large that terms involving x^{-4} are negligible, prove that approximately

$$\left[1 + \frac{1}{x} + \frac{3}{x^2}\right]^x = e\left[1 + \frac{5}{2x} + \frac{11}{24x^2} - \frac{93}{16x^3}\right].$$

Hence evaluate to four decimal places the value of the left-hand-side expression when $x = 36$.

Note It may be assumed that e = 2·718 282.

12 The following table gives the cranial measurements on eight characters $x_1, x_2, \ldots, x_8$ obtained from 60 Brahmins of the Kumaon region in Uttar Pradesh, India. For these data, calculate, as a continuous machine process,

(i) $\sum_{\nu=1}^{60} x_{i\nu} \quad (i = 1, 2, \ldots, 8)$

(ii) $\sum_{\nu=1}^{60} x_{i\nu}^2 \quad (i = 1, 2, \ldots, 8)$

Table for Exercise 12

x_{1v}	x_{2v}	x_{3v}	x_{4v}	x_{5v}	x_{6v}	x_{7v}	x_{8v}
11.4	18.5	14.6	10.5	13.0	9.8	4.8	3.1
11.5	19.1	14.1	10.5	13.5	10.2	4.8	3.5
12.1	19.5	14.0	10.0	13.3	10.6	5.1	3.7
11.0	19.8	13.5	10.2	12.8	10.0	5.3	3.7
12.3	20.1	14.0	10.5	13.6	11.0	5.0	3.8
10.7	19.1	14.6	10.0	13.8	10.6	5.2	3.0
12.1	19.0	14.3	10.1	13.4	9.5	4.5	3.4
11.4	18.3	14.3	10.4	12.9	9.7	4.9	3.4
11.5	19.1	14.2	10.6	13.5	10.3	4.6	3.4
12.1	19.1	13.9	10.0	13.2	10.5	5.2	3.9
10.9	20.0	13.6	10.1	12.9	10.1	5.2	3.8
11.3	19.0	14.0	10.7	13.9	9.8	4.7	3.1
12.5	19.7	13.8	10.3	13.3	9.8	4.9	3.6
13.3	20.7	14.2	10.5	13.8	11.1	4.9	4.1
12.8	19.4	14.6	10.0	13.8	9.7	4.8	3.3
13.6	18.8	14.1	10.4	13.4	9.9	5.4	3.6
12.2	19.2	13.4	10.1	14.2	9.7	4.6	3.1
13.1	19.0	14.4	10.4	14.0	10.8	5.2	3.2
11.7	19.2	14.3	10.5	12.5	10.0	5.2	3.6
11.9	19.4	14.7	10.5	12.7	10.8	4.9	3.3
12.7	19.3	14.1	11.2	12.5	10.3	4.9	3.2
11.7	19.8	15.4	10.7	13.7	11.0	4.8	3.6
12.8	19.2	14.1	10.6	12.3	10.1	4.8	3.7
12.6	20.6	15.8	10.6	13.0	10.8	4.7	3.5
11.8	19.6	14.7	10.6	12.8	11.1	4.9	3.5
12.1	18.9	13.8	10.5	12.7	9.9	4.8	3.6
12.5	18.9	14.6	10.5	12.5	11.3	4.9	3.2
11.6	18.9	13.8	11.5	12.5	10.1	4.9	3.4
12.9	19.4	15.3	10.7	14.1	11.2	5.1	3.6
11.8	19.4	14.2	10.5	12.9	10.8	5.2	3.5
12.2	19.2	14.4		12.7	11.3	4.9	3.2

$x_{1\nu}$	$x_{2\nu}$	$x_{3\nu}$	$x_{4\nu}$	$x_{5\nu}$	$x_{6\nu}$	$x_{7\nu}$	$x_{8\nu}$
12.5	20.0	15.6	11.6	14.2	10.8	4.8	3.5
12.1	19.3	14.1	11.0	12.9	10.3	4.4	3.3
11.7	19.5	14.5	10.5	12.7	9.8	5.0	3.4
12.8	19.2	13.8	10.7	12.7	10.2	4.8	3.3
11.1	19.3	14.7	11.3	13.9	11.5	5.0	3.7
12.9	18.6	14.3	10.6	12.3	10.5	5.5	3.2
11.7	19.9	15.5	11.7	14.4	11.3	4.8	3.4
11.8	19.7	14.8	11.5	14.6	11.5	4.9	3.5
12.0	19.2	14.6	10.5	12.7	11.1	5.8	3.0
10.6	19.7	14.1	10.5	12.1	10.5	5.9	3.5
13.9	19.8	14.8	11.2	13.3	10.7	4.7	3.7
12.4	19.4	14.0	10.9	13.4	10.3	5.0	3.7
12.5	21.4	15.4	10.7	13.6	10.7	4.6	3.6
12.0	20.1	14.9	10.7	13.4	10.2	5.3	3.5
12.5	20.0	16.1	10.8	13.5	10.6	5.0	3.8
12.2	19.9	16.0	10.6	13.6	10.4	5.0	3.8
13.2	19.4	14.6	11.0	13.4	10.1	5.1	3.7
11.6	20.2	15.7	11.5	13.7	11.1	4.6	3.4
12.1	20.1	16.3	10.8	13.5	10.3	5.1	3.8
13.7	19.4	15.8	10.7	13.4	10.2	4.8	3.6
12.4	19.0	14.8	11.4	13.0	10.2	4.8	3.5
12.5	19.4	14.9	10.8	13.2	10.6	4.9	3.6
13.5	18.5	14.3	11.1	12.5	10.2	5.0	3.5
13.3	19.7	15.4	11.4	13.4	11.0	4.8	3.7
12.9	20.4	16.3	11.1	13.2	10.4	5.0	3.7
13.5	19.5	14.6	11.5	13.5	11.0	5.1	3.8
11.8	19.5	15.8	11.2	12.9	11.1	5.8	3.6
12.4	19.7	15.8	10.4	13.5	10.7	4.8	3.6
12.8	19.3	14.5	10.5	13.0	10.2	5.2	3.5

x_1: Head height x_2: Head length x_3: Head breadth x_4: Minimum frontal diameter

x_5: Bizygomatic breadth x_6: Bigonial diameter x_7: Nasal height x_8: Nasal breadth

(iii) $\sum_{\nu=1}^{60} x_{i\nu}\, x_{j\nu}$ $(i > j = 1, 2, \ldots, 8)$

using the sum check for (ii) and (iii).

13 The following tables of a polynomial $f(x)$ and transcendental function $g(x)$ each contain one mistake. Correct them and form the corrected difference tables.

(i) Determine the explicit form of $f(x)$ and use it to evaluate the exact value of $f(x)$ for $x = 2\cdot3$.

(ii) Calculate, correct to four decimal places, the value of $g(x)$ for $x = 57\cdot25$.

x	$f(x)$	x	$g(x)$
0	326	50	0·8983
1	413	55	0·9933
2	514	60	1·0896
3	635	65	1·1869
4	803	70	1·2835
5	1201	75	1·3846
6	1898	80	1·4846
7	3119	85	1·5850
8	5110	90	1·6858

14 The following table of a function $f(x)$ contains two mistakes. Correct them and then form the corrected difference table of the function values. Use this table to determine the explicit form of $f(x)$ and then calculate exactly the value of $f(x)$ when $x = 7\cdot24$.

x	$f(x)$	x	$f(x)$
0	356	6	620
1	357	7	741
2	366	8	924
3	398	9	1157
4	432	10	1446
5	501	11	1797

15 From the following table of $f(x)$ calculate, correct to seven decimal places, the function values for $x = 1\cdot064$ and $x = 1\cdot156$. Also, find, correct to four decimal places, the value of x such that $f(x) = 0\cdot25$.

x	$f(x)$	x	$f(x)$
1·02	0·113 899 5	1·12	0·197 912 2
1·04	0·128 240 1	1·14	0·218 473 8
1·06	0·143 789 1	1·16	0·240 283 6
1·08	0·160 573 5	1·18	0·263 316 9
1·10	0·178 611 5	1·20	0·287 536 1

16 In a sample of size n from a normal distribution having variance σ^2, the mean value of the sample range is given for n = 90 (10) 150 by the following table:

n	Mean range/σ	n	Mean range/σ
90	4·939 40	130	5·199 96
100	5·015 19	140	5·251 18
110	5·082 95	150	5·298 49
120	5·144 17		

Calculate, correct to five decimal places, the mean range of a sample of size 95, and determine also the sample size which most nearly gives a mean range of 5σ.

17 Correct the mistakes, if any, in the following table of the polynomial function $f(x)$.

x	$f(x)$	x	$f(x)$
1·50	10·007 928 0	1·56	24·657 177 6
1·51	12·351 998 1	1·57	27·238 178 1
1·52	14·734 489 6	1·58	29·860 177 6
1·53	17·155 826 1	1·59	32·523 614 1
1·54	19·616 433 6	1·60	35·228 928 0
1·55	22·117 640 5	1·61	37·976 562 1

Calculate the value of $f(x)$, correct to seven decimal places, when $x = 1·525$, and find the solution, correct to five decimal places, of

$$[f(x) + 3]^2 = 221·890\ 816.$$

18 From the following table of $g(x) \equiv f(x) \times \sin \pi x$, where $f(x)$ is *not* a polynomial, calculate the value of $g(0·15)$, correct to five decimal places. By using this value and the given table of $\sin^{-1} x$, calculate $f(0·15)$ as accurately as possible, stating the accuracy obtained.

x	$g(x)$	x	$\sin^{-1} x$
0·14	0·441 128	0·45	0·466 765
0·16	0·477 845	0·46	0·477 995
0·18	0·512 298	0·47	0·489 291
0·20	0·544 604	0·48	0·500 655
0·22	0·574 835	0·49	0·512 090
0·24	0·603 029	0·50	0·523 599

Note It may be assumed that $\pi = 3·141\ 592\ 65$.

19 The table below gives the values of a polynomial function $y = f(x)$, for $0 \leqslant x(1) \leqslant 10$.

x	$f(x)$	x	$f(x)$
0	$-2\cdot670\ 33$	6	$14\cdot160\ 39$
1	$-1\cdot296\ 61$	7	$20\cdot045\ 51$
2	$0\cdot485\ 51$	8	$27\cdot348\ 23$
3	$2\cdot782\ 23$	9	$36\cdot340\ 35$
4	$5\cdot721\ 35$	10	$47\cdot329\ 67$
5	$9\cdot445\ 67$		

Check the table by differencing and correct if any erroneous entry is found.

Determine the exact relation between x and y, and use it to calculate, correct to seven decimal places, the value of $f(x)$ for $x = 3\cdot52$.

20 In the following table relating x to y there are two mistakes (of magnitude $+1$ in the last decimal place) in the values of y. Correct these mistakes. By introducing a new variable z, which takes the values 0, 1, 2, ... , 9, find the value of y when $x = 1\cdot1025$. Also, determine the relationship connecting x and y.

x	y	x	y
$1\cdot0816$	$0\cdot508\ 143$	$1\cdot2996$	$0\cdot914\ 823$
$1\cdot1236$	$0\cdot584\ 295$	$1\cdot3456$	$1\cdot004\ 175$
$1\cdot1664$	$0\cdot662\ 991$	$1\cdot3924$	$1\cdot096\ 311$
$1\cdot2100$	$0\cdot744\ 280$	$1\cdot4400$	$1\cdot191\ 279$
$1\cdot2544$	$0\cdot828\ 208$	$1\cdot4884$	$1\cdot289\ 127$

21 The $(r + 1)$th term of an infinite series is

$$u_r = q^r \bigg/ \binom{n + r}{r}, \quad \text{for } r \geqslant 0,$$

where q $(0 < q < 1)$ and n, a positive integer, are constants. Show that if the series is truncated after the $(m + 1)$th term, then the error in its sum is less than $(1 - q)^{-1} u_{m+2}$.

If $q = 0\cdot2$ and $n = 10$, evaluate, correct to six decimal places, the sum of the infinite series.

22 By assuming the logarithmic expansion

$$\log_e(1 + x) = \sum_{r=1}^{\infty} (-1)^{r-1} x^r/r, \quad \text{for } |x| < 1,$$

prove that

$$\log_e(y + 2) - \log_e(y + 1) = \frac{2}{2y + 3} \sum_{r=0}^{\infty} \left(\frac{1}{2y + 3}\right)^{2r} \bigg/ (2r + 1), \quad y \geqslant 0.$$

Use this expansion to calculate $\log_e 2$ and $\log_e 5$ correct to seven decimal places. Also, indicate how the accuracy of these computations can be evaluated.

23 The terms of an infinite series u_1, u_2, u_3, ... are such that for some integral value of k

$$\Delta^{k+1} u_r = bx^{r-1},$$

where $b \neq 0$ is a constant and x is independent of r. Prove that

$$u_r = \frac{bx^{r-1}}{(x-1)^{k+1}} + \sum_{j=0}^{k} a_j \, r^j,$$

where the a_j are arbitrary constants such that $a_k \neq 0$.

In particular, suppose that the first nine terms of the infinite series are 0·09, 0·75, 3·09, 10·73, 40·09, 171·59, 805·65, 3942·69, and 19583·13. Assuming that for some integral value of k, which is chosen to be the smallest possible, the $(k+1)$th differences of the u_r are in geometric progression, determine the general expression for u_r. Hence evaluate the sum of the infinite series to n terms. Verify that for $n = 1$ this sum correctly reduces to u_1.

24 The larger positive root of the cubic equation

$$10x^3 - 26x^2 - 78x + 21 = 0$$

is known to lie in the neighbourhood of $x = 4$. Show that a closer approximation for this root is $x = 4·35$. Determine this root correct to five decimal places. Hence also evaluate, correct to the same accuracy, the other two roots of the equation.

25 Tabulate the polynomial

$$f(x) \equiv 12·3x^5 - 25·2x^4 - 161·0x^3 + 173·0x^2 + 294·0x - 13·4$$

for values of $x = 1·8 \, (0·1) \, 4·2$.

Use these tabulated values to determine, correct to four decimal places, the two positive roots of the equation $f(x) = 0$ which lie in the neighbourhood of $x = 2$ and $x = 4$ respectively.

ANSWERS AND HINTS ON SOLUTIONS
Chapter 1

1 (i) 1438·205   (ii) 1482·451   (iii) 9298·283
   (iv) 0·941   (v) 0·235

2 (i) 2·4156   (ii) 4·9515   (iii) 0·8334
   (iv) 58·6690

3 (i) 0·043 45   (ii) 0·003 864   (iii) 0·8133
   (iv) 3·818   (v) 1·066   (vi) 4·639

4 (i) 6·939, 6·141   (ii) 1·583   (iii) 0·638
   (iv) 1·295   (v) −0·150   (vi) 1·162

5 (i) Here $\left(1 - \dfrac{3}{n}\right)^3 = 0.9465 = \dfrac{9465}{10\ 000}$. The cube roots are obtained

from *Barlow's Tables* whence, on division, $3/n = 0.018\ 16$. Hence $n = 165.2$ or 166 to the nearest integer containing n.

   (ii) Put $\left(1 - \dfrac{4}{n}\right)^2 = \theta$. The relevant root of the resulting quadratic

equation in θ is $\theta = 0.2128$. Hence $n = 7.4$ or 8 to the nearest integer containing n.

   (iii) Put $\left(1 - \dfrac{5}{n}\right)^3 = \theta$. The relevant root of the resulting quadratic

in θ is $\theta = 0.038\ 05 = \dfrac{3805 \times 10}{10^6}$. Hence, by using *Barlow's Tables*, we have

$$1 - \frac{5}{n} = \frac{15.6117 \times 2.1544}{100} = 0.3363.$$

Hence $n = 7.5$ or 8 to the nearest integer containing n.

6 To evaluate the accuracy use the method of Section 2.5, *Course* (pp. 15 *et seq.*).

(i) $7 \cdot 154 \pm 0 \cdot 006$ (ii) $-0 \cdot 6188 \pm 0 \cdot 0002$

(iii) $1 \cdot 2359 \pm 0 \cdot 0004$ (iv) $0 \cdot 5322 \pm 0 \cdot 0002$

7   Put $z_i = 1000 (x_i - 3 \cdot 4)$. Then $x_i = 10^{-3} z_i + 3 \cdot 4$. Therefore,

$$\sum x_i^2 = 10^{-6} \times \sum (z_i + 3400)^2$$

$$= 10^{-6} \times \left(\sum z_i^2 + 6800 \sum z_i + 20 \times 3400^2 \right).$$

But $\Sigma z_i^2 = 77\ 531$ and $\Sigma z_i = 1023$, whence $\Sigma x_i^2 = 238 \cdot 233\ 931$.

8   For $r = 0$,

$$P(0) = \binom{36}{5} \Big/ \binom{52}{5} = \frac{66}{455} = 0 \cdot 145\ 054\ 9.$$

Again, for $r \geqslant 1$,

$$\frac{P(r)}{P(r-1)} = \frac{(17 - r)(6 - r)}{r(31 + r)},$$

so that

$$P(r) = \frac{(17 - r)(6 - r)}{r(31 + r)} P(r - 1).$$

Use this recurrence relation to evaluate the values of $P(r)$. Hence, correct to five significant figures,

$P(0) = 0 \cdot 145\ 05$ $P(1) = 0 \cdot 362\ 64$ $P(2) = 0 \cdot 329\ 67$

$P(3) = 0 \cdot 135\ 75$ $P(4) = 10^{-1} \times 0 \cdot 252\ 10$

$P(5) = 10^{-2} \times 0 \cdot 168\ 07$

As a check, observe that the sum of these six values must add up to unity within the limits of rounding-off errors.

9   Note that $S_3 = 1 - S_1 - S_2$ and $e^{-2} = 0 \cdot 135\ 335$.

$$S_1 = e^{-2} \left(1 + 2 + 2 + \frac{4}{3} \right) = 0 \cdot 857\ 122$$

$$S_2 = e^{-2} \left(\frac{16}{24} + \frac{32}{120} + \frac{64}{720} + \frac{128}{5040} + \frac{256}{40\ 320} \right) = 0 \cdot 142\ 639$$

$S_3 = 0 \cdot 000\ 239.$ $X = 3 \cdot 286\ 234.$

10   Observe that a_{12} = the coefficient of x^{12} in $12!\,(e^x - 1)^8$

$$= \text{the coefficient of } x^{12} \text{ in } 12! \sum_{r=0}^{8} \binom{8}{r} (-1)^{8-r} e^{rx}.$$

Hence

$$a_{12} = 8^{12} - \binom{8}{1} 7^{12} + \binom{8}{2} 6^{12} - \binom{8}{3} 5^{12} + \binom{8}{4} 4^{12} - \binom{8}{3} 3^{12} + \binom{8}{2} 2^{12} - 8,$$

so that

$$8^{-12} \times a_{12} = 6\ 411\ 968\ 640/68\ 719\ 476\ 736 = 10^{-1} \times 0 \cdot 933\ 064.$$

11 We have

$$\left[1 + \frac{1}{x} + \frac{3}{x^2} \right]^x = \exp \left[x \log \left(1 + \frac{1}{x} + \frac{3}{x^2} \right) \right]$$

$$= \exp x \left[\left(\frac{1}{x} + \frac{3}{x^2} \right) - \frac{1}{2}\left(\frac{1}{x} + \frac{3}{x^2} \right)^2 + \frac{1}{3}\left(\frac{1}{x} + \frac{3}{x^2} \right)^3 - \frac{1}{4}\left(\frac{1}{x} + \frac{3}{x^2} \right)^4 + \cdots \right],$$

which, on simplification,

$$= e \times \exp \left(\frac{5}{2x} - \frac{8}{3x^2} - \frac{7}{4x^3} + \cdots \right)$$

$$= e \left[1 + \left(\frac{5}{2x} - \frac{8}{3x^2} - \frac{7}{4x^3} \right) + \frac{1}{2}\left(\frac{5}{2x} - \frac{8}{3x^2} - \frac{7}{4x^3} \right)^2 + \frac{1}{6}\left(\frac{5}{2x} - \frac{8}{3x^2} - \frac{7}{4x^3} \right)^3 + \cdots \right.$$

$$= e \left[1 + \frac{5}{2x} + \frac{11}{24x^2} - \frac{93}{16x^3} + \cdots \right], \text{ on reduction. Hence the result.}$$

For $x = 36$, using the approximation, we have

$$\left[1 + \frac{1}{x} + \frac{3}{x^2} \right]^x = e \left[1 + \frac{5}{72} + \frac{11}{31\ 104} - \frac{93}{746\ 496} \right]$$

$$= e \times 1 \cdot 069\ 673 = 2 \cdot 9077.$$

12

$\Sigma\, x_{1\nu} = 734 \cdot 5$	$\Sigma\, x_{2\nu} = 1168 \cdot 3$
$\Sigma\, x_{3\nu} = 878 \cdot 8$	$\Sigma\, x_{4\nu} = 642 \cdot 0$
$\Sigma\, x_{5\nu} = 794 \cdot 7$	$\Sigma\, x_{6\nu} = 629 \cdot 3$
$\Sigma\, x_{7\nu} = 298 \cdot 6$	$\Sigma\, x_{8\nu} = 210 \cdot 5$
$\Sigma\, x_{1\nu}^2 = 9024 \cdot 91$	$\Sigma\, x_{2\nu}^2 = 22\ 767 \cdot 13$
$\Sigma\, x_{3\nu}^2 = 12\ 903 \cdot 54$	$\Sigma\, x_{4\nu}^2 = 6881 \cdot 18$
$\Sigma\, x_{5\nu}^2 = 10\ 544 \cdot 13$	$\Sigma\, x_{6\nu}^2 = 6615 \cdot 71$
$\Sigma\, x_{7\nu}^2 = 1491 \cdot 16$	$\Sigma\, x_{8\nu}^2 = 741 \cdot 69$
$\Sigma\, x_{1\nu} x_{2\nu} = 14\ 303 \cdot 51$	$\Sigma\, x_{1\nu} x_{3\nu} = 10\ 764 \cdot 45$
$\Sigma\, x_{1\nu} x_{4\nu} = 7863 \cdot 60$	$\Sigma\, x_{1\nu} x_{5\nu} = 9730 \cdot 02$
$\Sigma\, x_{1\nu} x_{6\nu} = 7704 \cdot 61$	$\Sigma\, x_{1\nu} x_{7\nu} = 3653 \cdot 46$
$\Sigma\, x_{1\nu} x_{8\nu} = 2579 \cdot 49$	
$\Sigma\, x_{2\nu} x_{3\nu} = 17\ 123 \cdot 41$	$\Sigma\, x_{2\nu} x_{4\nu} = 12\ 504 \cdot 35$
$\Sigma\, x_{2\nu} x_{5\nu} = 15\ 480 \cdot 14$	$\Sigma\, x_{2\nu} x_{6\nu} = 12\ 259 \cdot 52$
$\Sigma\, x_{2\nu} x_{7\nu} = 5813 \cdot 26$	$\Sigma\, x_{2\nu} x_{8\nu} = 4102 \cdot 53$

$\Sigma\ x_{3\nu}\ x_{4\nu} = 9412{\cdot}99$ $\Sigma\ x_{3\nu}\ x_{5\nu} = 11\ 647{\cdot}26$

$\Sigma\ x_{3\nu}\ x_{6\nu} = 9226{\cdot}00$ $\Sigma\ x_{3\nu}\ x_{7\nu} = 4372{\cdot}74$

$\Sigma\ x_{3\nu}\ x_{8\nu} = 3085{\cdot}17$

$\Sigma\ x_{4\nu}\ x_{5\nu} = 8507{\cdot}62$ $\Sigma\ x_{4\nu}\ x_{6\nu} = 6739{\cdot}85$

$\Sigma\ x_{4\nu}\ x_{7\nu} = 3193{\cdot}66$ $\Sigma\ x_{4\nu}\ x_{8\nu} = 2253{\cdot}47$

$\Sigma\ x_{5\nu}\ x_{6\nu} = 8339{\cdot}35$ $\Sigma\ x_{5\nu}\ x_{7\nu} = 3951{\cdot}91$

$\Sigma\ x_{5\nu}\ x_{8\nu} = 2789{\cdot}00$

$\Sigma\ x_{6\nu}\ x_{7\nu} = 3133{\cdot}08$ $\Sigma\ x_{6\nu}\ x_{8\nu} = 2208{\cdot}67$

$\Sigma\ x_{7\nu}\ x_{8\nu} = 1047{\cdot}84$

13   The error in the fifth differences of $f(x)$ is $27\,(-1,\ 5,\ -10,\ 10,\ -5,\ 1)$. Therefore the incorrect value is $f(4)$, which is wrongly decreased by 27. Hence correct $f(4) = 830$.

Since $g(x)$ is a transcendental function its values are subject to rounding off. The error in the fourth differences of $g(x)$ is $18\,(-1,\ 4,\ -6,\ 4,\ -1)$. Therefore the erroneous value is $g(70)$ which is wrongly decreased by 18. Hence correct $g(70) = 1{\cdot}2835 + 0{\cdot}0018 = 1{\cdot}2853$.

(i) The corrected difference table of the $f(x)$ values is as follows:

x	$f(x)$	Δ	Δ^2	Δ^3	Δ^4
0	326				
		87			
1	413		14		
		101		6	
2	514		20		48
		121		54	
3	635		74		48
		195		102	
4	830		176		48
		371		150	
5	1201		326		48
		697		198	
6	1898		524		48
		1221		246	
7	3119		770		
		1991			
8	5110				

Since the fourth differences are constant, $f(x)$ is a fourth-degree polynomial in x. Therefore, for any $x = a + \theta h$, where h is the interval of the argument and $0 < \theta < 1$, we have by the Gregory–Newton formula

$$f(a + \theta h) = \sum_{r=0}^{4} \frac{\theta^{(r)}}{r!} \Delta^r f(a).$$

Hence for $h = 1$, $a = 0$, $x = \theta$ and so

$$f(x) = \sum_{r=0}^{4} \frac{x^{(r)}}{r!} \Delta^r f(0)$$

$$= 326 + 87x + 7x^{(2)} + x^{(3)} + 2x^{(4)}$$

$$= 326 + 70x + 26x^2 - 11x^3 + 2x^4, \text{ on reduction.}$$

Therefore $f(2 \cdot 3) = 546 \cdot 6712$.

(ii) The corrected difference table of the $g(x)$ values is as follows. The fourth differences indicate rounding-off errors only.

x	$g(x)$	Δ	Δ^2	Δ^3	Δ^4
50	0·8983				
		950			
55	0·9933		13		
		963		−3	
60	1·0896		10		4
		973		1	.
65	1·1869		11		−3
		984		−2	
70	1·2853		9		0
		993		−2	
75	1·3846		7		−1
		1000		−3	
80	1·4846		4		3
		1004		0	
85	1·5850		4		
		1008			
90	1·6858				

To obtain $g(x)$ for $x = 57 \cdot 25$, observe that the argument of the leading diagonal is $a = 55$, the interval of the argument is $h = 5$ and $\theta = 0 \cdot 45$, since $a + \theta h = 57 \cdot 25$. Forward interpolation gives $g(57 \cdot 25) = 1 \cdot 0365$.

14   The mistakes in the fourth differences of $f(x)$ are $9(1, -4, 6, -4, 1)$ and $18(1, -4, 6, -4, 1)$ with overlap. Therefore the incorrect function values are $f(3)$ and $f(6)$. The corrected values are $f(3) = 398 - 9 = 389$ and $f(6) = 620 - 18 = 602$.

The corrected difference table of the $f(x)$ values is as at top of p. 15. Since the third differences are constant, $f(x)$ is a third-degree polynomial in x. The interval of the argument is unity and so, using the leading diagonal of the table, we have

$$f(x) = \sum_{r=0}^{3} \frac{x^{(r)}}{r!} \Delta^r f(0)$$

$$= 356 + x + 4x^{(2)} + x^{(3)}$$

$$= 356 - x + x^2 + x^3.$$

Hence $f(7 \cdot 24) = 780 \cdot 681 \, 024$.

x	$f(x)$	Δ	Δ^2	Δ^3
0	356			
		1		
1	357		8	
		9		6
2	366		14	
		23		6
3	389		20	
		43		6
4	432		26	
		69		6
5	501		32	
		101		6
6	602		38	
		139		6
7	741		44	
		183		6
8	924		50	
		233		6
9	1157		56	
		289		6
10	1446		62	
		351		
11	1797			

15 Difference the function values up to fifth differences. The forward difference table is as follows:

x	$f(x)$	Δ	Δ^2	Δ^3	Δ^4	Δ^5
1·02	0·113 899 5					
		143 406				
1·04	0·128 240 1		12 084			
		155 490		270		
1·06	0·143 789 1		12 354		− 88	
		167 844		182		− 3
1·08	0·160 573 5		12 536		− 91	
		180 380		91		−18
1·10	0·178 611 5		12 627		− 109	
		193 007		− 18		0
1·12	0·197 912 2		12 609		− 109	
		205 616		− 127		−11
1·14	0·218 473 8		12 482		− 120	
		218 098		−247		− 9
1·16	0·240 283 6		12 235		− 129	
		230 333		− 376		
1·18	0·263 316 9		11 859			
		242 192				
1·20	0·287 536 1					

To obtain $f(x)$ for $x = 1.064$, observe that the argument of the leading diagonal is $a = 1.06$, the interval of the argument is $h = 0.02$, and $\theta = 0.2$, where $a + \theta h = 1.064$. Forward interpolation then gives $f(1.064) = 0.147\ 046\ 5$.

To determine $f(x)$ for $x = 1.156$, use backward differences – see the *Course* (Section 3.9, p. 41). In this case, $a = 1.18$, $h = -0.02$ and $\theta = 0.2$, where $a + \theta h = 1.156$. Hence $f(1.156) = 0.235\ 822\ 8$.

Let $x_0 = a + \theta h$ be the root of the equation $f(x) = 0.25$. Again use backward differences with $a = 1.18$ and $h = -0.02$. Inverse interpolation for the determination of θ is carried out by the equation

$$-0.023\ 033\ 3\ \theta = 0.25 - 0.263\ 316\ 9 - \frac{\theta^{(2)}}{2!} \times 0.001\ 223\ 5 - \frac{\theta^{(3)}}{3!}$$

$$\times 0.000\ 024\ 7 + \frac{\theta^{(4)}}{4!} \times 0.000\ 012\ 0 - \frac{\theta^{(5)}}{5!} \times 0.000\ 001\ 1.$$

Clearly, the contributions arising from the third and higher differences are negligible for the required accuracy in the determination of θ. Iteration with the second difference correction only gives a stable value of $\theta = 0.5717$. Hence $x_0 = 1.1686$.

16   Leicester, 1960.

The difference table of the function values is as follows:

n	Mean range/σ	Δ	Δ^2	Δ^3	Δ^4	Δ^5
90	4.939 40					
		7579				
100	5.015 19		−803			
		6776		149		
110	5.082 95		−654		−38	
		6122		111		13
120	5.144 17		−543		−25	
		5579		86		5
130	5.199 96		−457		−20	
		5122		66		
140	5.251 18		−391			
		4731				
150	5.298 49					

To determine the mean range/σ for $n = 95$, observe that the argument of the leading diagonal is $a = 90$, the interval of the argument is $h = 10$, and $\theta = 0.5$, where $a + \theta h = 95$. Then forward interpolation gives mean range/$\sigma = 4.978\ 411$ for $n = 95$.

To evaluate the sample size such that the corresponding mean range is 5σ use inverse interpolation. The required sample size $n_0 = 90 + 10\theta$, where the equation for θ is

$$0 \cdot 075\ 79\ \theta = 0 \cdot 060\ 60 + 0 \cdot 008\ 03\ \frac{\theta^{(2)}}{2!} - 0 \cdot 001\ 49\ \frac{\theta^{(3)}}{3!} +$$

$$+ 0 \cdot 000\ 38\ \frac{\theta^{(4)}}{4!} - 0 \cdot 000\ 13\ \frac{\theta^{(5)}}{5!} \ .$$

For the required accuracy of θ, the contributions from the third and higher differences are negligible. Iteration then gives $\theta = 0 \cdot 7909$, whence $n_0 = 97 \cdot 9$ or 98 to the nearest integer.

17   Leicester, 1961.
The incorrect function value is $f(1 \cdot 55)$ since the fourth differences should be 24. The error in the fourth differences is $9000(1, -4, 6, -4, 1)$. Hence correct $f(1 \cdot 55) = 22 \cdot 117\ 640\ 5 - 0 \cdot 000\ 900\ 0 = 22 \cdot 116\ 740\ 5$. The correct forward difference table is as follows:

x	$f(x)$	Δ	Δ^2	Δ^3	Δ^4
1·50	10·007 928 0				
		2·344 070 1			
1·51	12·351 998 1		384 214		
		2·382 491 5		4236	
1·52	14·734 489 6		388 450		24
		2·421 336 5		4260	
1·53	17·155 826 1		392 710		24
		2·460 607 5		4284	
1·54	19·616 433 6		396 994		24
		2·500 306 9		4308	
1·55	22·116 740 5		401 302		24
		2·540 437 1		4332	
1·56	24·657 177 6		405 634		24
		2·581 000 5		4356	
1·57	27·238 178 1		409 990		24
		2·621 999 5		4380	
1·58	29·860 177 6		414 370		24
		2·663 436 5		4404	
1·59	32·523 614 1		418 774		24
		2·705 313 9		4428	
1·60	35·228 928 0		423 202		
		2·747 634 1			
1·61	37·976 562 1				

To obtain $f(x)$ for $x = 1 \cdot 525$, observe that the argument of the leading diagonal is $a = 1 \cdot 52$, the interval of the argument is $h = 0 \cdot 01$ and $\theta = 0 \cdot 5$, where $a + \theta h = 1 \cdot 525$. Forward interpolation then gives $f(1 \cdot 525) = 15 \cdot 940\ 275\ 7$.

The given quadratic equation gives $f(x) = 11 \cdot 896$ exactly. Let $x_0 = a + \theta h$ be the required root, where $a = 1 \cdot 50$, $h = 0 \cdot 01$. The equation for the determination of θ by inverse interpolation is

$$2 \cdot 344\ 070\ 1\ \theta = 11 \cdot 896 - 10 \cdot 007\ 928\ 0 - 0 \cdot 038\ 421\ 4\ \frac{\theta^{(2)}}{2!} -$$

$$- 0 \cdot 000\ 423\ 6\ \frac{\theta^{(3)}}{3!} - 0 \cdot 000\ 002\ 4\ \frac{\theta^{(4)}}{4!}\ .$$

To the required accuracy, the contributions from the third and fourth differences are negligible and may be ignored. On iteration, the stable value of $\theta = 0 \cdot 8067$. Hence $x_0 = 1 \cdot 508\ 07$.

18   Leicester, 1961.

The forward difference table of the $g(x)$ values is as follows:

x	$g(x)$	Δ	Δ^2	Δ^3	Δ^4	Δ^5
0·14	0·441 128					
		36 717				
0·16	0·477 845		− 2264			
		34 453		117		
0·18	0·512 298		− 2147		− 45	
		32 306		72		11
0·20	0·544 604		− 2075		− 34	
		30 231		38		
0·22	0·574 835		− 2037			
		28 194				
0·24	0·603 029					

To obtain $g(x)$ for $x = 0 \cdot 15$, observe that the argument of the leading diagonal is $a = 0 \cdot 14$, the interval of the argument is $h = 0 \cdot 02$ and $\theta = 0 \cdot 5$ where $a + \theta h = 0 \cdot 15$. Forward interpolation then gives $g(0 \cdot 15) = 0 \cdot 459\ 78$.

If we set $x_0 = \sin 0 \cdot 15\pi$, then $\sin^{-1} x_0 = 0 \cdot 15\pi = 0 \cdot 471\ 239$. Hence use inverse interpolation to determine $x_0 = a + \theta h$, with $a = 0 \cdot 45$ and $h = 0 \cdot 01$. The appropriate forward difference table of the $\sin^{-1} x$ values is as follows:

x	$\sin^{-1} x$	Δ	Δ^2	Δ^3
0·45	0·466 765			
		11 230		
0·46	0·477 995		66	
		11 296		2
0·47	0·489 291		68	
		11 364		3
0·48	0·500 655		71	
		11 435		3
0·49	0·512 090		74	
		11 509		
0·50	0·523 599			

To determine θ by inverse interpolation use the relation

$$0\cdot011\ 230\ \theta = 0\cdot471\ 239 - 0\cdot466\ 765 - 0\cdot000\ 066\ \frac{\theta^{(2)}}{2!} - 0\cdot000\ 002\ \frac{\theta^{(3)}}{3!}.$$

The contribution from the third difference is negligible and may be ignored. The stable value of $\theta = 0\cdot399\ 109$, whence $x_0 = 0\cdot453\ 991$. Hence

$$f(0\cdot15) = 0\cdot459\ 78/0\cdot453\ 991 = 1\cdot012\ 75,$$

and maximum value of $f(0\cdot15) = 0\cdot459\ 785/0\cdot453\ 990\ 5 = 1\cdot012\ 76$.

Therefore $\qquad f(0\cdot15) = 1\cdot012\ 75 \pm 0\cdot000\ 01$.

19 Leicester, 1964.

The error in the sixth differences of the function values $f(x)$ is $900\,(-1,\ 6,\ -15,\ 20,\ -15,\ 6,\ -1)$. The incorrect value is $f(5)$ which should be $9\cdot445\ 67 + 0\cdot009\ 00 = 9\cdot454\ 67$.

The corrected forward difference table is as follows:

x	$f(x)$	Δ	Δ^2	Δ^3	Δ^4	Δ^5
0	$-2\cdot670\ 33$					
		1·373 72				
1	$-1\cdot296\ 61$		40 840			
		1·782 12		10 620		
2	0·485 51		51 460		2160	
		2·296 72		12 780		240
3	2·782 23		64 240		2400	
		2·939 12		15 180		240
4	5·721 35		79 420		2640	
		3·733 32		17 820		240
5	9·454 67		97 240		2880	
		4·705 72		20 700		240
6	14·160 39		117 940		3120	
		5·885 12		23 820		240
7	20·045 51		141 760		3360	
		7·302 72		27 180		240
8	27·348 23		168 940		3600	
		8·992 12		30 780		
9	36·340 35		199 720			
		10·989 32				
10	47·329 67					

Since the fifth differences are constant, $f(x)$ is a fifth-degree polynomial in x. The argument of the leading diagonal is $a = 0$, and the interval of the argument is $h = 1$. Hence by the Gregory–Newton formula

$$f(x) = \sum_{r=0}^{5} \frac{x^{(r)}}{r!}\,\Delta^r f(0),$$

so that

$$10^5 f(x) = -267\ 033 + 137\ 372\ x + 20\ 420\ x^{(2)} + 1770 x^{(3)} + 90 x^{(4)} + 2 x^{(5)}$$

$$= 2x^5 + 70 x^4 + 1300 x^3 + 1600 x^2 + 120\ 000\ x - 267\ 033,$$

on reduction. Hence $f(3 \cdot 52) = 4 \cdot 221\ 392\ 1$.

20   The errors in the fourth differences of y are $(1, -4, 6, -4, 1)$ and $(1, -4, 6, -4, 1)$ with overlap. Therefore the incorrect y values correspond to $x = 1 \cdot 2100$ and $x = 1 \cdot 2544$. The corrected y values are $0 \cdot 744\ 279$ and $0 \cdot 828\ 207$.

If $x = f(z)$ and $y = g(z)$, then the difference tables of the given function values are as follows:

z	$x = f(z)$	Δ	Δ^2	$y = g(z)$	Δ	Δ^2	Δ^3
0	1·0816			0·508 143			
		420			76 152		
1	1·1236		8	0·584 295		2544	
		428			78 696		48
2	1·1664		8	0·662 991		2592	
		436			81 288		48
3	1·2100		8	0·744 279		2640	
		444			83 928		48
4	1·2544		8	0·828 207		2688	
		452			86 616		48
5	1·2996		8	0·914 823		2736	
		460			89 352		48
6	1·3456		8	1·004 175		2784	
		468			92 136		48
7	1·3924		8	1·096 311		2832	
		476			94 968		48
8	1·4400		8	1·191 279		2880	
		484			97 848		
9	1·4884			1·289 127			

Clearly, x is a quadratic in z. Hence for any $z = a + \theta h$, where h is the interval of the argument and $0 < \theta < 1$, the Gregory–Newton formula gives

$$x = \sum_{r=0}^{2} \frac{\theta^{(r)}}{r!} \, \Delta^r f(a).$$

In particular, since $h = 1$, we have for $a = 0$, $z = \theta$. Therefore

$$1 \cdot 1025 = 1 \cdot 0816 + 0 \cdot 0420 z + 0 \cdot 0008 \frac{z^{(2)}}{2!},$$

which reduces to $4z^2 + 416z - 209 = 0$ or $(2z - 1)(2z + 209) = 0$.
Therefore, $z = 0 \cdot 5$ when $x = 1 \cdot 1025$.

For $z = \theta = 0 \cdot 5$ forward interpolation gives $y = g(0 \cdot 5) = 0 \cdot 545\ 904$.

In general,
$$x = f(z) = 1{\cdot}0816 + 0{\cdot}0420z + 0{\cdot}0008\frac{z^{(2)}}{2!} \qquad (1)$$
and
$$y = g(z) = 0{\cdot}508\ 143 + 0{\cdot}076\ 152\ z + 0{\cdot}002\ 544\ \frac{z^{(2)}}{2!} + 0{\cdot}000\ 048\ \frac{z^{(3)}}{3!}. \qquad (2)$$

Now (1) can be rewritten as $10^4 x = 4(z + 52)^2$ with the solution $z = 50\sqrt{x} - 52$.

Elimination of z from (2) gives after some reduction
$$10^6 y = -1\ 136\ 721 + 5 \times 10^5\sqrt{x} + 10^6 x^{3/2}$$

or
$$y = x^{3/2} + 0{\cdot}5\sqrt{x} - 1{\cdot}136\ 721.$$

21   The remainder after $m + 1$ terms of the series
$$= \sum_{r=m+1}^{\infty} q^r \bigg/ \binom{n+r}{r} < \sum_{r=m+1}^{\infty} q^r \bigg/ \binom{n+m+1}{m+1} = q^{m+1}(1-q)^{-1} \bigg/ \binom{n+m+1}{m+1}$$
$$= (1-q)^{-1}\, u_{m+2}.$$

Again, for $r \geqslant 1$, $\quad u_r = \dfrac{qn}{n+r}\, u_{r-1}$ and $u_0 = 1$.

For $n = 10$, $q = 0{\cdot}2$, $qn = 2$. Use the recurrence relation to evaluate the terms of the series up to u_8. The sum of the series up to u_7 is $1{\cdot}217\ 550\ 4$, and the truncation error is
$$< 10^{-5} \times 0{\cdot}145\ 098\ 1/0{\cdot}8 = 10^{-6} \times 0{\cdot}1814.$$

Hence, correct to six decimal places, the sum of the series is $1{\cdot}217\ 550$.

22   We have
$$\log_e (1 + x) - \log_e (1 - x) = 2 \sum_{r=0}^{\infty} x^{2r+1}/(2r + 1).$$

If we now set $x = 1/(2y + 3)$, then $(1 + x)/(1 - x) = (y + 2)/(y + 1)$.

Hence $\log_e (y + 2) - \log_e (y + 1) = \dfrac{2}{2y + 3} \displaystyle\sum_{r=0}^{\infty} \dfrac{1}{(2r + 1)(2y + 3)^{2r}}.$

By putting $y = 3$ and $y = 0$ successively in this formula, we obtain
$$\log_e 5 = 2 \log_e 2 + \frac{2}{9} \sum_{r=0}^{\infty} \frac{1}{(2r + 1)9^{2r}} \qquad (1)$$

and
$$\log_e 2 = \frac{2}{3} \sum_{r=0}^{\infty} \frac{1}{(2r + 1)3^{2r}}. \qquad (2)$$

Compute successively the first seven terms of the series in (2), and their sum gives $\log_e 2 = 0 \cdot 693\ 147\ 17$. The truncation error is

$$< \frac{2}{3} \times \frac{1}{15} \times \frac{1}{3^{14}} \sum_{r=0}^{\infty} \left(\frac{1}{3} \right)^{2r} = 10^{-7} \times 0 \cdot 1045.$$

Hence, correct to seven decimal places, $\log_e 2 = 0 \cdot 693\ 147\ 2$.

Similarly, compute the first four terms of the series in (1). Then $\log_e 5 = 1 \cdot 609\ 437\ 89$, and the truncation error is

$$< 10^{-7} \times 0 \cdot 2090 + \frac{2}{9} \times \frac{1}{9} \times \frac{1}{9^8} \sum_{r=0}^{\infty} \left(\frac{1}{9} \right)^{2r} = 10^{-7} \times 0 \cdot 2148.$$

Hence, correct to seven decimal places, $\log_e 5 = 1 \cdot 609\ 437\ 9$.

23 The general solution of the homogeneous difference equation $\Delta^{k+1} u_r = 0$ is $u_r = a_0 + a_1 r + \ldots + a_k r^k$. To obtain a particular solution of the given difference equation, try $u_r = A x^{r-1}$, where A is a constant to be determined. Now $\Delta^{k+1} u_r = (E - 1)^{k+1} u_r$, where $E = 1 + \Delta$ is the shift operator,

$$= \left[E^{k+1} - \binom{k+1}{1} E^k + \ldots + (-1)^{k+1} \right] u_r$$

$$= u_{r+k+1} - \binom{k+1}{1} u_{r+k} + \ldots + (-1)^{k+1} u_r.$$

Hence, on substituting the trial solution, we have

$$A \left[x^{r+k} - \binom{k+1}{1} x^{r+k-1} + \ldots + (-1)^{k+1} x^{r-1} \right] = b x^{r-1},$$

which gives $A = b/(x-1)^{k+1}$. The required general solution of the given difference equation is the sum of the particular solution and the general solution of $\Delta^{k+1} u_r = 0$.

The fourth differences of the given nine terms of the series show that

$$\Delta^4 u_r = 12 \cdot 80 \times 5^{r-1}.$$

Hence $$u_r = \frac{12 \cdot 80 \times 5^{r-1}}{(5-1)^4} + a_0 + a_1 r + a_2 r^2 + a_3 r^3$$

$$= 0 \cdot 05 \times 5^{r-1} + c_0 + c_1 (r-1) + c_2 (r-1)^{(2)} + c_3 (r-1)^{(3)},$$

where the c's are some constants. Put $r = 1, 2, 3, 4$ successively, and by using the given values of u_r derive four linear equations for the c's as follows:

$0 \cdot 09 = 0 \cdot 05 + c_0 \cdot$ $\qquad$ $0 \cdot 75 = 0 \cdot 25 + c_0 + c_1$

$3 \cdot 09 = 1 \cdot 25 + c_0 + 2c_1 + 2c_2$ $\qquad$ $10 \cdot 73 = 6 \cdot 25 + c_0 + 3c_1 + 6c_2 + 6c_3.$

Hence $c_0 = 0 \cdot 04; \; c_1 = 0 \cdot 46; \; c_2 = 0 \cdot 44; \; c_3 = 0 \cdot 07.$

Therefore $u_r = 0 \cdot 05 \times 5^{r-1} + (0 \cdot 07 r^3 + 0 \cdot 02 r^2 - 0 \cdot 09 r + 0 \cdot 04)$

and $\displaystyle\sum_{r=1}^{n} u_r = \frac{1}{80} (5^n - 1) + \frac{n}{12} (0 \cdot 21 n^3 + 0 \cdot 50 n^2 - 0 \cdot 21 n - 0 \cdot 02).$

24   To obtain the closer approximation put $x = 4 + \epsilon$ in the equation. A first approximation then gives $194\epsilon - 67 = 0$ or $\epsilon = 0 \cdot 35.$
      Tabulate the cubic for $4 \cdot 30 \leqslant x(0 \cdot 01) \leqslant 4 \cdot 35$. The forward difference table formed from these function values is as follows:

x	$10x^3 - 26x^2 - 78x + 21$	Δ	Δ^2	Δ^3
4·30	$- 0 \cdot 070 \ 00$			
		2·541 31		
4·31	2·471 31		2066	
		2·561 97		6
4·32	5·033 28		2072	
		2·582 69		6
4·33	7·615 97		2078	
		2·603 47		6
4·34	10·219 44		2084	
		2·624 31		
4·35	12·843 75			

The root x_0 lies between $4 \cdot 30$ and $4 \cdot 31$. If we set $x_0 = a + \theta h$, where $a = 4 \cdot 30$ and $h = 0 \cdot 01$ is the interval of the argument, then inverse interpolation gives the following equation for θ.

$$2 \cdot 541 \ 31 \ \theta = 0 \cdot 070 \ 00 - 0 \cdot 020 \ 66 \ \frac{\theta^{(2)}}{2!} - 0 \cdot 000 \ 06 \ \frac{\theta^{(3)}}{3!}.$$

The stable value of $\theta = 0 \cdot 0277$, whence $x_0 = 4 \cdot 300 \ 277$. If x_1 and x_2 are the other two roots of the cubic, then

$$x_1 + x_2 = 2 \cdot 6 - 4 \cdot 300 \ 277 = -1 \cdot 700 \ 277$$

and $\qquad x_1 x_2 = -2 \cdot 1/4 \cdot 300 \ 277 = -0 \cdot 488 \ 341.$

Hence $(x_1 - x_2)^2 = 4 \cdot 844 \ 306$, so that $x_1 - x_2 = 2 \cdot 200 \ 978.$
Hence $x_1 = 0 \cdot 250 \ 350$ and $x_2 = -1 \cdot 950 \ 627.$

25   Tabulate $f(x)$ by the method of differences; see *Course* (Section 3.5, pp. 27–8). The accompanying table gives the differences of $f(x)$.

x	$f(x)$	Δ	Δ^2	Δ^3	Δ^4	Δ^5
1·8	105·245 344					
		− 63·663 087				
1·9	41·582 257		− 8·919 170			
		− 72·582 257		662 730		
2·0	− 31·000 000		− 8·256 440		234 720	
		− 80·838 697		897 450		14 760
2·1	− 111·838 697		− 7·358 990		249 480	
		− 88·197 687		1·146 930		14 760
2·2	− 200·036 384		− 6·212 060		264 240	
		− 94·409 747		1·411 170		14 760
2·3	− 294·446 131		− 4·800 890		279 000	
		− 99·210 637		1·690 170		14 760
2·4	− 393·656 768		− 3·110 720		293 760	
		−102·321 357		1·983 930		14 760
2·5	− 495·978 125		− 1·126 790		308 520	
		−103·448 147		2·292 450		14 760
2·6	− 599·426 272		1·165 660		323 280	
		−102·282 487		2·615 730		14 760
2·7	− 701·708 759		3·781 390		338 040	
		− 98·501 097		2·953 770		14 760
2·8	− 800·209 856		6·735 160		352 800	
		− 91·765 937		3·306 570		14 760
2·9	− 891·975 793		10·041 730		367 560	
		− 81·724 207		3·674 130		14 760
3·0	− 973·700 000		13·715 860		382 320	
		− 68·008 347		4·056 450		14 760
3·1	−1041·708 347		17·772 310		397 080	
		− 50·236 037		4·453 530		14 760
3·2	−1091·944 384		22·225 840		411 840	
		− 28·010 197		4·865 370		14 760
3·3	−1119·954 581		27·091 210		426 600	
		− 0·918 987		5·291 970		14 760
3·4	−1120·873 568		32·383 180		441 360	
		31·464 193		5·733 330		14 760
3·5	−1089·409 375		38·116 510		456 120	
		69·580 703		6·189 450		14 760
3·6	−1019·828 672		44·305 960		470 880	
		113·886 663		6·660 330		14 760
3·7	− 905·942 009		50·966 290		485 640	
		164·852 953		7·145 970		14 760
3·8	− 741·089 056		58·112 260		500 400	
		222·965 213		7·646 370		14 760
3·9	− 518·123 843		65·758 630		515 160	
		228·723 843		8·161 530		14 760
4·0	− 229·400 000		73·920 160		529 920	
		362·644 003		8·691 450		
4·1	133·244 003		82·611 610			
		445·255 613				
4·2	578·499 616					

Let x_1 and x_2 be the required roots such that $1 \cdot 9 < x_1 < 2 \cdot 0$ and $4 \cdot 0 < x_2 < 4 \cdot 1$.

To determine $x_1 = a + \theta h$, where $a = 1 \cdot 9$, $h = 0 \cdot 1$, use forward differences for inverse interpolation. The equation for θ is

$$-72 \cdot 582\ 257\ \theta = -41 \cdot 582\ 257 + 8 \cdot 256\ 440\ \frac{\theta^{(2)}}{2!} - 0 \cdot 897\ 450\ \frac{\theta^{(3)}}{3!} -$$

$$- 0 \cdot 249\ 480\ \frac{\theta^{(4)}}{4!} - 0 \cdot 014\ 760\ \frac{\theta^{(5)}}{5!}.$$

The stable value of $\theta = 0 \cdot 587$, whence $x_1 = 1 \cdot 9587$.

To determine $x_2 = a + \theta h$, where $a = 4 \cdot 1$, $h = -0 \cdot 1$, use backward differences for inverse interpolation. The equation for θ is in this case

$$-362 \cdot 644\ 003\ \theta = -133 \cdot 244\ 003 - 73 \cdot 920\ 160\ \frac{\theta^{(2)}}{2!} + 8 \cdot 161\ 530\ \frac{\theta^{(3)}}{3!} -$$

$$- 0 \cdot 515\ 160\ \frac{\theta^{(4)}}{4!} + 0 \cdot 014\ 760\ \frac{\theta^{(5)}}{5!}.$$

The stable value of $\theta = 0 \cdot 343$, whence $x_2 = 4 \cdot 0657$.

2 Statistical mathematics and probability

1 The following extract is from *Equality* by Richard Henry Tawney (1880–1962).

The conception of class is, therefore, at once more fundamental and more elusive than that of the division between different types of occupation. It is elusive because it is comprehensive. It relates, not to this or that specific characteristic of a group, but to a totality of conditions by which several sides of life are affected. The classification will vary, no doubt, with the purpose for which it is made, and with the points which accordingly are selected for emphasis. Conventional usage, which is concerned, not with the details of the social structure, but with its broad outlines and salient features, makes a rough division of individuals according to their resources and manner of life, the amount of their income and the source from which it is derived, their ownership of property or their connection with those who own it, the security or insecurity of their economic position, the degree to which they belong by tradition, education and association to social strata which are accustomed, even on a humble scale, to exercise direction, or, on the other hand, to those whose normal lot is to be directed by others. It draws its class lines, in short, with reference partly to consumption, partly to production; partly by standards of expenditure, partly by the position which different individuals occupy in the economic system. Though its criteria change from generation to generation, and are obviously changing today with surprising rapidity, its general tendency is clear. It sets at one end of the scale those who can spend much, or who have what is called an independent income, because they are dependent for it on persons other than themselves, and at the other end those who can spend little and live by manual labour. It places at a point between the two those who can spend more than the second but less than the first, and who own a little property or stand near to those who own it.

Thus conventional usage has ignored, in its rough way, the details, and has emphasized the hinges, the nodal points, the main watersheds. And in so doing, it has come nearer, with all its crudity, to grasping certain significant sides of the reality than have those who would see in the idea of class merely the social expression of

26

the division of labour between groups engaged in different types of economic activity. For, though differences of class and differences of occupation may often have sprung from a common source, they acquire, once established, a vitality and momentum of their own, and often flow in distinct, or even divergent, channels. The essence of the latter is difference of economic function: they are an organ of co-operation through the division of labour. The essence of the former is difference of status and power: they have normally been, in some measure at least, the expression of varying degrees of authority and subordination. Class systems, in fact, in the historical forms which they most commonly have assumed, have usually been associated — hence, indeed, the invidious suggestion which the word sometimes conveys — with differences, not merely of economic *métier*, but of social position, so that different groups have been distinguished from each other, not only, like different professions, by the nature of the service they render. Even today, indeed, though somewhat less regularly than in the past, class tends to determine occupation rather than occupation class.

Public opinion has in all ages been struck by this feature in social organization, and has used terms of varying degrees of appropriateness to distinguish the upper strata from the lower, describing them sometimes as the beautiful and good, sometimes as the fat men, sometimes as the twice-born, or the sons of gods and heroes, sometimes merely, in nations attached to virtue rather than beauty, as the best people. Such expressions are not terms of precision, but they indicate a phenomenon which has attracted attention, and which has certainly deserved it. The note of most societies has been, in short, not merely vertical differentiation, as between partners with varying tasks in a common enterprise, but also what, for want of a better term, may be called horizontal stratification, as between those who occupy a position of special advantage and those who do not.

Reproduced from *Equality* by R.H. Tawney (paperback edition, 1964, pp. 60–1) by kind permission of George Allen and Unwin Ltd, London

(i) Count the number of letters occurring in each word of the passage, assuming hyphenated words to be single words.

(ii) Make a frequency distribution of word length and evaluate, correct to two decimal places, the mean and standard deviation of the distribution.

(iii) Plot the frequency distribution as a histogram and as a frequency-polygon.

2 The following extract is from an essay entitled *Lay Morals* by Robert Louis Stevenson (1850–94).

Money gives us food, shelter, and privacy; it permits us to be clean in person, opens for us the doors of the theatre, gains us books for study or pleasure, enables us to help the distresses of others, and puts us above necessity so that we can choose the best in life. If we love, it enables us to meet and live with the loved one, or even to prolong her health and life; if we have scruples, it gives us an opportunity to be honest; if we have any bright designs, here is what will smooth the way to their accomplishment. Penury is the worst slavery, and will soon lead to death.

But money is only a means; it presupposes a man to use it. The rich can go where he pleases, but perhaps please himself nowhere. He can buy a library or visit the whole world, but perhaps has neither patience to read nor intelligence to see. The table may be loaded and the appetite wanting; the purse may be full and the heart empty. He may have gained the world and lost himself; and with all his wealth around him, in a great house and spacious and beautiful demesne, he may live as blank a life as any tattered ditcher. Without an appetite, without an aspiration, void of appreciation, bankrupt of desire and hope, there, in his great house, let him sit and look upon his fingers. It is perhaps a more fortunate destiny to have a taste for collecting shells than to be born a millionaire. Although neither is to be despised, it is always better policy to learn an interest than to make a thousand pounds; for the money will soon be spent, or perhaps you may feel no joy in spending it; but the interest remains imperishable and ever new. To become a botanist, a geologist, a social philosopher, an antiquary, or an artist, is to enlarge one's possessions in the universe by an incalculably higher degree, and by a far surer sort of property, than to purchase a farm of many acres. You had perhaps two thousand a year before the transaction; perhaps you have two thousand five hundred after it. That represents your gain in the one case. But in the other, you have thrown down a barrier which concealed significance and beauty. The blind man has learned to see. The prisoner has opened up a window in his cell and beholds enchanting prospects; he will never again be a prisoner as he was; he can watch clouds and changing seasons, ships on the river, travellers on the road, and the stars at night; happy prisoner ! his eyes have broken gaol! And again he who has learned to love an art or science has wisely laid up riches against the day of riches; if prosperity come, he will not enter poor into his inheritance; he will not slumber and forget himself in the lap of money, or spend his hours in counting idle treasures, but be up and briskly doing; he will have the true alchemic touch, which is not that of Midas, but which transmutes dead money into living delight and satisfaction. *Etre et pas avoir* — to be, not to possess — that is the problem of life. To be wealthy, a rich nature is the first requisite and money but the second. To be of a quick and healthy blood, to share in all honourable curiosities, to be rich in admiration and free from envy, to rejoice greatly in the good of others, to love with such generosity of heart that your love is still a dear possession in absence or unkindness — these are the gifts of fortune which money cannot buy and without which money can buy nothing. For what can a man possess, or what can he enjoy, except himself ? If he enlarges his nature, it is then he enlarges his estates. If his nature be happy and valiant, he will enjoy the universe as if it were his park and orchard.

(i) Excluding the personal name Midas, but including the foreign words, count the number of letters occurring in each of the other words of the passage. Hyphenated words are to be regarded as single words.

(ii) Make a frequency distribution of word length and evaluate, correct to two decimal places, the mean and standard deviation of the distribution.

(iii) Plot the distribution as a histogram and as a frequency-polygon.

3 The accompanying passage is an essay entitled *Of Studies* by Francis Bacon (1561–1626).

Studies serve for delight, for ornament, and for ability. Their chief use for delight is in privateness and retiring; for ornament is in discourse; and for ability, is in the judgement and disposition of business. For expert men can execute, and perhaps judge of particulars, one by one; but the general counsels, and the plots and marshalling of affairs, come best from those that are learned. To spend too much time in studies is sloth; to use them too much for ornament is affectation; to make judgement wholly by their rules is the humour of a scholar. They perfect nature, and are perfected by experience; for natural abilities are like natural plants, that need proyning by study; and studies themselves do give forth directions too much at large, except they be bounded in by experience. Crafty men contemn studies; simple men admire them; and wise men use them: for they teach not their own use; but that is a wisdom without them and above them, won by observation. Read not to contradict and confute; nor to believe and take for granted; nor to find talk and discourse; but to weigh and consider. Some books are to be tasted, others to be swallowed, and some few to be chewed and digested: that is, some books are to be read only in parts; others to be read, but not curiously; and some few to be read wholly, and with diligence and attention. Some books also may be read by deputy, and extracts made of them by others; but that would be only in the less important arguments, and the meaner sort of books; else distilled books are like common distilled waters, flashy things. Reading maketh a full man; conference a ready man; and writing an exact man. And, therefore, if a man write little, he had need have a great memory; if he confer little, he had need have a present wit; and if he read little, he had need have much cunning, to seem to know that he doth not. Histories make men wise; poets witty; the mathematics subtile; natural philosophy deep; moral grave; logic and rhetoric able to contend. *Abeunt studia in mores.* Nay, there is

no stond or impediment in the wit, but may be wrought out by fit
studies: like as diseases of the body may have appropriate exercises.
Bowling is good for the stone and reins; shooting for the lungs and
breast; gentle walking for the stomach; riding for the head; and the
like. So if a man's wit be wandering, let him study the mathematics;
for in demonstrations, if his wit be called away never so little, he must
begin again; if his wit be not apt to distinguish or find differences,
let him study the schoolmen; for they are *cymini sectores*: if he be
not apt to beat over matters, and to call one thing to prove and illustrate another, let him study the lawyers' cases: so every defect of
the mind may have a special receipt.

(i) Make a frequency distribution of the number of letters occurring
in each word of the essay, including the foreign words used.

(ii) Calculate, correct to two decimal places, the mean and standard
deviation of the distribution of word length.

(iii) Plot the distribution as a histogram and as a frequency-polygon.

4   The following passage on the definition of a gentleman is from *The
Idea of a University* by John Henry Newman (1801–90).

Hence it is that it is almost a definition of a gentleman to say
he is one who never inflicts pain. This description is both refined
and, as far as it goes, accurate. He is mainly occupied in merely
removing the obstacles which hinder the free and unembarrassed ac-
tion of those about him; and he concurs with their movements rather
than takes the initiative himself. His benefits may be considered as
parallel to what are called comforts or conveniences in arrangements
of a personal nature: like an easy chair or a good fire, which do their
part in dispelling cold and fatigue, though nature provides both means
of rest and animal heat without them. The true gentleman in like
manner carefully avoids whatever may cause a jar or a jolt in the
minds of those with whom he is cast; – all clashing of opinion, or
collision of feeling, all restraint, or suspicion, or gloom, or resent-
ment; his great concern being to make every one at their ease and at
home. He has his eyes on all his company; he is tender towards the
bashful, gentle towards the distant, and merciful towards the absurd;
he can recollect to whom he is speaking; he guards against unseason-
able allusions, or topics which may irritate; he is seldom prominent
in conversation, and never wearisome. He makes light of favours while
he does them, and seems to be receiving when he is conferring. He
never speaks of himself except when compelled, never defends himself by a mere retort, he has no ears for slander or gossip, is scrup-
ulous in imputing motives to those who interfere with him, and inter-
prets every thing for the best. He is never mean or little in his dis-
putes, never takes unfair advantage, never mistakes personalities or
sharp sayings for arguments, or insinuates evil which he dare not say

out. From a long-sighted prudence, he observes the maxim of the ancient sage, that we should ever conduct ourselves towards our enemy as if he were one day to be our friend. He has too much good sense to be affronted at insults, he is too well employed to remember injuries, and too indolent to bear malice. He is patient, forbearing, and resigned, on philosophical principles; he submits to pain, because it is inevitable, to bereavement, because it is irreparable, and to death, because it is his destiny. If he engages in controversy of any kind, his disciplined intellect preserves him from the blundering discourtesy of better, perhaps, but less educated minds, who, like blunt weapons, tear and hack instead of cutting clean, who mistake the point in argument, waste their strength on trifles, misconceive their adversary, and leave the question more involved than they find it. He may be right or wrong in his opinion, but he is too clear-headed to be unjust; he is as simple as he is forcible, and as brief as he is decisive. Nowhere shall we find greater candour, consideration, indulgence: he throws himself into the minds of his opponents, he accounts for their mistakes. He knows the weakness of human reason as well as its strength, its province and its limits. If he be an unbeliever, he will be too profound and large-minded to ridicule religion or to act against it; he is too wise to be a dogmatist or fanatic in his infidelity. He respects piety and devotion; he even supports institutions as venerable, beautiful, or useful, to which he does not assent; he honours the ministers of religion, and it contents him to decline its mysteries without assailing or denouncing them. He is a friend of religious toleration, and that, not only because his philosophy has taught him to look on all forms of faith with an impartial eye, but also from the gentleness and effeminacy of feeling, which is the attendant on civilization.

(i) Count the number of letters occurring in each word of the passage, hyphenated words being treated as single words.

(ii) Make a frequency distribution of word length and calculate, correct to two decimal places, the mean and standard-deviation of the distribution.

(iii) Plot the distribution as a histogram and as a frequency-polygon.

5　The accompanying table gives the last digit in each of 930 consecutive telephone numbers obtained from a directory.

(i) Form a frequency distribution of the integers 0, 1, 2, ... , 9.

(ii) Evaluate, correct to two decimal places, the mean and standard deviation of the distribution.

(iii) Determine, correct to four decimal places, the observed relative frequency of the odd integers 1, 3, 5, 7, and 9.

(iv) Plot the distribution as a histogram and as a frequency-polygon.

Table for Exercise 5

4	7	9	1	8	2	3	4	7	7	4	4	2	2	4	9	5	6	5	9	1	5	8	2	4	6	8	4	9	2
4	8	5	2	4	1	6	2	6	0	3	4	4	6	3	5	9	3	0	1	4	7	7	8	5	3	8	6	1	6
6	1	3	8	8	6	2	0	8	1	0	1	1	0	0	2	4	8	0	2	8	4	6	3	9	1	0	5	0	2
3	0	1	8	0	3	2	9	0	8	2	2	3	0	3	2	5	6	5	5	5	9	8	2	0	4	6	6	9	2
0	0	9	3	8	9	4	2	8	3	9	4	7	4	6	2	0	5	7	1	1	3	8	2	7	1	1	6	6	6
2	4	2	7	7	5	7	8	2	3	3	5	2	1	1	5	2	3	2	7	7	2	9	8	5	0	7	7	7	4
0	7	0	3	2	9	1	6	0	9	6	2	9	0	4	8	4	3	6	6	7	6	1	2	3	6	5	9	1	8
7	7	0	8	4	0	9	6	4	7	2	7	4	8	2	6	7	7	6	3	3	1	9	0	1	4	8	5	6	2
7	5	1	2	4	7	6	3	5	5	4	2	4	6	3	4	1	2	3	6	2	7	0	4	7	3	2	8	8	4
9	6	2	0	1	5	4	4	0	4	3	1	7	4	7	5	4	2	7	6	8	7	3	3	3	8	8	0	3	9
0	0	9	4	0	4	9	3	2	2	2	0	4	8	6	0	9	8	8	4	7	1	1	6	1	5	8	1	2	4
9	4	4	9	9	3	9	5	3	6	2	1	1	2	8	5	4	6	6	7	3	8	8	1	5	2	9	2	7	3
3	2	7	4	8	2	4	8	7	1	0	9	7	6	7	1	9	8	2	9	4	7	1	2	8	1	6	8	5	9
2	8	8	9	0	3	2	0	4	0	8	5	8	8	4	7	7	8	2	6	6	8	8	3	3	8	7	1	7	0
5	1	1	5	7	8	2	1	8	2	7	7	8	1	5	5	7	7	3	1	6	4	4	1	3	0	4	1	0	2
4	5	2	2	8	5	0	1	2	4	6	5	2	2	7	2	0	0	6	0	8	3	7	5	5	6	6	5	2	1
8	7	4	3	9	1	1	6	0	9	2	5	1	7	7	5	5	5	9	2	4	9	4	5	5	7	6	9	4	3
6	1	4	8	0	5	3	3	1	5	0	4	5	5	8	1	9	9	7	5	9	9	8	3	4	3	9	1	6	5
9	9	5	5	6	0	5	9	7	3	4	0	4	7	7	9	9	5	3	7	2	8	9	3	8	3	6	8	5	3
4	6	7	3	9	7	2	2	4	2	9	5	4	4	0	8	3	7	5	4	9	6	2	6	5	3	6	4	1	7
5	2	4	5	3	4	3	6	1	5	6	5	9	3	2	0	6	8	1	3	4	3	5	4	8	6	7	9	0	1
6	3	0	5	2	9	4	4	9	3	2	4	8	5	7	1	5	7	9	2	2	4	2	3	1	4	9	5	6	8
4	1	9	4	0	8	0	9	2	3	1	3	3	3	6	1	0	3	4	4	5	7	5	6	9	1	1	4	6	0
9	3	5	8	9	4	1	0	7	0	7	3	9	3	6	4	1	6	1	0	0	2	1	3	8	0	1	3	4	6
8	1	4	7	7	1	0	0	1	2	8	2	1	4	5	9	3	7	8	1	6	7	8	2	9	7	4	3	0	8
3	4	0	7	1	4	1	1	3	0	7	0	7	3	6	4	9	4	5	0	0	1	9	0	7	4	8	8	0	1
0	5	4	5	6	3	4	8	4	6	9	8	9	4	7	4	7	5	4	5	8	3	6	0	3	5	4	0	6	4
9	0	9	6	5	5	1	2	0	7	9	7	9	5	6	7	4	7	6	7	8	9	9	6	9	3	0	4	8	1
7	2	7	3	4	7	9	5	2	0	0	7	1	9	9	6	6	0	6	7	0	3	0	7	4	5	4	3	7	2
3	4	8	9	6	7	9	9	3	9	9	9	8	5	4	4	9	2	7	8	0	2	2	5	8	9	7	1	2	7
0	9	2	8	0	0	6	0	7	4	4	1	8	8	7	5	1	7	3	6	6	6	9	2	6	3	1	5	3	5

6   Each of the integers from 0 to 36 occurs once on a standard roulette wheel, and when the wheel is spun any one of these numbers is equally likely to turn up. The accompanying table gives the outcomes of 720 such independent trials with the wheel.

(i) Make a frequency distribution of the 37 numbers and then calculate, correct to three decimal places, the mean and standard deviation of the distribution.

(ii) Plot the distribution as a histogram and as a frequency-polygon.

Table for Exercise 6

12	1	28	36	22	0	24	6	11	7	14	20	1	7	22	3	3	11	34	34	31	29	30	34
33	4	14	30	11	6	22	28	30	25	29	14	17	21	34	0	22	15	30	11	1	9	26	25
7	24	1	11	8	15	13	6	13	18	24	1	29	20	10	15	23	22	19	0	2	22	24	13
26	31	5	2	5	0	17	2	33	28	3	5	8	2	6	34	15	23	18	21	32	34	23	12
7	11	28	35	4	10	1	28	11	9	1	25	13	36	20	5	0	29	2	25	34	33	19	33
3	10	31	5	19	31	8	17	6	17	18	36	20	34	26	19	15	4	8	5	5	9		
0	30	10	4	17	15	21	24	12	8	7	5	2	21	34	1	7	34	23	34	33	21	5	36
15	35	1	6	12	1	26	15	2	20	23	0	0	27	6	9	25	6	8	33	11	27	1	6
7	24	11	2	34	27	4	18	5	11	10	26	24	23	0	35	16	31	36	4	4	7	9	22
20	35	8	8	24	20	25	15	7	24	9	27	25	1	25	3	36	14	25	31	34	30	30	4
2	33	8	13	10	23	16	14	6	19	34	7	0	33	29	13	11	32	15	5	10	13	22	8
21	19	24	13	16	20	16	35	7	24	12	35	33	7	7	5	16	25	11	23	1	8	7	21
7	34	7	18	2	12	20	29	9	24	15	24	29	22	7	15	29	6	11	30	30	31	18	32
6	14	22	32	1	4	32	25	24	17	1	15	21	26	7	17	24	3	4	1	36	13	5	9
31	4	16	3	19	33	18	34	36	6	12	31	9	16	13	11	34	23	6	27	29	31	32	22
27	18	31	23	23	5	6	23	16	6	7	3	12	18	9	23	35	22	16	15	20	20	17	19
31	25	19	28	1	5	18	8	2	4	36	19	28	18	36	0	20	21	30	16	35	21	10	18
27	6	33	30	22	19	13	19	0	36	19	4	17	30	25	34	26	25	20	30	17	33	10	19
11	33	23	18	20	19	35	18	19	4	8	25	10	0	33	16	18	0	20	13	26	0	10	2
30	13	19	1	5	28	0	0	31	10	15	33	12	13	9	5	1	9	7	36	11	25	12	28
15	30	31	31	17	11	26	13	18	7	13	18	24	10	22	21	0	7	15	14	14	36	34	34
9	2	14	19	14	7	3	2	4	34	0	7	29	26	32	9	23	11	25	1	12	31	32	13
4	3	23	28	28	18	3	32	8	27	0	16	20	7	28	10	4	11	31	16	22	25	36	27
0	27	9	10	7	33	27	36	36	26	31	28	17	14	2	2	9	33	22	18	4	5	28	13
12	1	13	2	11	33	34	26	8	32	28	8	8	31	11	9	10	21	3	19	29	8	9	29
12	8	23	10	5	29	18	20	21	28	16	21	7	24	31	23	19	24	2	8	0	26	19	33
36	35	26	24	21	9	6	15	21	16	34	19	12	16	26	9	2	22	29	1	5	13	35	1
11	36	5	22	21	8	18	13	30	25	26	13	0	23	31	22	0	19	25	31	18	17	4	27
31	1	18	16	14	1	23	25	26	6	9	22	19	25	20	7	9	1	10	5	21	20	27	9
6	7	21	36	23	1	20	30	0	9	34	14	35	8	3	4	4	26	23	25	26	33	18	0

Table for Exercise 7

Observations

Row no.																														
1	S	F	S	S	F	S	F	S	S	S	F	S	S	S	F	S	S	S	F	F	S	S	F	S	S	S	S	F	S	S
2	S	S	S	F	F	F	S	F	S	F	S	S	S	S	F	F	S	S	F	F	S	S	S	S	S	S	S	S	S	F
3	S	S	F	F	S	S	S	S	S	S	S	S	F	S	S	S	S	F	S	S	S	F	S	S	S	S	S	F	S	F
4	S	S	F	S	S	F	S	S	S	S	S	S	S	S	S	S	S	S	S	S	F	S	F	S	F	S	S	F	S	F
5	S	S	S	S	S	S	S	S	F	F	S	S	S	S	S	S	S	S	S	S	S	S	S	S	S	S	S	S	S	S
6	S	S	S	S	F	S	F	S	F	S	S	S	F	F	F	F	S	S	S	S	S	S	F	S	S	S	F	S	S	S
7	S	S	F	S	F	S	F	F	S	S	S	S	F	S	F	S	F	S	S	S	S	S	S	S	S	F	S	S	S	S
8	F	F	S	F	S	F	S	F	S	S	S	F	S	S	S	S	S	S	S	S	F	S	F	S	S	S	S	F	S	S
9	S	F	S	F	S	F	S	S	S	S	S	S	S	S	S	S	S	S	S	S	F	S	S	S	S	S	F	S	S	S
10	S	S	S	S	S	S	S	S	S	S	S	S	S	S	S	S	S	S	F	S	S	S	S	F	F	S	S	S	S	S
11	S	F	S	S	S	S	S	F	S	S	S	S	F	S	S	S	S	S	S	S	S	S	F	S	S	F	S	S	S	S
12	S	S	F	S	S	S	S	S	S	S	S	S	S	S	S	S	S	F	S	S	S	F	S	S	F	S	S	S	F	S
13	S	S	S	F	S	S	F	S	S	F	F	S	S	S	S	S	S	S	S	S	F	S	S	F	S	S	S	F	S	S
14	S	F	S	S	S	S	S	F	F	S	S	S	S	S	S	S	F	S	S	S	S	F	S	S	S	S	S	S	S	S
15	S	S	S	F	S	S	S	S	S	S	S	S	S	S	S	F	S	S	S	S	F	S	S	S	F	S	S	S	S	S
16	S	F	S	S	F	S	F	S	F	S	S	S	S	S	S	F	S	S	F	S	S	F	S	S	S	S	S	S	S	S
17	S	S	F	F	S	F	S	S	S	F	F	S	S	S	S	S	S	S	S	S	S	S	F	S	S	F	S	S	S	S
18	S	F	S	F	S	S	S	S	S	S	S	S	S	S	S	S	F	S	S	S	S	F	S	F	S	S	S	S	S	S
19	S	S	S	F	S	S	S	S	S	S	S	S	S	S	S	F	S	S	F	S	S	S	F	S	S	S	F	S	S	S
20	S	S	F	S	F	F	S	S	S	F	S	S	S	F	S	S	S	S	S	S	F	F	S	S	F	S	S	F	S	S
21	S	S	S	S	S	S	S	S	F	F	S	S	S	S	S	S	F	S	S	S	F	S	S	S	S	S	S	S	S	S
22	S	S	S	S	S	S	F	F	F	S	S	S	S	S	S	S	S	S	F	S	S	S	F	S	S	F	S	S	S	S
23	S	F	S	S	F	S	S	F	F	S	S	S	S	S	S	S	S	S	F	S	S	S	F	S	S	S	S	F	S	S
24	S	F	S	S	S	F	S	S	S	S	S	F	S	S	S	S	S	S	S	F	S	S	S	S	S	F	S	S	S	S
25	S	F	S	S	S	F	S	F	S	S	S	S	F	S	S	S	S	S	S	S	S	F	S	S	S	S	S	F	S	S

Table for Exercise 7 (concluded)

Observations

Row no.																									
26	S	S	S	S	S	S	S	S	F	F	S	S	S	S	S	S	S	F	S	S	F	S	S	S	S	F
27	S	S	S	S	S	S	S	S	S	F	F	F	S	S	S	S	S	S	S	S	S	S	F	S	S
28	S	S	S	S	S	S	F	S	S	S	S	F	S	S	S	S	S	F	S	S	S	F	F	S	S
29	S	S	S	S	F	S	S	S	S	F	S	S	S	S	F	S	F	S	S	S	F	S	S	F	S
30	S	S	S	S	S	F	S	F	S	S	S	S	F	S	F	S	S	S	S	F	S	S	S	F	S
31	F	S	F	S	S	S	F	S	S	S	S	S	F	S	S	S	S	S	S	S	S	F	F	F	S
32	S	S	S	S	F	S	F	S	S	S	F	S	S	S	F	S	F	S	S	F	S	S	F	S	F
33	S	F	S	S	S	F	S	S	S	S	F	S	S	S	F	S	F	S	S	F	S	F	F	S	S
34	F	F	S	S	F	F	S	S	S	S	S	F	S	S	F	S	S	F	S	S	S	S	F	F	S
35	S	F	S	S	S	S	S	F	S	S	S	S	F	S	S	S	F	S	S	F	S	S	F	S	S
36	F	F	S	F	S	S	F	S	S	S	S	F	S	S	F	S	S	S	S	S	S	S	F	S	S
37	S	F	S	S	F	S	S	F	S	S	S	F	S	S	S	S	F	S	S	F	S	S	S	S	S
38	S	F	S	S	S	S	S	S	F	S	S	S	S	S	F	S	S	S	F	S	S	S	S	S	S
39	S	S	S	S	S	S	F	S	S	S	S	S	S	S	S	F	F	S	S	S	S	S	S	S	F
40	S	S	S	F	F	S	S	F	S	S	S	S	S	S	S	S	F	S	S	S	S	S	S	F	S
41	S	S	S	S	S	S	S	S	S	S	S	S	S	S	S	F	F	S	S	S	S	S	F	F	S
42	S	F	S	S	S	S	F	F	S	S	S	S	S	F	S	F	S	F	S	S	S	S	F	F	S
43	S	S	S	S	S	F	S	S	S	F	S	S	S	S	S	S	F	S	S	S	S	S	S	F	F
44	S	S	F	S	S	S	S	F	S	S	S	S	F	S	S	S	F	S	S	S	S	S	S	F	S
45	S	F	S	S	F	S	S	S	S	S	F	S	S	S	S	S	F	S	S	S	S	F	S	S	S
46	F	S	S	S	S	S	F	S	S	F	S	S	S	S	S	F	S	S	S	S	F	S	S	S	S
47	S	F	S	S	S	S	F	S	F	S	S	S	F	S	S	S	F	S	S	S	S	F	S	S	S
48	S	F	S	S	S	S	S	S	S	S	S	S	S	S	S	S	S	S	F	S	S	F	S	S	S
49	S	S	S	F	S	S	S	S	S	S	S	S	S	S	S	S	S	F	F	S	S	S	F	S	S
50	F	S	S	S	F	S	S	F	S	S	S	S	S	S	S	S	F	S	S	S	S	S	S	S	S

7   In industrial quality control of a product, 2000 items were tested consecutively, a non-defective being denoted as a success (S) and a defective as a failure (F). The results obtained are given in the table on pages 34–35, the sequence being considered row-wise, from left to right.

A run of r ($\geqslant 1$) successes is said to have occurred whenever (a) r consecutive successes either precede or follow at least one failure; or (b) r consecutive successes are preceded and followed by at least one failure.

(i) Determine the lengths of the runs of successes observed in the complete series, and then make a frequency distribution of the run lengths observed.

(ii) Calculate, correct to two decimal places, the mean and standard deviation of the distribution.

(iii) Plot the distribution as a histogram and as a frequency-polygon.

8   The table on page 37 gives the yield (in grams) of 400 square-yard plots of barley.

(i) Make a frequency distribution and histogram for the barley yields using a class-interval of 21, and let the class-marks of the first class-interval be 14-34.

(ii) Make a frequency distribution and histogram by doubling the length of the class-interval.

(iii) Calculate, correct to four decimal places, the mean and standard deviations of the frequency distributions (i) and (ii), and compare these values with those obtained from the individual 400 observations.

(iv) Comment on the results obtained.

9   The table on page 38 gives the areas in arbitrary units of 500 bull sperms. The data are due to A. Savage, Department of Animal Pathology, University of Manitoba, Canada.

(i) Make a frequency distribution and histogram of the areas, using a class-interval of 3, the first class-interval having the class-marks 123-125.

(ii) Make a frequency distribution and histogram by doubling the length of the class-interval.

(iii) Calculate, correct to four decimal places, the mean and standard deviation of the frequency distributions (i) and (ii), and compare these values with those obtained from the individual 500 observations.

(iv) Comment on the results obtained.

Table for Exercise 8

185	162	136	157	141	130	129	176	171	190	157	147	176	126	175	134	169	189	180	128
169	205	129	117	144	125	165	170	153	186	164	123	165	203	156	182	164	176	176	150
216	154	184	203	166	155	215	190	164	204	194	148	162	146	174	185	171	181	158	147
165	157	180	165	127	186	133	170	134	177	109	169	128	152	165	139	146	144	178	188
133	128	161	160	167	156	125	162	128	103	116	87	123	143	130	119	141	174	157	168
195	180	158	139	139	168	145	166	118	171	143	132	126	171	176	115	165	147	186	157
187	174	172	191	155	169	139	144	130	146	159	164	160	122	175	156	119	135	116	134
157	182	209	136	153	160	142	179	125	149	171	186	196	175	189	214	169	166	164	195
189	108	118	149	178	171	151	192	127	148	158	174	191	134	188	248	164	206	185	192
147	178	189	141	173	187	167	128	139	152	167	131	203	231	214	177	161	194	141	161
124	130	112	122	192	155	196	179	166	156	131	179	201	122	207	189	164	131	211	172
170	140	156	199	181	181	150	184	154	200	187	169	155	107	143	145	190	176	162	123
189	194	146	22	160	107	70	84	112	162	124	156	138	101	138	141	143	135	163	183
99	118	150	151	83	136	171	191	155	164	98	136	115	168	130	111	136	129	122	120
179	172	192	171	151	142	193	174	146	180	140	137	138	194	109	120	124	126	126	147
115	148	195	154	149	139	163	118	126	127	139	174	167	175	179	172	174	167	142	169
122	163	144	147	123	160	137	161	122	101	158	103	119	164	112	57	94	106	132	122
164	142	155	147	115	143	68	184	183	167	160	138	191	133	160	156	122	111	153	148
103	131	180	142	191	175	146	181	111	110	154	176	168	175	175	146	148	167	106	123
121	154	148	91	93	74	113	79	131	119	96	80	97	98	106	107	69	86	94	129

Reproduced from *Methods of Statistical Analysis* by C.H. Goulden, (1939 edition, p. 18) by kind permission of the author and John Wiley and Sons, Inc., New York.

Table for Exercise 9

146	138	145	138	146	138	146	138	146	139	149	140	149	140	149	140	149	138	149	140
129	145	135	145	136	145	134	139	136	139	137	138	137	140	136	140	136	132	137	133
152	160	152	159	150	159	151	160	152	155	152	149	152	149	152	147	139	147	139	147
145	146	145	144	145	145	146	145	146	141	144	141	145	142	137	142	136	143	136	139
158	149	157	149	157	148	157	148	158	148	158	149	135	147	155	147	153	148	153	140
151	149	152	149	150	149	150	149	151	141	151	155	150	153	150	153	150	153	150	153
141	161	141	161	142	159	143	148	141	147	130	147	129	148	134	149	133	149	133	149
143	142	143	143	141	143	136	143	125	143	128	141	137	141	127	157	132	157	134	158
144	136	144	137	144	155	149	161	162	159	162	138	165	139	169	141	147	141	147	141
146	146	146	144	146	145	144	145	145	146	146	144	144	144	145	145	145	144	146	144
126	153	127	153	135	162	135	162	132	148	134	148	127	148	137	149	137	149	135	138
151	143	150	141	149	146	149	145	149	145	147	145	147	146	148	144	148	144	151	144
142	134	143	123	143	125	143	137	143	136	144	132	146	134	144	124	144	124	145	134
133	139	135	146	137	140	136	140	137	138	137	138	137	138	137	139	156	146	157	146
151	149	152	149	152	151	152	152	152	150	152	150	152	150	152	152	150	139	150	139
140	149	140	149	140	149	138	155	142	153	142	154	142	154	143	149	141	149	141	149
139	154	140	154	146	155	146	154	144	154	144	154	144	135	144	159	143	160	143	161
153	141	138	142	139	142	140	142	138	142	145	141	145	141	145	141	146	141	144	142
145	137	146	137	145	137	145	135	145	135	146	136	146	137	146	135	146	136	146	136
156	140	157	138	157	138	157	140	157	140	156	140	156	145	157	146	135	146	134	146
152	155	150	153	150	153	150	153	150	153	150	153	150	153	151	153	151	154	151	155
143	150	141	152	142	150	142	150	142	147	141	147	150	142	142	142	141	142	142	143
138	134	139	134	139	129	139	129	149	130	151	131	150	129	133	130	133	131	135	134
158	133	156	133	147	132	147	134	148	137	153	127	153	139	140	139	140	139	138	140
139	147	139	148	139	149	143	147	141	147	142	147	141	147	141	147	152	148	158	148

Reproduced from *Methods of Statistical Analysis* by C.H. Goulden (1939 edition, p. 19) by kind permission of the author and John Wiley and Sons, Inc., New York.

Table for Exercise 10

H T T H H	T H T H T	H H T T T	T H H T H	H T T T H
T T H T H	T H H H T	T T T T T	T H T H T	T H H H H
T T H H T	T T H T H	T T H H T	H T H H H	T T T H H
H T T T H	H H T H H	T T H T H	H H H T H	T T H H T
T H H H H	T H T T H	H T T H H	H H T T T	T T T H T
H T T T T	T H T T T	H T H H H	T H T H T	T H H H T
T T T T H	T H T H H	H T T H T	T H T H T	T H H T T
T H T H H	H T H H T	T T T T T	T H H H T	H T H T H
T H H T H	T H H T T	T H T H H	H H T T T	H T T T T
T H H H T	T H T T H	T H H T T	H H T H T	H H H T H
T H H H T	T H H H T	T H H H H	T H T H T	H T H T H
H T H H H	T H T T T	T H H H H	H H H H T	T T H H T
T H T T H	H H T H T	T H H H T	T H H T T	T H H T H
T T H H H	H T T H T	T T H H H	H H H T H	H H H H H
T H T T T	T T T T H	H T T T T	T T H H H	T T T H H
H T T T T	T H T H T	T T T H H	H H T H H	H T H H H
T T T H T	H H H H H	H H H H H	T T T T H	T H T T H
T T H T T	T H H T T	H H T T T	H T T T T	H T T H T
T T H T H	T T T T H	T H T T H	T T H H H	H T H T T
T H H H H	T H T T T	T H H H T	T T T T T	T H T H H
T T T H T	H H T H H	T T H H T	T H T T T	T H T T T
T H H H H	T T T T H	H T T T H	H T T H T	T T T T H
T T T T T	H H H H H	T T T H H	H T H H H	H H T H H
T H H T T	T T H H T	T H H T T	H H T H T	H H T H T
T T H H T	H H T T T	T H T T T	T T H T H	T H H T T
H T H T T	T T H H H	T H T H T	H T H H H	T H H T T
T H H T H	T T T T T	T T T T H	T H T H H	T T T H H
H T T H H	T T H T T	H H T H H	H H H H H	T T T H T
T H H T H	H H H T H	H T H T T	H T H H H	T H H T H
H T H H T	H H T T T	H H H H T	H H H H T	H T T T T
T H T T H	H H H T H	H T H H T	H H T T T	H T T H T
T T H T T	T T T T H	T T H H T	T T H H T	T T T T T
H H H T T	H T T H H	T H T T T	T H T T H	T H H T H

Table for Exercise 10 (concluded)

T T H H H	T T T T T	T H H T H	T H T H H	H T H T H
T T T T H	H H T H H	H H T H H	H H H T H	H T T T H
H T H T H	T T H H T	H H H H T	T H H T H	H H T T H
H H T H T	T H H T T	T T T H T	T T H H T	T T H H H
T H T T H	H T T T T	H T H H T	T T T T T	H H T H H
T T T T H	T T H T T	T H T T H	H H H H T	H T H T H
H H H T H	T H H T T	H T H T H	H T H T T	T T T H H
T T T T T	H T H H T	T T H H T	H T H T H	T T T H H
T H T T T	H T H H T	T H T T H	T T H H T	T H H T T
H H H T H	H T H T T	H H T T T	H H H T T	T T T H T
H T H T T	T H H T H	H H H H T	H T H T H	T T T H T
T H T H T	H H H H H	H T T T T	H H H H H	T H H H H
H H T T T	H T H T H	H H T T T	T T H T T	H T H H T
T H H T T	H T H T T	T T T T T	H T H T H	T T H T T
H H T H H	H T T T T	H T T H H	T T H H T	T H T H T
H H H H T	T T H H H	T H T H H	T H T T T	H H T T H
T H H H H	H T H T H	H T H H H	H H T T H	T H H T H
T H H T H	H T H T T	H T H T H	T H H H T	T T T H T
T H H H T	T H T H T	T H H T H	T T T T T	T T T H T
T H T T T	H T T T T	T H H T T	H T H T H	T H H T T
H T H H H	T T H T T	H T T T T	H H T H H	T T T T T
T H H T H	T T T H H	T H H T T	H H T H T	H H T H H
H T H H T	H T H H H	H H T H H	T H H H H	H H H T H
H T T T T	H H H T T	H T H H H	T H H T T	T T H H T
H H T H H	T H H H T	T T H H H	H T T H H	H H T H H
H T T T H	H H H T H	H T H T T	T H H T H	H T T H T
H H T H T	T H T T T	H T T H T	H H T H T	T H H T T
T T H H H	H H T T T	H T H T T	H T H T T	H T H T H
H T H H T	T H T T H	H T T H H	H H H T T	H H H T H
H T T T T	H T H H H	T H H H T	H T H T T	H H H T T
H T T T T	T T H T T	T T T H H	H T T T H	H H T H H
T H T H T	H T H T T	H T H H H	H H T H H	H H H H T
T H T T T	H T H T H	H H T T H	T H H H T	T T H T T

10   In an experiment a penny is tossed five times and the outcomes —
heads (H) or tails (T) — of the trials are observed. The table on pages
39–40 gives the results of 330 independent repetitions of the experiment.

(i) Count the number of heads obtained in each of the experiments,
and then make a frequency distribution of the number of heads.

(ii) Calculate, correct to four decimal places, the mean and standard
deviation of the observed frequency distribution, and compare these
values with the corresponding ones obtained on the hypothesis that the
penny used in the experiment is unbiased.

(iii) Compare the observed frequency distribution with the distribution
expected on the hypothesis of unbiasedness of the penny.

(iv) Comment on the results obtained.

11   The following table gives the frequency distribution of the weights
(in pounds) of adult males born in England, Scotland, Wales and Ireland.
The weights were taken correct to the nearest pound and so the class-
limits of the class-intervals are 89·5–99·5, 99·5–109·5, ... , 279·5–289·5.

Weight (class-mark)	Frequency	Weight (class-mark)	Frequency
90 –	2	190 –	263
100 –	34	200 –	107
110 –	152	210 –	85
120 –	390	220 –	41
130 –	867	230 –	16
140 –	1623	240 –	11
150 –	1559	250 –	8
160 –	1326	260 –	1
170 –	787	270 –	0
180 –	476	280 –	1
Total			7749

Source   *Final Report of the Anthropometric Committee to the British Association*
(Report, 1883).

(i) Calculate, correct to four decimal places, the mean and standard
deviation of the frequency distribution.

(ii) Calculate, correct to four significant figures, g_1 and g_2, the
Fisher coefficients of skewness and kurtosis of the frequency distribution.

(iii) Calculate, correct to four decimal places, the three quartiles
and the semi-interquartile range of the frequency distribution.

12   The following table gives the frequency distribution of heights in
inches (without shoes) of adult males born in England, Scotland, Wales
and Ireland. The heights were measured correct to one-sixteenth of an
inch and so the class-limits of the class-intervals used in tabulation are
56·9375–57·9375, 57·9375–58·9375, ... , 76·9375–77·9375.

Height (class-mark)	Frequency	Height (class-mark)	Frequency
57 –	2	68 –	1230
58 –	4	69 –	1063
59 –	14	70 –	646
60 –	41	71 –	392
61 –	83	72 –	202
62 –	169	73 –	79
63 –	394	74 –	32
64 –	669	75 –	16
65 –	990	76 –	5
66 –	1223	77 –	2
67 –	1329		
Total			8585

Source   *Final Report of the Anthropometric Committee to the British Association* (Report, 1883).

(i) Calculate, correct to four decimal places, the mean and standard deviation of the frequency distribution.

(ii) Calculate, correct to four significant figures, g_1 and g_2, the Fisher coefficients of skewness and kurtosis of the frequency distribution.

(iii) Calculate, correct to four decimal places, the three quartiles, the semi-interquartile range and the ninth decile of the frequency distribution.

13   The following table gives the frequency distributions of the lengths and breadths (in millimetres) of 243 cuckoo's (*Cuculus canorus*) eggs from the collections in the Charterhouse Museum and the British Museum of Natural History, London. Of the eggs whose measurements are included in the distributions, 223 were known to have been deposited in the nests of 42 different species of birds, while the foster-parents of the remaining 20 were not ascertainable.

For each of the two frequency distributions calculate, correct to four decimal places,

(i) the mean and standard deviation;

(ii) g_1 and g_2, the Fisher coefficients of skewness and kurtosis; and

(iii) the three quartiles, the semi-interquartile range, and the five and ninety-five per cent points.

Length of egg (class-limit)	Frequency	Breadth of egg (class-limit)	Frequency
18·75 –	1	13·75 –	1
19·25 –	1	14·25 –	1
19·75 –	7	14·75 –	5
20·25 –	3	15·25 –	9
20·75 –	29	15·75 –	73
21·25 –	13	16·25 –	51
21·75 –	54	16·75 · ·	80
22·25 –	38	17·25 –	15
22·75 –	47	17·75 –	7
23·25 –	22	18·25 –	0
23·75 –	21	18·75 –	1
24·25 –	5		
24·75 –	2		
Total	243	Total	243

Source   O.H. Latter (1901), *Biometrika*, Vol. 1, p. 164.

14   The following table gives the frequency distribution of the estimated intensities of cloudiness at Greenwich during the years 1890–1904 (excluding 1901) for the month of July. The class-marks used in the table are the midpoints of the class-intervals.

Degrees of cloudiness (midpoint of interval)	Frequency	Degrees of cloudiness (midpoint of interval)	Frequency
10	676	4	45
9	148	3	68
8	90	2	74
7	65	1	129
6	55	0	320
5	45		
Total			1715

Source   G.E. Pearse (1928), *Biometrika*, Vol. 20A, p. 336.

Calculate, correct to four decimal places, the following statistical measures of the frequency distribution:

(i) the mean and standard deviation; and

(ii) g_1 and g_2, the Fisher coefficients of skewness and kurtosis.

15   The following table gives the frequency distribution of sentences according to their length in number of words in two passages selected from *Biographia Literaria* (1817) by Samuel Taylor Coleridge (1772–1834). The variable (length of sentence) is discrete and so, for example, the first class-interval includes sentences whose lengths do not exceed five words.

Length of sentence	Frequency	Length of sentence	Frequency
1 –	11	96 –	8
6 –	58	101 –	10
11 –	90	106 –	4
16 –	95	111 –	2
21 –	131	116 –	6
26 –	120	121 –	5
31 –	112	126 –	2
36 –	103	131 –	2
41 –	101	136 –	0
46 –	76	141 –	2
51 –	53	146 –	3
56 –	45	151 –	1
61 –	39	156 –	1
66 –	37	161 –	1
71 –	29	166 –	0
76 –	16	171 –	1
81 –	15	. . .	. . .
86 –	14	196 –	1
91 –	13		
Total			1207

Source   G.U. Yule (1939), *Biometrika*, Vol. 30, p. 363.

Calculate, correct to four decimal places, the following measures of the frequency distribution:

(i) the mean and standard deviation;

(ii) g_1 and g_2, the Fisher coefficients of skewness and kurtosis; and

(iii) the three quartiles, the semi-interquartile range, and the five and ninety-five per cent points.

16   Trypanosomes are flagellate protozoa which cause sleeping-sickness in vertebrates, and they are transmitted to the animals by their host, the tsetse fly (*Glossina morsitans*). In an experiment, five different strains of tsetse fly were fed on rats and in each case 500 trypanosomes from a single rat were used for purposes of measurement. The following table gives the combined frequency distribution of the lengths of the trypanosomes (in microns) measured in the experiment.

Calculate, correct to four decimal places, the following measures of the frequency distribution:

(i) the mean and standard deviation;

(ii) g_1 and g_2, the Fisher coefficients of skewness and kurtosis; and

(iii) the three quartiles, the semi-interquartile range, and the five and ninety-five per cent points.

Length (midpoint of interval)	Frequency	Length (midpoint of interval)	Frequency
15	7	26	110
16	31	27	127
17	148	28	133
18	230	29	113
19	326	30	96
20	252	31	54
21	237	32	44
22	184	33	11
23	143	34	7
24	115	35	2
25	130		
Total			2500

Source   K. Pearson (1914–15), *Biometrika*, Vol. 10, p. 85.

17   In September, 1926, human bones, possibly dating from Roman times, were discovered during excavations conducted by the City of London Corporation preparatory to building the westward extension of Spitalfields Market. The following table gives the frequency distribution of the lengths (in millimetres) of 600 skulls found in the remains. The class-marks used in tabulation are, in fact, class-limits and so, for example, the midpoint of the first class-interval is 155.

Length	Frequency	Length	Frequency
154 –	1	184 –	50
156 –	0	186 –	37
158 –	0	188 –	37
160 –	3	190 –	40
162 –	2	192 –	34
164 –	21	194 –	:26
166 –	16	196 –	21
168 –	18	198 –	8
170 –	23	200 –	7
172 –	36	202 –	3
174 –	22	204 –	2
176 –	38	206 –	4
178 –	36	. . .	. . .
180 –	47	216 –	1
182 –	67		
Total			600

Source   G.M. Morant (1931), *Biometrika*, Vol. 23, p. 191 (adapted).

For this frequency distribution calculate, correct to four decimal places, the following statistical measures.

(i) the mean and standard deviation;

(ii) g_1 and g_2, the Fisher coefficients of skewness and kurtosis; and

(iii) the three quartiles, the semi-interquartile range and the five and ninety-five per cent points.

Repeat the above calculations using a grouping interval of four instead of two millimetres. Comment on any differences observed.

18 The performance of a cow in the production of butter-fat is not constant from day to day, but varies considerably. Records of the Advanced Register of the Holstein—Friesian Association of America are in pounds of butter-fat made in seven consecutive days, and are given along with the age of the cow at the time of the test. This register admits cows subject to their satisfying certain minimum requirements of butter-fat production in relation to their age.

The following frequency distribution of butter-fat production is based on the performance of such a select group of pure-bred Holstein—Friesian cows. All the cows included in the sample were under sixteen years of age. The class-marks used are the midpoints of the class-intervals.

Butter-fat	Frequency	Butter-fat	Frequency
7·0	9	15·5	167
7·5	137	16·0	157
8·0	170	16·5	113
8·5	215	17·0	120
9·0	260	17·5	73
9·5	243	18·0	49
10·0	257	18·5	39
10·5	231	19·0	27
11·0	219	19·5	35
11·5	259	20·0	28
12·0	297	20·5	15
12·5	359	21·0	21
13·0	349	21·5	6
13·5	298	22·0	4
14·0	264	22·5	5
14·5	254	23·0	1
15·0	183		
Total			4864

Source H.L. Rietz (1909), *Biometrika*, Vol. 7, p.106.

Calculate, correct to four decimal places, the following measures for the above frequency distribution:

(i) the mean and standard deviation;

(ii) g_1 and g_2, the Fisher coefficients of skewness and kurtosis; and

(iii) the three quartiles, the semi-interquartile range and the five and ninety-five per cent points.

Table for Exercise 19

No. of stamens	Frequency			
	Bordighera	Guernsey	Surrey	Dorset
13				1
14		1		0
15		0		0
16	1	0		0
17	1	2		0
18	8	3	1	0
19	4	1	1	0
20	9	11	0	2
21	18	15	0	1
22	21	29	6	1
23	49	40	11	5
24	42	29	2	5
25	90	38	14	7
26	65	53	29	8
27	67	45	20	8
28	86	43	35	14
29	51	25	29	17
30	28	31	27	23
31	28	31	42	20
32	23	24	31	35
33	8	24	30	48
34	7	22	41	36
35	5	18	33	36
36	4	10	24	26
37	4	6	23	29
38	1	6	23	40
39	1	8	20	22
40	0	3	13	29
41	0	1	8	18
42	0	0	11	16
43	1	1	1	18
44	0		5	8
45	0		4	6
46	0		2	4
47	1		5	2
48	1		4	9
49			2	3
50			2	3
51			0	2
52			1	1
...				...
59				1
60				0
61				1
Total	624	520	500	505

Source   K. Pearson *et al.* (1903), *Biometrika*, Vol. 2, p. 145.

19   In a comparative study of the regional variations in the flowers of the lesser celandine (*Ficaria ranunculoides*), samples were obtained from the following four places:

(a) 624 heads gathered in a vineyard at Bordighera, Italy, between 4th and 7th March, 1902;

(b) 520 heads gathered on roadside banks at St Peter Port, Guernsey, between 25th and 30th April, 1902;

(c) 500 heads gathered in two lanes and a meadow near Thursley, on the north side of Hindhead, Surrey, between 12th and 17th April, 1902; and

(d) 505 heads gathered in a lane at Studland, Dorsetshire, on 7th April, 1902.

The accompanying table gives the frequency distributions of the number of stamens per flower based on the samples from the four regions.

Calculate, correct to four decimal places, the mean and standard deviation for each of the above four frequency distributions.

Pool the data to determine the frequency distribution of the number of stamens per flower for all the four regions taken together. Use a two-unit class-interval for the pooled distribution, and let the first class-interval be 13–14.

Calculate, correct to four decimal places, the following measures for the pooled frequency distribution:

(i) the mean and standard deviation; and

(ii) g_1 and g_2, the Fisher coefficients of skewness and kurtosis.

Make the cumulative distribution based on the pooled data and represent it graphically.

20   During the autumn of 1902 an investigation was conducted at the West of Scotland Marine Biological Station, Millport, to determine some of the external features of the brittle-star (*Ophiocoma nigra*, Müller). The specimens studied were dredged from near the Tan Buoy between the Cumbraes at a depth of from five to seven fathoms, from gravelly ground well known to be frequented by brittle-stars. Among several measurements made on the specimens, one pertained to the length (measured in milli-metres) of the longest arm. The following table gives the frequency dis-tribution of arm-length based on 1000 five-rayed specimens which were part of the collection obtained by dredging on 27th September, 1902.

Calculate, correct to four decimal places, the following measures for the above frequency distribution:

(i) the mean and standard deviation;

(ii) g_1 and g_2, the Fisher coefficients of skewness and kurtosis; and

(iii) the three quartiles, the semi-interquartile range and the five and ninety-five per cent points.

Arm-length	Frequency	Arm-length	Frequency
20 –	7	53 –	112
23 –	10	56 –	94
26 –	22	59 –	81
29 –	16	62 –	54
32 –	39	65 –	44
35 –	35	68 –	35
38 –	61	71 –	13
41 –	62	74 –	13
44 –	69	77 –	2
47 –	116	80 –	3
50 –	112		
Total			1000

Source   D.C. McIntosh (1903), *Biometrika*, Vol. 2, p. 463 (adapted).

21   A farmer uses an automatic machine loader to fill sacks of wheat, each of which contains a nominal 112 lb of grain. In fact, because of random fluctuations in the weighing mechanism of the loader, the mean weight of grain in a sack is 112·375 lb with standard deviation 0·226 lb. Find the probability that a randomly selected sack contains less than the nominally stipulated amount of grain, assuming that the weight of grain in the sacks is normally distributed.

The farmer supplies the wheat to a miller on the understanding that not more than five per cent of the sacks will be underweight. Determine the lowest value of the mean weight of grain per sack which would, on an average, just satisfy the requirement for underweight sacks. Hence calculate, correct to the nearest lb, the expected saving to the farmer in 10 000 sacks filled by the new method.

On the other hand, suppose there is a loss of weight due to evaporation during the period of transit of the wheat from the farm to the mill. If this loss is, on an average, $\frac{1}{4}$ lb per sack but otherwise the weight distribution remains unchanged, determine the adjustment in the mean weight to ensure meeting the miller's stipulation for underweight sacks. Hence determine the additional amount of grain required for the 10 000 sacks supplied.

22   A consultant surgeon lives on the outskirts of a city, and the direct route to the hospital passes through the centre of the city. The journey by car takes, on an average, 28 minutes with standard deviation 7·5 minutes. The surgeon can bypass the congestion of the city centre either by going through an adjoining village or by taking the ring-road round the city. This journey through the village takes, on an average, 32 minutes with standard deviation 4·3 minutes, and the corresponding figures for the journey via the ring-road are 33 minutes and 1·2 minutes. Assuming that all three journey times are normally distributed, determine which is the best route to take if the surgeon has (i) 30 minutes; and (ii) 35 minutes to reach the hospital for an appointment.

23 A factory produces festoon lamps for motor-cycle tail-lights by two different processes A and B. The lamps produced by process A have a mean lifetime of 3675 hours with standard deviation 430 hours, whereas those produced by process B have a mean lifetime of 4500 hours with standard deviation 980 hours. According to the specifications of a motor-cycle manufacturer, lamps with a lifetime of less than 3000 hours are to be regarded as definitely bad. In view of this, determine which of the two processes will be the better one to use, assuming that the lifetime distribution of lamps produced by A and B are both normal. How is this decision affected if the customer's specification for bad lamps is not 3000 but 3500 hours?

  Also, calculate the probability that a randomly selected lamp produced by A will have a lifetime greater than that of a similarly selected lamp produced by B.

24 Cylindrical steel rods are mass-produced at a factory. According to the specifications of a customer, the diameter of a rod should be 4 inches, but the rods will be acceptable if their diameters lie within the limits 3·995 and 4·005 inches. On testing the rods supplied to him by the factory, the customer finds that 5 per cent are under-size and 12 per cent over-size. If the diameters are assumed to have a normal distribution, determine its mean and standard deviation.

  Also, evaluate the percentage reduction in the standard deviation of the distribution which would ensure that not more than 2 per cent of the rods were over-size.

25 A manufacturer uses a machine which packs automatically 20 oz cartons of a detergent. When the machine is in operation, the cartons emerge from it in a steady stream ready for dispatch to retailers. In order to check that the machine weighs correctly, a random sample of cartons is taken and weighed accurately. It is observed that the average weight of detergent per carton is 20·37 oz with standard deviation 0·5634 oz. Assuming that the weight distribution is normal, calculate the proportion of cartons which contain more detergent than the nominally stipulated amount.

  Also, evaluate the mean and standard deviation of the weight distribution which would halve the proportion of overweight cartons but at the same time ensure that, on an average, not more than one per cent of the cartons will be less than 19·9 oz in weight. Hence determine the percentage saving in the amount of detergent used per carton.

26 A firm manufactures carbon rods, and it is known that, measured at a constant temperature, the mean resistance of the rods is 319·98 ohms with standard deviation 1·7645 ohms. Find the proportion of rods which, on an average, have resistances (i) less than 318 ohms and (ii) greater than 321 ohms, assuming that the resistances have a normal distribution.

A prospective customer is not satisfied with the existing limits of control and requires that the production process be so modified that the proportions in (i) and (ii) are halved. Calculate, correct to four decimal places, the mean and standard deviation of the resistance distribution which would meet the customer's specifications. Also, determine whether this new distribution of resistance will meet the requirements of another customer who requires that, on an average, not more than 0·5 per cent of the rods have resistances less than 317 ohms and not more than 5 per cent have resistances greater than 322 ohms.

27 A company has a country-wide network of petrol stations which are all fitted with automatic petrol pumps. A pump set to deliver a gallon of petrol gives, on an average, 1·004 gallon with standard deviation 0·007 568 gallon. A rival company installs a similar network of stations but with the difference that its pumps deliver, on an average, 1·0038 gallon for each one metered with standard deviation 0·007 249 gallon. Determine, by using three different criteria, which of the two companies offers a better arrangement for customers, assuming that the amount of petrol delivered per gallon metered by both types of pumps is normally distributed and the price of petrol is the same for both companies.

28 Each of two detergents, A and B, is marketed in cartons which nominally contain 25 oz of powder. However, it is observed that 15 per cent of the cartons of A contain less than 24·5 oz and 6 per cent more than 26 oz of detergent. On the other hand, 5 per cent of the cartons of B contain less than 24 oz and 8 per cent more than 26·5 oz. Assuming that the weight distributions for both detergents are normal, calculate, correct to four decimal places, their means and standard deviations. Hence, assuming that the price of the detergents is the same, determine

(i) which detergent gives, on an average, better value;

(ii) the probability that detergent B gives better value than detergent A; and

(iii) the probabilities that the contents of randomly selected cartons of A and B will be less than the nominal weight.

How is the conclusion based on (i)–(iii) affected if the price of detergent B is 4 per cent more than that of detergent A?

29 A commercial traveller has a car which averages 28·75 miles per gallon on the standard grade of petrol with standard deviation 3·28 miles, whereas its performance on superfine petrol is 29·62 miles per gallon with standard deviation 4·25 miles. The price of the standard grade of petrol is 25p per gallon and that of superfine petrol 26½p. The traveller plans a sales trip of 1150 miles and his petrol allowance is £9.

Determine which of the two grades of petrol is better value for the trip, assuming that the mileages per gallon are normally distributed for both grades of petrol.

30   Four undergraduates plan a 2000 miles touring holiday during the summer vacation to visit certain historic places. Each undergraduate contributes £80 towards the price of a second-hand car and its running costs in terms of petrol and oil. The undergraduates have a choice between two cars offered for sale at a garage. The first is a mini which is priced at £306, and the second a bigger but older saloon costing £297. The average petrol consumption of the mini is 39·65 miles per gallon with standard deviation 4·58 miles, whereas that of the saloon is 23·95 miles per gallon with standard deviation 2·89 miles. The garage mechanic also estimates that the cost of oil for the trip will be £2 for the mini and £3 for the saloon. The price of petrol is 25p per gallon. If the mileage per gallon for both cars is normally distributed, determine which car is the better purchase for the undergraduates.

31   An automatic machine produces bolts whose diameters should be 0·5 in., but according to a customer's specifications the bolts will be serviceable if their diameters lie within the range 0·485 in. to 0·515 in. The production process is so organised that the diameters of the bolts are normally distributed with mean 0·500 24 in. and standard deviation 0·009 45 in.

Calculate the proportion of bolts which fall below and above the specification limits of the customer. Determine the effect on these proportions if the mean diameter of the bolts is reduced to 0·499 76, the standard deviation remaining unchanged.

If the mean of the distribution remains 0·500 24 in., calculate the reduced standard deviation which would ensure that not more than 2·5 per cent of the bolts have diameters exceeding the upper specification limit 0·515 in. Also, with this new standard deviation, determine the proportion of bolts having diameters less than the lower specification limit 0·485 in.

32   The weights of cakes of soap produced by a large manufacturer are normally distributed with standard deviation 0·078 oz. The nominal weight of a cake is 6·5 oz, but because of random variations the production process is so arranged that only 0·4 per cent of cakes produced fall below the stipulated weight. Calculate the mean weight of the cakes of soap produced.

The weekly output of soap is 450 000 cakes, and the cost in new pence of producing a cake of x oz in weight is

$$8·56 + 0·1834x,$$

where x is the random variable denoting the actual weight (in oz) of a cake of soap whose nominal weight is 6·5 oz. If by a modification of the production process the standard deviation of the weight distribution of the cakes is halved but the proportion of underweight cakes remains the same, determine, correct to the nearest pound sterling, the average saving in weekly cost to the manufacturer.

33   A factory produces packets of margarine, and the nominal weight of the contents of each packet is 16 oz. However, due to random fluctuations in the automatic weighing machine, the actual weight of the contents of a packet varies, and it is observed that 10 per cent of the packets are underweight, whereas 5 per cent weigh not less than 16·25 oz. Calculate, correct to six decimal places, the mean and standard deviation of the weight distribution, assuming that this distribution is normal.

The daily output of the factory is 375 000 packets, and the cost in new pence of producing a packet containing x oz of margarine is

$$16 \cdot 75 + 0 \cdot 2689x - 0 \cdot 000\ 54\ x^2,$$

where x is the random variable denoting the actual amount (in oz) of margarine in a 16 oz packet. Because of an increase in the world demand for edible oils, the cost of production of the margarine rises. To offset a part of this increase, the manufacturer makes certain small economies in his basic expenses and also establishes checks to lessen waste. As a result of these measures, the cost function becomes

$$16 \cdot 734 + 0 \cdot 2795x - 0 \cdot 000\ 39\ x^2.$$

Evaluate, correct to the nearest pound sterling, the change in the expected daily cost of production of the 375 000 packets.

34   At a milk pasteurisation plant one-pint bottles are filled and sealed automatically. The amount of milk in a bottle varies, and it is observed that, on an average, 8 per cent of the bottles contain less than one pint of milk, whereas 30 per cent contain not less than 1·05 pint. If the amount of milk in the bottles is normally distributed, calculate the mean and standard deviation of the distribution.

The plant processes 500 000 pint bottles a day, and the total cost in new pence of producing x gallons of milk is

$$22 \cdot 67 + 0 \cdot 1675x - 0 \cdot 000\ 478\ x^2,$$

where $x = y/8$ and y is the random variable denoting the amount (in pints) of milk in a pint bottle. Evaluate, correct to the nearest pound sterling, the total expected daily cost of production of the plant.

A change in government policy reduces the milk subsidy to dairy farmers, and so the plant authority is obliged to pay more for the unpro-cessed milk. This increase cannot be wholly passed on to the consumers, and so the plant authority considers a reduction in processing costs. It is suggested that certain improvements in the bottling department could halve the proportion of bottles containing not less than 1·05 pint without affect-ing the proportion of bottles which contain less milk than the nominal amount. These improvements require capital outlay for new machinery, and it is estimated that the new cost function for x gallons of milk would be

$$22 \cdot 78 + 0 \cdot 1784x - 0 \cdot 000\ 596\ x^2.$$

Calculate, correct to the nearest pound sterling, the difference in the total expected daily cost of production of the plant.

35 The table below gives the heights (measured to the nearest centimetre) of 694 girls in Berne aged nine years.

Height	Frequency	Height	Frequency
117 –	8	137 –	148
121 –	28	141 –	69
125 –	82	145 –	15
129 –	140	149 –	16
133 –	188		
Total			694

(i) Calculate, correct to four decimal places, the mean and standard deviation of the frequency distribution.

(ii) Assuming that these data follow a normal distribution, calculate, correct to three decimal places, the expected frequencies in each height group. Also, determine to the same accuracy the expected frequencies outside the range of the observed distribution.

36 The following table gives the frequency distribution of the weights (measured to the nearest lb) of 1000 schoolchildren.

Weight	Frequency	Weight	Frequency
28 –	1	48 –	263
32 –	14	52 –	156
36 –	56	56 –	67
40 –	172	60 –	23
44 –	245	64 –	3
Total			1000

(i) Calculate, correct to four decimal places, the first four moments of the distribution. Hence evaluate, also to the same accuracy, g_1 and g_2, the Fisher coefficients of skewness and kurtosis of the distribution.

(ii) If the observed distribution can be assumed to be reasonably normal, calculate, correct to two decimal places, the expected frequency in each class-interval.

(iii) Determine, correct to four decimal places, the fractional errors made in extending the fitted normal curve from $-\infty$ to $+\infty$ instead of the range given by the observed frequency distribution.

(iv) Calculate, correct to five decimal places, the expected proportion of children whose weight is

(a) less than 35 lb;

(b) greater than 50 lb;

(c) greater than 40 lb and less than 48 lb.

(v) Calculate, correct to five decimal places, the expected probability that if two children are taken randomly from the population sampled

(a) both have weights less than 35 lb;

(b) one has weight less than 35 lb and the other greater than 50 lb;

(c) both have weights lying between 40 and 48 lb;

(d) both have weights greater than 50 lb;

(e) one has weight greater than 35 lb and the other less than 50 lb.

37 A factory produces cartridge electrical fuses by an automatic process. The fuses are packed in boxes, each box containing 24 fuses, for dispatch to customers. If the production is under statistical control and it is known from past experience that three per cent of the fuses produced are defective, find, correct to five decimal places, the probability that a randomly selected box contains at least one defective fuse.

Also, evaluate, correct to the same accuracy, the probability that of four randomly selected boxes inspected

(i) each contains at least one defective fuse;

(ii) each contains exactly one defective fuse;

(iii) each contains less than four defective fuses;

(iv) some three contain not more than one defective fuse each, and the fourth box has at least two defective fuses;

(v) some two contain less than four defective fuses each, and the other two have at least three defective fuses each.

(vi) If an electrician buys 100 boxes of fuses, determine, correct to the nearest integer, the expected number of boxes, each of which contains (a) exactly two defective fuses; (b) at least two defective fuses; and (c) not more than two defective fuses.

38 The experience of births at a maternity hospital over a long period of time indicates that 51 per cent of the babies born are boys and the rest girls. If in a particular week 20 pregnant women are admitted to the hospital, determine, correct to four significant figures, the probability that

(i) there will be an equal number of boys and girls;

(ii) the number of boys will be less than the number of girls;

(iii) there will be at least eight boys;

(iv) there will not be more than twelve girls.

Also, if the first five babies born are all boys, evaluate, correct to four significant figures, the probability that

(v) in all there will be an equal number of boys and girls; and

(vi) in all there will be at least six boys more than the number of girls.

Note   It may be assumed that there were no multiple births or miscarriages amongst the 20 women admitted to the hospital.

39   Under normal conditions, a building society provides, on an average, financial assistance for house mortgages to 90 per cent of the applicants. If in a particular week 15 applications are received by the society, calculate, correct to four significant figures, the probability that

(i) all applicants will receive financial assistance;

(ii) at least ten of the applications will be turned down;

(iii) not more than seven applicants will receive financial assistance;

(iv) the number of applications refused will not be less than one and a half times the number approved.

Also, calculate the expected amount of capital advance made by the society if, on an average, a single mortgage application needs a sum of £2750.

In a period of financial stringency, the building society is required to reduce its approved applications to 80 per cent of the number received. Calculate, correct to four significant figures, the probabilities (i), (ii), (iii) and (iv) under this restriction. If the average amount advanced for each mortgage is reduced by ten per cent, evaluate the percentage expected saving on the transactions made.

40   At a factory, motor-cycle tyre tubes are manufactured by an automatic process, and the production is so organised that the outgoing product includes, on an average, six per cent of defective tubes. The tubes are packed in batches of 24 each for dispatch to retailers. Calculate, correct to four decimal places, the probability that a randomly selected batch contains

(i) exactly three defective tubes;

(ii) at least four defective tubes;

(iii) not more than three defective tubes.

A retailer orders regularly 12 batches of tubes per month. Determine the expected number of defective tubes received by the retailer in a month. Also, if the manufacturer allows a refund of 37½p on each defective tube received by the retailer, evaluate, correct to the nearest new penny, the expected amount of the refund payable to the retailer in a month.

To reduce the cost of correspondence, the manufacturer agrees to refund 50p on each defective tube supplied, but on the condition that the refund will not be payable unless a batch contains at least two defective

tubes. Calculate, correct to four decimal places, the probability of the
manufacturer paying a refund to the retailer on a batch. Also, evaluate
the expected amount of refund payable on such a batch. Hence, determine,
correct to the nearest new penny, the expected amount of refund payable
to the retailer on a month's supply under the new system.

41   An undergraduate lives in a hall of residence on the outskirts of a
city, and during a lecturing term of nine weeks he has to go to the university in the city on six days per week. He normally catches a bus to the
university in the morning but walks back to the hall in the evening. If the
undergraduate catches a city transport bus which leaves the stop near the
hall at 8.40 a.m., then the fare for the journey up to the university is 6p.
However, if the city bus is late by at least five minutes, then the undergraduate is obliged to take the fast long-distance bus, and the fare to the
university is 9p. On a weekday, the probability of the city bus being late
by at least five minutes is 0·16, and it may be assumed that the undergraduate can always catch the long-distance bus when the city bus is late
by at least five minutes. Calculate, correct to four significant figures,
the probability that during any given week the undergraduate will make

    (i) all six journeys by the city bus;

    (ii) some five journeys by the city bus;

    (iii) at least two of the journeys by the long-distance bus;

    (iv) not more than three journeys by the city bus;

    (v) more journeys by the city than by the long-distance bus.

    (vi) Calculate the expected cost of a single journey to the university,
and hence determine, correct to the nearest new penny, the total expected
cost of travel for a whole term. Verify that this total expected cost is
almost the same as that calculated on the assumption that in each week
of the term the undergraduate is obliged to make one journey by the long-
distance bus.

    (vii) Calculate, correct to four significant figures, the probability
that during the nine weeks of a term, the undergraduate will travel in each
of some four weeks at least four times by the city bus, and in each of the
remaining weeks not more than twice by the city bus.

42   A pipe-smoker normally buys tobacco of one of two brands A and B.
Both brands are sold in two-ounce tins, and the price of a tin of A is 65p
and that of B 67½p. On an average, the smoker uses one pound weight of
tobacco a month, and he buys this on eight different occasions. He pre-
fers brand A, but if this is not available at his usual tobacconist, then
he buys a tin of brand B. If the probability of the smoker obtaining a tin
of brand A on any visit to the tobacconist is 0·85, calculate

    (i) the expected cost of a month's supply of tobacco, correct to the
nearest new penny;

(ii) the probability that the monthly cost will be at least 5p more than the minimum, correct to four significant figures; and

(iii) the probability that the monthly cost will not be more than 10p above the minimum, correct to four significant figures.

Also, determine, correct to four significant figures, the probability that in twelve calendar months the smoker will spend

(iv) in each of some six months at least 5p more than the minimum;

(v) in at least six months at least 5p per month more than the minimum;

(vi) in some eight months exactly 10p per month more than the minimum;

(vii) in not more than one month exactly 10p per month more than the minimum.

43   A psychological test consists of twenty items of the yes/no type. Each correct response scores one mark and an incorrect one zero. If the probability is 0·46 that a candidate will give a correct response on any one item, find, correct to four significant figures, the probability that the candidate's score on the test is

(i) more than fifteen;

(ii) not less than twelve;

(iii) not more than five;

(iv) at least ten.

Suppose a battery of tests consists of six separate ones such that each test is similar to the one considered above, and that the probability of a correct response on an item is the same for all the tests. Find, correct to four significant figures, the probability that on each of the tests in the battery the candidate scores

(v) at least half the maximum possible score;

(vi) not more than one-fourth the maximum possible score;

(vii) more than three-fourths the maximum possible score;

(viii) on some two tests at least half the maximum possible score, and on the others more than three-fourths the maximum possible score;

(ix) on some three tests at least half the maximum possible score, on two others not more than one-fourth the maximum possible score, and on the remaining test more than three-fourths the maximum possible score.

44   A farmer buys regularly a bulk supply of seed-potatoes from a seed-store. The potatoes are supplied in bags, each bag containing 320 potatoes. Because of natural deterioration in quality and transportation hazards, there is, on an average, a 1·25 per cent chance of a potato being damaged in a bag. According to the farmer's agreement with the seed-store, he accepts a 2·5 per cent damage in a bag, but compensation is payable to the farmer for greater damage. Calculate, correct to five significant figures, the probability that the seed-store will have to pay compensation on a bag of seed-potatoes supplied.

The annual requirement of the farmer is 120 bags, and it is known that, on an average, compensation on a bag amounts to 47½p. Evaluate

(i) the probability of the expected annual compensation payable by the seed-store being at least £2·85; and

(ii) the expected compensation for a year.

45   A factory mass-produces a standard type of bulb for electric torches, and production is so arranged that, on an average, only 1·5 per cent of the bulbs produced are defective. The bulbs produced are packed in batches of 144 each for dispatch to retailers. According to a standard agreement between the manufacturer and the retailers, a retailer accepts the risk of receiving at most two defective bulbs in a batch. If a batch contains more than two but not more than five defective bulbs, then the manufacturer pays a refund of 15p to the retailer. For a batch containing more than five defective bulbs the refund is 25p.

(i) Calculate, correct to four significant figures, the probability that on three randomly selected batches received by a retailer, the total refund will be exactly 40p.

Also, evaluate, correct to the nearest new penny,

(ii) the expected refund on a randomly selected batch; and

(iii) the expected refund on a batch, if it is known that the batch contains at least three defective bulbs.

Explain the difference between the expectations calculated in (ii) and (iii).

46   A twenty-four-hour service station is located at a point just before the start of a motorway. It is observed over a period of time that the number of cars arriving at the station for petrol in any half-hour has a Poisson distribution with mean 3·85, and that, on an average, the amount of petrol purchased by any one customer is 4·25 gallons. Calculate, correct to the nearest gallon, the expected amount of petrol sold by the station during a week.

The petrol station has a storage capacity of 5800 gallons, and it receives its supplies from the distributors once a week.

(i) Assuming that the average amount of petrol purchased by a customer remains unchanged, calculate, correct to two decimal places, the percentage increase in the mean half-hourly arrival rate of customers which would ensure that, on an average, the weekly stock of petrol would be just adequate to meet demand.

(ii) Hence, by using a suitable approximation, calculate, correct to four decimal places, the probability that in any one week the average half-hourly rate of arrival of customers will be large enough to exhaust the week's supply.

(iii) Also, evaluate, correct to four significant figures, the probability that the event in (ii) will happen at least four times in a year, it being assumed that a year has 52 weeks exactly.

47   The records of a Ministry of Transport test centre for learner drivers show that the average number of test appearances of an examinee for passing the driving test is 1·86, and the frequency distribution of the examinees according to the number of tests taken to pass is Poisson with the zero class missing. It may be assumed that a learner driver continues to take the test until success is achieved. If X is a discrete random variable denoting the number of test appearances required by a learner driver to pass the driving test, calculate, correct to six decimal places, the probabilities for $1 \leqslant X(1) \leqslant 4$ and $X \geqslant 5$. Hence evaluate, correct to four significant figures, the probability that a learner driver will pass the driving test

(i) on the first attempt;

(ii) in exactly three attempts;

(iii) after the fourth unsuccessful attempt.

Also, determine the conditional probability of a learner driver passing the test on the second attempt given that he has failed in the first examination.

A driving school presented its pupils regularly at the test centre for examination. The school's experience indicated that, on an average, pupils passing the test on the first attempt had had 15 hours of tuition, those passing on the second attempt 25 hours of tuition, those passing on the third attempt 40 hours of tuition, and those passing in more than three attempts 70 hours of tuition. If the fee for a one-hour lesson at that time was £1 and this was also the fee for a test examination, calculate, correct to the nearest pound, the expected cost of passing the driving test for a pupil entering the driving school.

48   A large fashion store plans to order in advance a stock of the latest style in ladies' nylon stockings which would be sold during the next winter season. The order can be placed in number of gross units, and the cost per gross in £54. The store hopes to sell the stockings in season at one and half times the cost price, but the residual stock will have to be disposed of in the spring sale at a loss of 15 per cent. The past records of the store indicate that the seasonal sales of a new line in nylons are 175·5x gross, where x is a random variable having a Poisson distribution with mean 4·96.

(i) Determine the optimum size of the stock which the store must order to maximise its expected profit.

(ii) Hence evaluate, correct to the nearest pound sterling, the maximum expected profit.

Also, determine, correct to four significant figures,

(iii) the maximum expected profit as a percentage of the total outlay; and

(iv) the probability that the seasonal demand will exceed the stock bought to maximise the expected profit.

49   If Z is a Poisson variable with mean μ, and k and ρ are positive integers such that $k \leqslant \rho$, prove that

$$\sum_{r=\rho}^{\infty} r^{(k)} \times P(Z = r) = \mu^k \times P(Z \geqslant \rho - k).$$

A factory produces small batteries by an automatic process, and it is known that, on an average, the outgoing product contains 1·28 per cent defective batteries. The batteries are packed in cartons of 75 for dispatch to a retailer. The cost of a battery to the retailer is 9p. However, if there is a single defective battery in a batch supplied, the retailer gets a refund of 12p, and for two defectives in a batch the refund is 15p per battery. If a batch contains r $(r \geqslant 3)$ defective batteries then the refund is $(12 + 2r)$ new pence per battery.

Evaluate, correct to seven decimal places, the probabilities that a batch contains 0, 1, 2, and $\geqslant 3$ defective batteries. Hence, calculate, correct to four decimal places, the mean and variance of the amount of refund payable to the retailer for a battery in a batch.

50   The net price of a book is £1·25, but the publishers give trade discounts to retailers for bulk purchases. If a retailer buys r $(2 \leqslant r \leqslant 10)$ copies at a time, then the publishers charge £1·25 − r new pence per copy, and for 11 or more copies purchased at a time the price per copy is £1·12. Assuming that the number of copies purchased at a time by individual retailers has a Poisson distribution with mean 4·6 and the zero class missing, calculate, correct to four decimal places, the expected discount per copy paid by the publishers. Hence determine, correct to two decimal places, the expected discount as a percentage of the net price of the book. Also, evaluate, correct to four decimal places, the variance of the discount to the retailers offered by the publishers.

The edition of the book consists of 3750 copies of which 50 are distributed free for publicity purposes. Calculate, correct to the nearest pound sterling, the expected financial return to the publishers on all copies of the book sold.

51   If Z is a Poisson variable with mean μ, prove that

$$\frac{1}{\mu} \times P(a + 2 \leqslant Z \leqslant a + 1 + N) < \sum_{k=1}^{N} P(Z = a + k)/(a + k)$$

$$< \left(\frac{a + 2}{a + 1}\right) \times \frac{1}{\mu} \times P(a + 2 \leqslant Z \leqslant a + 1 + N),$$

where $a > 0$ is a constant and N is a fixed integer.

According to post office regulations, the charge for a local telephone call in a certain area is 6p, irrespective of its duration. As a concession to offices and other frequent users of the telephone, the post office agrees to a different system of charging for calls, depending upon the hourly intensity of demand on the telephone service. Accordingly, it is agreed that in any hour the first ten calls from the same office telephone will be charged at 3p per call, the next ten at 6p per call, and thereafter all calls made in the hour will be charged at a flat rate of 9p per call. If the hourly intensity of calls emanating from an office telephone has a Poisson distribution with mean 15, prove that, according to the new system of charging, the percentage reduction to the office in the average cost per call lies between 32·59 and 35·15 approximately.

Also, show that it is practically certain that the average cost per call to the office according to the new system will not be greater than the standard charge of 6p.

52   Two gamblers A and B play a game of chance as follows. From an ordinary well-shuffled pack of 52 cards (13 cards of each suit), three cards are dealt to A and then from the remainder of the pack five cards are dealt to B. If A and B have the same number of spades, or if neither of them has any spades, A wins; otherwise B wins. Calculate, correct to four decimal places, the probability of A winning a game, and hence show that if A wins a succession of three games, it is reasonable to doubt the randomness of the card distribution in the pack.

53   A hand of 13 cards is distributed randomly to a bridge player from a standard pack of 52 cards. If $P(r)$ is the probability that the hand contains r spades, prove that

$$P(r) = \binom{13}{r}\binom{39}{13 - r} \bigg/ \binom{52}{13}, \quad \text{for } 0 \leqslant r \leqslant 13.$$

Determine $P(0)$ correct to six decimal places. Also, show that $P(r)$ satisfies the recurrence relation

$$P(r + 1) = \frac{(13 - r)^2}{(r + 1)(27 + r)} \cdot P(r), \quad \text{for } 0 \leqslant r \leqslant 12.$$

Hence determine, correct to six decimal places, the probabilities $P(r)$ for $1 \leqslant r \leqslant 9$, and $r \geqslant 10$.

Find, correct to four decimal places, the probability that in three consecutive deals the player has

(i) exactly three spades in each deal;

(ii) exactly two spades in each of any two deals and five spades in the third.

54 A discrete random variable X has the negative binomial distribution with parameters p and r such that its probability-generating function is

$$G(\theta) = p^r (1 - q\theta)^{-r}, \quad p + q = 1.$$

Prove that $P(X = x) = \dfrac{q(x + r - 1)}{x} \times P(X = x - 1)$ for $X \geqslant 1$.

(i) If $p = 0 \cdot 062\,88$ and $r = 0 \cdot 051\,69$, calculate $P(X = x)$ for $0 \leqslant X \leqslant 5$ and $P(X \geqslant 6)$ correct to five decimal places.

(ii) It is known that at a grocery store the monthly sales of the number of packets of a certain brand of breakfast cereal has a negative binomial distribution with the above values of p and r. Find the probability, correct to four decimal places, that in a month, of three independent and randomly selected persons, one buys at least three packets, another exactly four packets, and the third not more than five packets respectively.

55 The experience of a motor insurance company shows that of the claims made in any year 60 per cent are from comprehensive policy-holders and the rest from insurers having third-party cover only. If the average amount paid out by the company on a comprehensive policy claim is £85 and that on a third-party risk cover policy claim is £35, find the average amount paid on a claim. Also, if the total number of policy-holders is 178 540 and, on an average, 20 per cent of these make claims, calculate the expected amount paid out in claims by the company during a year.

With the increase in vehicular traffic in the country due to a new road development programme, the insurance company expects that the claims from comprehensive insurance policy-holders will increase to 65 per cent of the total number received in a year. Furthermore, it also estimates that due to rising prices, the average cost per claim on comprehensive and third-party risk cover policies will increase to £108 and £52 respectively. If the number of policy-holders increases by 10 per cent and of these 25 per cent make claims, calculate the total increase in the expected cost of claims in a year and also the percentage increase in the cost per claim.

Also, if the average premium for an insurance policy is £15 originally, determine, correct to two decimal places, the percentage increase in it which would maintain the existing margin between expected receipts and payments in the future.

56 The production of electric-light bulbs is so organised at a factory that five per cent of the bulbs produced are defective and the rest non-defective. The bulbs are inspected in large batches, and a batch is accepted if a random sample of 20 bulbs from it contains not more than two defectives but otherwise it is rejected. Find, correct to four significant figures, the probability that

(i) a batch will be rejected on inspection;

(ii) of 30 batches inspected, ten will be rejected;

(iii) of 30 batches inspected, at least five will be rejected.

Also, determine the probabilities (i), (ii) and (iii) if it is known that three-fourths of all batches inspected have five per cent defective bulbs and the rest eight per cent defective, the condition for the rejection of a batch remaining the same as before.

Calculate, correct to one decimal place, the percentage increase in the proportion of batches rejected.

57 A standard make of razor blade is sold in packets of five. The probability that a blade in a packet is defective is 0·32. Calculate, correct to four decimal places, the probability that a customer buying a packet of blades will find that

(i) not more than one blade is defective;

(ii) at least three of the blades are defective;

(iii) at least two of the blades are non-defective.

If, on an average, the customer uses twelve packets of blades in a year, find, correct to four significant figures, the probability that he receives

(iv) at least two defective blades in each packet;

(v) more than half the packets containing at least one defective blade in each;

(vi) not more than four packets containing not more than two defective blades in each.

58 A nursery offers an assortment of pink and deep red potted geranium plants. The individual plants cannot be identified for colour definitely, but the nurseryman maintains that 25 per cent of the plants are of the deep red variety. A lady wishes to obtain six pots of the deep red plants, and she has enough money to purchase ten plants. Determine, correct to four significant figures, the probability that the lady will obtain

(i) exactly six deep red plants;

(ii) at least six deep red plants;

(iii) exactly one deep red plant.

Also, evaluate, correct to four significant figures, the probabilities of the events (i), (ii) and (iii) when the lady knows that

(a) one specified plant of her selection of ten is definitely of the deep red variety; and

(b) at least one of her ten plants is of the deep red variety.

Finally, suppose the lady has unrestricted funds and she is willing to buy in all 40 plants. In this case, calculate, correct to four significant figures, the probability that she will obtain less than the required six deep red plants.

59   At a holiday resort the probability of rain on any one day of a week during the summer season is 0·42. Calculate, correct to four significant figures, the probability that

(i) the first four days of a given week will be wet and the rest fine;

(ii) rain will fall on exactly four days of a given week;

(iii) rain will fall on at least four days of a given week;

(iv) rain will fall on not more than four days of a given week.

It may be assumed that the weather on any one day of a week is independent of the weather on the other days of the week.

Also, determine, correct to four significant figures, the probability that in a given random selection of five independent weeks during the season

(v) in each of two or more weeks rain will fall on at least four days;

(vi) none will have rain for more than four days;

(vii) at least two will have rain for not more than four days.

60   The experience of a house-agent indicates that he can provide suitable accommodation for 75 per cent of the clients who come to him. If on a particular occasion 18 clients approach him independently, calculate, correct to four significant figures, the probability that

(i) not more than five clients will get satisfactory accommodation;

(ii) exactly eight clients will get satisfactory accommodation;

(iii) at least seven clients will get satisfactory accommodation.

If the first four clients who approached the agent obtained satisfactory accommodation, evaluate, correct to four significant figures, the probability that, in all,

(iv) ten clients will obtain satisfactory accommodation;

(v) not less than eight clients will obtain satisfactory accommodation;

(vi) not more than five clients will obtain satisfactory accommodation;

(vii) at least four clients will not obtain satisfactory accommodation.

61   A landlady runs a lodging-house for commercial travellers which consists of ten single bed-sitting rooms. During a week of normal business, the probability that a given room will be occupied for a night is 0·85. Calculate, correct to four significant figures, the probability that during a week a given room will be occupied for

   (i) exactly four nights;

   (ii) at least five nights;

   (iii) not more than three nights;

   (iv) all seven nights.

Hence, determine, correct to the nearest integer, the expected number of rooms in the lodging-house which are occupied for the durations (i) to (iv), it being assumed that the rooms are rented out to lodgers independently.

   Also, calculate, correct to four significant figures, the probability that during a week

   (v) each of some five rooms is occupied for exactly four nights;

   (vi) each of some five rooms is occupied for not more than three nights, and each of the remaining rooms for at least four nights;

   (vii) all the rooms are occupied for all the seven nights;

   (viii) at least three of the rooms are occupied for all the seven nights;

   (ix) one room is occupied for exactly four nights, each of some six rooms for at least five nights, and each of the remaining rooms for not more than three nights.

62   A golfer plays on a standard course of eighteen holes, and it is known from his past performance that the probability of his obtaining a par or a better score for a hole is 0·43. Calculate, correct to four significant figures, the probability that in one round the player will attain a par or a better score for

   (i) exactly nine holes;

   (ii) at least four holes;

   (iii) less than six holes;

   (iv) not more than eight holes;

   (v) more than eight holes.

   If the golfer plays three complete rounds, find, correct to four significant figures, the probability that

   (vi) in some two rounds he will attain a par or a better score for less than six holes, and in the remaining round he will achieve the same kind of score for more than eight holes;

   (vii) in some two rounds he will attain a par or a better score for exactly nine holes, and in the remaining round he will achieve the same kind of score for not more than eight holes;

   (viii) in some one round he will attain a par or a better score for at least four holes, in another round the same kind of score for at least six holes, and in the remaining round the same kind of score for more than eight holes.

63   In a singles tennis match the player who first wins six games wins a set, but if the state of play reaches five games in favour of each of the two players, then the winner of the set is the player who first wins two more games than his opponent.

In a match between two players, A and B, the probability that A will win a game is $0 \cdot 48$. Calculate, correct to four significant figures, the probability that A wins a set with the score

(i) eight games to six;

(ii) seven games to five;

(iii) six games to three;

(iv) six games to two.

Hence calculate, correct to the same accuracy, the probability that A wins the match in three straight sets with the *ordered* scores

(v) eight games to six, six games to two, seven games to five;

(vi) seven games to five, six games to three, six games to three.

Also, determine the probabilities when the set scores in (v) and (vi) are not ordered.

64   To promote sales, a manufacturer of a breakfast cereal runs a gift coupon scheme. In each packet of the cereal there is a coupon which may be one of two kinds, A and B. Ten A coupons can be exchanged for a free packet of the cereal, whereas fifteen B coupons can be exchanged for a free packet of the cereal and a two-ounce tin of instant coffee. The distribution of the coupons is so organised that 60 per cent of the packets contain A coupons and the rest B coupons.

If a housewife buys 25 packets of the cereal, find, correct to four significant figures, the probability that she will obtain

(i) at least ten A coupons;

(ii) not more than ten A coupons;

(iii) not more than five A coupons;

(iv) exactly five A coupons.

If each of two housewives buys 25 packets of the cereal and there is no exchange of coupons between them, calculate, correct to four significant figures, the probability that

(v) not more than one of them will get gifts on both A and B coupons;

(vi) both of them will get just one packet of the cereal each;

(vii) some one of them will get one packet of the cereal and the tin of coffee only, and the other a packet of the cereal only; and

(viii) some one of them will get two packets of the cereal only, and the other one packet of cereal and a tin of coffee only.

65   The board of directors of a nationalised industry consists of seventeen members including the chairman, who has a casting vote only. At a meeting of the board, a controversial proposal is put to the vote. If it is known that each member of the board is equally likely to vote either way, find, correct to four significant figures, the probability that the chairman will have to exercise his casting vote. Also, determine this probability if it is known that

(i) some one of the directors is against the proposal and some other is in favour of it, and each of the rest is equally likely to vote either way;

(ii) some two of the directors will abstain, but each of the others is equally likely to vote either way.

Suppose that, after the first vote results in a tie, the chairman is unwilling to force a decision by his casting vote. He appeals to the members to give a clear decision on a second vote. If one of the directors abstains in the second vote, but the others are equally divided on the proposal, find the probability that the chairman's proposal is carried by the narrowest majority of the directors actually voting.

In the course of a long agenda, six independent but equally controversial proposals are presented to the board. Find, correct to four significant figures, the probability that, apart from the casting vote of the chairman,

(iii) on exactly four proposals the first vote will result in a tie;

(iv) on at least four proposals the first vote will result in a tie;

(v) on not more than three proposals the first vote will result in a tie;

(vi) on all the six proposals the first vote will result in a tie.

66   In a coal-mining district there are 10 000 miners, and it is known from past records of the area that, on an average, in a year one miner out of 2500 loses his life in a colliery accident. Evaluate, correct to four decimal places, the probability that in a year there will be

(i) at least one death in an accident;

(ii) exactly two deaths in accidents;

(iii) not more than three deaths in accidents;

(iv) more than three deaths in accidents.

If, on an average, the compensation paid by the Coal Board to the dependents of a miner killed in a colliery accident is £3250, determine the expected amount of compensation for the district in any one year. Also, calculate, correct to four decimal places, the probability that the compensation paid in a year will exceed the expected amount.

67   An electrical dealer has a stock of six similar television sets which he rents out to customers on a weekly basis. It is known from the past experience of the dealer that the weekly demand for the sets has a Poisson distribution with mean 3·56. Calculate, correct to four significant figures, the probability that in any week

(i) at least two sets remain unused;

(ii) the demand is not fully met;

(iii) not more than three sets remain unused.

(iv) Evaluate, correct to four significant figures, the probability that, on an average, exactly one set is not used in exactly two out of any four weeks.

(v) Also, calculate correct to two decimal places, the percentage of the expected demand which the dealer cannot meet with his available stock, and the expected percentage of the available sets which remain unused.

(vi) If one of the sets is temporarily withdrawn for repairs, prove that with five sets

(a) the probability of not satisfying the total demand in any week is 0·1504;

(b) the percentage of unsatisfied demand is 7·47; and

(c) the percentage of the available stock in excess of expected demand is 34·12.

68 A local newspaper normally charges 10p per line for a single insertion of an advertisement, irrespective of the number of lines required. As a concession to commercial advertisers, the newspaper management agrees to a different price system which depends upon the length of the advertisements. Accordingly, for each insertion, it is agreed that an advertisement whose length does not exceed eight lines will cost 60p; an advertisement exceeding eight lines but not exceeding twelve lines in length £1; an advertisement exceeding twelve lines but not exceeding sixteen lines in length £1·25; and thereafter all advertisements exceeding sixteen lines in length will be charged at a flat rate of 9p per line.

A cinema regularly advertises its forthcoming programmes in the newspaper, and it is observed that, over a period of time, the length of its advertisements averages twelve lines. If the distribution of the length of the cinema's advertisements may be regarded as Poisson, calculate, by using a suitable approximation, the percentage reduction in its average advertisement charges due to the concessional rates of the newspaper.

69 A manufacturer produces a standard quality of woollen blankets. Before leaving the factory, the blankets are inspected for minor blemishes such as oil stains or irregularities in the weaving. A blanket found to have no defects is classed as grade A, a blanket having one to three defects as grade B, and the remaining blankets as grade C. If it is known from past experience of the production process that the number of defects per blanket is a random variable having a Poisson distribution with mean 2·48, calculate, correct to four significant figures, the probabilities that a randomly selected blanket will be classified as A, B or C.

Also, evaluate to the same accuracy the probability that a randomly selected blanket will be of grade B, if it is known that it has at least one defect.

If the retail price of a grade A blanket is £4·50, that of a grade B blanket £4, and that of a grade C blanket £3·75, determine, correct to the nearest new penny, the average price of

(i) a randomly selected blanket; and

(ii) a blanket which is known to have at least one defect.

The manufacturer wishes to tender for a bulk order from a hospital for ungraded blankets at a uniform price. If the cost of inspection of a blanket for grading is 3p, determine, correct to the nearest new penny, the price that the manufacturer should quote in his tender so that, on an average, his profit remains the same as on graded blankets.

70   A large city transport organisation has a fleet of 1000 buses, and it is observed that, on an average, there is a 0·24 per cent chance that a bus will break down on any one day.

(i) If the maintenance department of the transport organisation has sufficient strength to cope with six breakdowns on any day, find, correct to five significant figures, the probability that on any day there will be inadequate staff to attend to all the breakdowns occurring on that day.

(ii) Hence, assuming that a year has 365 days, calculate, correct to four decimal places, the probability that on at least four days in the year, there will be inadequate maintenance staff to attend to all the breakdowns on the day of their occurrence.

(iii) Also, determine how large should be the strength of the maintenance department to ensure approximately a 99·9 per cent chance that all daily breakdowns will be attended to on the very day of their occurrence.

(iv) With this increased strength of the maintenance department, evaluate the probability of the event (ii).

71   It is known from the past experience of a publishing house that a small proportion of typographical errors remains undetected by the usual methods of proof-reading. In order to improve upon the existing procedure, the publishing house agrees to an additional proof-reading with the proposal that the proof-reader will be paid

(i) 25p for a page on which he detects no error;

(ii) £1 for a page on which he detects a single error; and

(iii) £1·25 for a page on which he detects more than one error.

If the probability that a word on a page is typographically incorrect is 1/1000, and, on an average, there are 400 words on a page, determine the probability that there are exactly $r\ (\geqslant 0)$ errors on a page. Hence, assuming that the proof-reader detects the errors present, calculate, correct to the nearest new penny, the expected payment to the proof-reader for proof-reading a book of 250 pages.

The proof-reader is unwilling to accept the suggested rates of payment because he realises that it is unlikely that many pages will contain one or more errors per page. As a counter-proposal, the proof-reader asks for an arrangement which would give him an expected payment of 75p per page. Calculate, correct to the nearest new penny, the increased payment that the publishing house should make for each page on which the proof-reader detects no errors, in order to meet his demand.

Also, evaluate the percentage increase in the expected payment to the proof-reader if his proposal is accepted by the publishing house.

72   A garage regularly keeps a stock of a standard make of car battery. It is known from past experience that the monthly demand for this type of battery has a Poisson distribution with mean 5·28, and the stock is made up to a fixed number of batteries at the beginning of every month. Calculate the number of batteries the garage should stock in order that the probability of being out of stock at the end of the month is as near as possible to 0·025.

During a period of limited supplies the garage begins each month with only six batteries. Calculate, correct to four decimal places, the probability that during six months of limited supplies, the garage will not lose any custom.

Also, evaluate the percentage monthly loss of custom.

73   A factory normally employs a labour force of 1500 men, and it is observed that, over a period of time, there is a 0·25 per cent chance of a worker being absent on any one day. Assuming that absentee workers are statistically independent, calculate the average daily rate of absenteeism at the factory. If X is a random variable denoting the number of absentee workers on any day, determine, correct to six significant figures, the probabilities of X for $0 \leqslant X(1) \leqslant 10$ and $X \geqslant 11$. Hence evaluate, correct to four decimal places, the probability that there will be not more than ten absentees on any one day.

A full working day at the factory is of eight hours duration, and in a year there are 300 working days. Calculate, correct to the nearest hour, the expected number of man-hours lost during a full working year. How is this expectation altered if it is assumed that not more than ten workers are absent on any one day in the working year?

Also, evaluate, by using a suitable approximation, the probability that the rate of absenteeism in a year will be at least five per cent more than the average.

74   A commercial secretarial organisation charges 15p for cutting a standard foolscap stencil. The price includes the cost of the stencil and the remuneration of the typist, and the residual profit for the organisation is 4p per stencil, provided the typist makes no typographical errors. However, if an error is made, then the profit of the organisation is

diminished by the additional cost of correction. On empirical considerations, it is known that the cost in new pence of correcting r errors on a
stencil is $2r(3r + 2)/(r + 1)$, for $r \geqslant 1$. A stencil, on an average, contains
420 words, and it is observed that a typist has a chance of $1/1400$ of
making a typographical error in a word.

Calculate, correct to six significant figures,

   (i) the probability that the typist makes no error on a stencil; and

   (ii) the expected profit that the organisation makes on a stencil.

Hence evaluate, correct to the nearest new penny, the expected profit of
the organisation on the typing of a manuscript covering 500 stencils.

Also, evaluate, correct to four decimal places, the probability that

   (iii) on none of the 500 stencils the typist makes more than two
         typographical errors; and

   (iv) in typing the 500 stencils the typist makes, on an average, ten
         per cent errors more than in the past.

75   The accounts department of a large county council receives bills for
payment daily. The amounts of the bills vary considerably, from less than
50p to several thousand pounds sterling. Nevertheless, according to the
standard procedure, all bills have to be checked before payment is sanc-
tioned. This scrutiny of individual bills is a long and often expensive
procedure because different agencies are involved in the making of the
orders and the receipt of the goods ordered, apart from the suppliers.

An analysis of the bills received over a period of time and paid after
scrutiny showed that, on an average, 40 per cent of them were individually
for amounts not greater than £5, and of these only 2·5 per cent were either
partly or wholly incorrect. In order to reduce the costs of payment, it is
decided that all bills presented to the accounts department for payment,
and which are individually for amounts not exceeding £5, should be paid
straightaway without any scrutiny. However, other bills should be checked
as before.

On an average, the accounts department receives 485 bills a day; and
of those which are individually for amounts not exceeding £5, the number
of partly or wholly incorrect bills have a Poisson distribution with mean
as determined by the analysis of the earlier records. It is also estimated
that under the new system the average excess payment per incorrect bill
would be £3·39. If a working year consists of 300 days and the average
cost of scrutiny of a bill not exceeding £5 in amount is 12½p, evaluate,
correct to the nearest pound sterling, the expected annual saving in the
payment of bills under £5 in amount by the new system.

Also, assuming that the average excess payment on an unchecked bill
remains the same, determine, correct to four decimal places, the increased
daily rate of incorrect bills for amounts not exceeding £5, which would
ensure that, on an average, no saving is made by the new system of payment. Hence show that it is practically impossible that no saving will be
made by the new system of payment over a period of one working year.

ANSWERS AND HINTS ON SOLUTIONS

Chapter 2

1 (ii) The frequency distribution of word-length is as follows:

Word-length (x_i)	Frequency (f_i)	Word-length (x_i)	Frequency (f_i)
1	12	9	38
2	137	10	27
3	141	11	17
4	97	12	3
5	81	13	4
6	70	14	4
7	43	15	1
8	44		
Total			719

$\Sigma\ f_i\,x_i = 3533$ $\qquad\qquad$ $\Sigma\ f_i\,x_i^2 = 22\ 801$

Mean = 4·91 $\qquad\qquad$ Standard deviation = 2·75

2 (ii) The frequency distribution of word-length is as follows:

Word-length (x_i)	Frequency (f_i)	Word-length (x_i)	Frequency (f_i)
1	22	8	28
2	161	9	9
3	147	10	13
4	99	11	8
5	76	12	6
6	43	13	0
7	59	14	1
Total			672

$\Sigma\ f_i\,x_i = 2841$ $\qquad\qquad$ $\Sigma\ f_i\,x_i^2 = 15\ 761$

Mean = 4·23 $\qquad\qquad$ Standard deviation = 2·36

3 (i) The frequency distribution of word-length is as follows:

Word-length (x_i)	Frequency (f_i)	Word-length (x_i)	Frequency (f_i)
1	9	8	14
2	96	9	19
3	136	10	9
4	81	11	11
5	49	12	0
6	37	13	0
7	40	14	1
Total			502

(ii) $\Sigma f_i x_i = 2188$ $\Sigma f_i x_i^2 = 12\ 292$

  Mean = 4·36   Standard deviation = 2·35

4 (ii) The frequency distribution of word-length is as follows:

Word-length (x_i)	Frequency (f_i)	Word length (x_i)	Frequency (f_i)
1	10	8	44
2	180	9	32
3	109	10	21
4	75	11	10
5	61	12	6
6	38	13	4
7	51		
Total			641

$\Sigma f_i x_i = 2971$ $\Sigma f_i x_i^2 = 18\ 561$

Mean = 4·63   Standard deviation = 2·74

5 (i) The frequency distribution of the digits is as follows:

Digit (x_i)	Frequency (f_i)	Digit (x_i)	Frequency (f_i)
0	88	5	88
1	87	6	86
2	94	7	100
3	93	8	89
4	114	9	91
Total			930

(ii) $\Sigma f_i x_i = 4197$ $\Sigma f_i x_i^2 = 26\ 387$

  Mean = 4·51   Standard deviation = 2·83

(iii) The relative frequency of 1, 3, 5, 7, 9 is 459/930 = 0·4935.

6 (i) The frequency distribution of the numbers is as follows:

Number (x_i)	Frequency (f_i)	Number (x_i)	Frequency (f_i)	Number (x_i)	Frequency (f_i)
0	26	13	22	26	19
1	27	14	13	27	13
2	19	15	18	28	16
3	13	16	17	29	14
4	21	17	13	30	17
5	21	18	24	31	23
6	20	19	25	32	10
7	29	20	21	33	20
8	22	21	20	34	24
9	23	22	20	35	12
10	18	23	22	36	20
11	22	24	19		
12	14	25	23		
Total					720

$\Sigma f_i x_i = 12\ 402$ $\Sigma f_i x_i^2 = 296\ 480$

Mean = 17·225   Standard deviation = 10·735

7 (i) The frequency distribution of run-length of successes is as follows:

Length (x_i)	Frequency (f_i)	Length (x_i)	Frequency (f_i)	Length (x_i)	Frequency (f_i)
1	68	10	5	19	6
2	56	11	5	20	0
3	61	12	3	21	2
4	33	13	7	22	0
5	17	14	4	23	0
6	23	15	3	24	0
7	11	16	5	25	0
8	11	17	2	26	1
9	7	18	1		
Total					331

(ii) $\Sigma f_i x_i = 1593$ $\Sigma f_i x_i^2 = 14\ 517$

Mean = 4·81   Standard deviation = 4·55

8 (i) The frequency distribution of the barley yields is as follows:

Yield (Midpoint)	Frequency (f_i)	Yield (Midpoint)	Frequency (f_i)	Yield (Midpoint)	Frequency (f_i)
24	1	108	36	192	50
45	0	129	82	213	13
66	5	150	93	234	1
87	12	171	106	255	1
Total					400

(ii) This frequency distribution is obtained by pooling the frequencies in (i) of consecutive pairs of class-intervals.

Yield (Midpoint)	Frequency (f_i)	Yield (Midpoint)	Frequency (f_i)	Yield (Midpoint)	Frequency (f_i)
35	1	119	118	203	63
77	17	161	199	245	2
Total					400

(iii) For the ungrouped observations,

Raw S.S. = 9 597 372   Grand total = 60 698
Mean = 151·7450   Standard deviation = 31·1337.

If x_i denotes the midpoint of the ith class-interval in (i), put $x_i = 150 + 21z_i$. Then,

$$\sum f_i z_i = 38 \qquad \sum f_i z_i^2 = 914.$$

Hence, in terms of the units of the given observations,

Mean = 151·9950   Standard deviation = 31·7210.

If x_i denotes the midpoint of the ith class-interval in (ii), put $x_i = 161 + 42z_i$. Then,

$$\sum f_i z_i = -88 \qquad \sum f_i z_i^2 = 266.$$

Hence, in terms of the units of the given observations,

Mean = 151·7600   Standard deviation = 33·0213.

9 (i) The frequency distribution of the bull sperm areas is as follows:

Area (Midpoint)	Frequency (f_i)	Area (Midpoint)	Frequency (f_i)	Area (Midpoint)	Frequency (f_i)
124	5	142	68	160	12
127	6	145	83	163	4
130	10	148	69	166	1
133	25	151	54	169	1
136	44	154	35		
139	61	157	22		
Total					500

(ii) This frequency distribution is obtained by pooling the frequencies in (i) of consecutive pairs of class-intervals.

Area (Midpoint)	Frequency (f_i)	Area (Midpoint)	Frequency (f_i)	Area (Midpoint)	Frequency (f_i)
125·5	11	143·5	151	161·5	16
131·5	35	149·5	123	167·5	2
137·5	105	155·5	57		
Total					500

(iii) Subtract 100 from each of the ungrouped observations. For the ungrouped transformed observations,

Raw S.S. = 1 023 144   Grand total = 22 288.

Hence, for the original ungrouped observations,

Mean = 144·5760   Standard deviation = 7·7063.

If x_i denoted the midpoint of the ith class-interval in (i), put $x_i = 145 + 3z_i$. Then,

$$\sum f_i z_i = -74 \qquad \sum f_i z_i^2 = 3328.$$

Hence, in terms of the units of the given observations,

Mean = 144·5560   Standard deviation = 7·7348.

If x_i denotes the midpoint of the ith class-interval in (ii), put $x_i = 143·5 + 6z_i$. Then,

$$\sum f_i z_i = 85 \qquad \sum f_i z_i^2 = 871.$$

Hence, in terms of the units of the given observations,

Mean = 144·5200   Standard deviation = 7·8610.

10 (i) The observed frequency distribution of the number of heads in five tosses of a penny is as follows:

No. of heads (x_i)	Frequency (f_i)	No. of heads (x_i)	Frequency (f_i)	No. of heads (x_i)	Frequency (f_i)
0	12	2	106	4	57
1	54	3	94	5	7
Total					330

$\sum f_i x_i = 811$ $\qquad\qquad\qquad\qquad$ $\sum f_i x_i^2 = 2411$

Mean = 2·4576 $\qquad\qquad\qquad\qquad$ Standard deviation = 1·1270.

For an unbiased penny the expected number of heads in five trials is 2·5 and the standard deviation is $\frac{1}{2}\sqrt{5} = 1·1180$.

The expected frequency distribution is $\frac{330}{32}$ (1, 5, 10, 10, 5, 1) which has the frequencies

10·3125, 51·5625, 103·1250, 103·1250, 51·5625, 10·3125

corresponding to the six classes of the observed frequency distribution.

11 (i) Let x_i be the midpoint of the ith class-interval with frequency f_i. Make the transformation $x_i = 144·5 + 10z_i$. Then,

$$\sum f_i z_i = 9475; \quad \sum f_i z_i^2 = 46\ 871; \quad \sum f_i z_i^3 = 197\ 281; \quad \sum f_i z_i^4 = 1\ 304\ 075.$$

Hence, in terms of the original units of measurement,

Mean $= 156 \cdot 7274$. Standard deviation $= 21 \cdot 3405$.

(ii) The first four moments of z about the origin are:

$m_1'(z) = 1 \cdot 2227$; $m_2'(z) = 6 \cdot 0487$; $m_3'(z) = 25 \cdot 4589$; $m_4'(z) = 168 \cdot 2895$.
Hence the central moments

$$m_2(z) = 4 \cdot 5537; \quad m_3(z) = 6 \cdot 9276; \quad m_4(z) = 91 \cdot 3268;$$

whence $\quad\quad\quad\quad\quad g_1 = 0 \cdot 7129; \quad\quad g_2 = 1 \cdot 4042.$

(iii) Linear interpolation gives the quartiles

$$Q_1 = 142 \cdot 5330; \quad Q_2 = 154 \cdot 6732; \quad Q_3 = 158 \cdot 4348;$$

and $\quad\quad\quad\quad\quad\quad\quad$ S.I.Q.R. $= 7 \cdot 9509$.

12 (i) Let x_i be the midpoint of the ith class-interval with frequency f_i. Make the transformation $x_i = 67 \cdot 4375 + z_i$. Then,

$$\sum f_i z_i = 179; \quad \sum f_i z_i^2 = 56\ 809; \quad \sum f_i z_i^3 = 1769; \quad \sum f_i z_i^4 = 1\ 182\ 061.$$

Hence, in terms of the original units of measurement,

Mean $= 67 \cdot 4584$ Standard deviation $= 2 \cdot 5725$.

(ii) The first four moments of z about the origin are:

$$m_1'(z) = 0 \cdot 020\ 850; \quad m_2'(z) = 6 \cdot 617\ 239; \quad m_3'(z) = 0 \cdot 206\ 057;$$

$$m_4'(z) = 137 \cdot 689\ 109.$$

Hence the central moments

$$m_2(z) = 6 \cdot 6168; \quad m_3(z) = -0 \cdot 207\ 83; \quad m_4(z) = 137 \cdot 6892;$$

whence $\quad\quad\quad\quad\quad g_1 = -0 \cdot 012\ 21; \quad\quad g_2 = 0 \cdot 1449.$

(iii) Linear interpolation gives the quartiles

$$Q_1 = 65 \cdot 7155; \quad Q_2 = 67 \cdot 4668; \quad Q_3 = 69 \cdot 2110;$$

$$\text{S.I.Q.R.} = 1 \cdot 7478; \quad\quad \text{9th decile} = 70 \cdot 7355.$$

13 *Length of egg*

(i) Let x_i be the midpoint of the ith class-interval with frequency f_i. Make the transformation $x_i = 22 \cdot 0 + 0 \cdot 5 z_i$. Then,

$$\sum f_i z_i = 200; \quad \sum f_i z_i^2 = 1286; \quad \sum f_i z_i^3 = 2294; \quad \sum f_i z_i^4 = 18\ 098.$$

Hence, in terms of the original units of measurement,

Mean $= 22 \cdot 4115$ Standard deviation $= 1 \cdot 0763$.

(ii) The first four moments of z about the origin are:

$m'_1(z) = 0.823\ 045;$ $m'_2(z) = 5.2922;$ $m'_3(z) = 9.4403;$ $m'_4(z) = 74.4774.$

Hence the central moments

$$m_2(z) = 4.6148;\quad m_3(z) = -2.5118;\quad m_4(z) = 66.5313;$$

whence $g_1 = -0.2534;$ $g_2 = 0.1241.$

(iii) Linear interpolation gives the quartiles

$$Q_1 = 21.8125;\quad Q_2 = 22.4276;\quad Q_3 = 22.6356;$$

S.I.Q.R. $= 0.4116;$ 5 per cent point $= 20.7759;$ 95 per cent point $= 24.1274.$

Breadth of egg

(i) Let x_i be the midpoint of the ith class-interval with frequency f_i. Make the transformation $x_i = 16.5 + 0.5z_i$. Then,

$$\sum f_i z_i = 21;\quad \sum f_i z_i^2 = 423;\quad \sum f_i z_i^3 = 45;\quad \sum f_i z_i^4 = 3015.$$

Hence, in terms of the original units of measurement,

Mean $= 16.5432$ Standard deviation $= 0.6596.$

(ii) The first four moments of z about the origin are:

$m'_1(z) = 0.086\ 420;$ $m'_2(z) = 1.7407;$ $m'_3(z) = 0.185\ 185;$ $m'_4(z) = 12.4074.$

Hence the central moments

$$m_2(z) = 1.7332;\quad m_3(z) = -0.2648;\quad m_4(z) = 12.4212;$$

whence $g_1 = -0.1160;$ $g_2 = 1.1349.$

(iii) Linear interpolation gives the quartiles

$$Q_1 = 16.0565;\quad Q_2 = 16.5686;\quad Q_3 = 17.0156;$$

S.I.Q.R. $= 0.4796;$ 5 per cent point $= 15.5361;$ 95 per cent point $= 17.6117.$

14 (i) Let x_i be the midpoint of the ith class-interval with frequency f_i. Make the transformation $x_i = 5 + z_i$. Then,

$$\sum f_i z_i = -1908;\ \sum f_i z_i^2 = 31\ 440;\ \sum f_i z_i^3 = -46\ 134;\ \sum f_i z_i^4 = 708\ 924.$$

Hence, in terms of the original units of measurement,

Mean $= 3.8875$ Standard deviation $= 4.1358.$

(ii) The first four moments of z about the origin are:

$m'_1(z) = -1.1125;$ $m'_2(z) = 18.3324;$ $m'_3(z) = -26.9003;$ $m'_4(z) = 413.3668.$

Hence the central moments

$$m_2(z) = 17.0946;\quad m_3(z) = 31.5303;\quad m_4(z) = 425.2004;$$

whence $g_1 = 0.4461;$ $g_2 = -1.5450.$

15   (i) Let x_i be the midpoint of the ith class-interval with frequency f_i. Make the transformation $x_i = 43 + 5z_i$. Then,

$$\sum f_i z_i = -643; \quad \sum f_i z_i^2 = 33\ 033; \quad \sum f_i z_i^3 = 214\ 223; \quad \sum f_i z_i^4 = 5\ 456\ 253.$$

Hence, in terms of the original units of measurement,

Mean $= 40 \cdot 3364$ Standard deviation $= 26 \cdot 0320$.

(ii) The first four moments of z about the origin are:

$$m'_1(z) = -0 \cdot 532\ 726; \quad m'_2(z) = 27 \cdot 3679; \quad m'_3(z) = 177 \cdot 4838; \quad m'_4(z) = 4520 \cdot 5079.$$

Hence the central moments

$$m_2(z) = 27 \cdot 0841; \quad m_3(z) = 220 \cdot 9202; \quad m_4(z) = 4945 \cdot 0688;$$

whence $g_1 = 1 \cdot 5674; \qquad g_2 = 3 \cdot 7413.$

(iii) Linear interpolation gives the quartiles

$$Q_1 = 22 \cdot 3225; \quad Q_2 = 34 \cdot 8973; \quad Q_3 = 51 \cdot 2783;$$

S.I.Q.R. $= 14 \cdot 4779$; 5 per cent point $= 9 \cdot 7543$; 95 per cent point $= 91 \cdot 1346$.

16   (i) Let x_i be the midpoint of the ith class-interval with frequency f_i. Make the transformation $x_i = 22 + z_i$. Then,

$$\sum f_i z_i = 1738; \quad \sum f_i z_i^2 = 47\ 646; \quad \sum f_i z_i^3 = 205\ 252; \quad \sum f_i z_i^4 = 2\ 353\ 602.$$

Hence, in terms of the original units of measurement,

Mean $= 22 \cdot 6952$ Standard deviation $= 4 \cdot 3107$.

(ii) The first four moments of z about the origin are:

$$m'_1(z) = 0 \cdot 6952; \quad m'_2(z) = 19 \cdot 0584; \quad m'_3(z) = 82 \cdot 1008; \quad m'_4(z) = 941 \cdot 4408.$$

Hence the central moments

$$m_2(z) = 18 \cdot 5751; \quad m_3(z) = 43 \cdot 0246; \quad m_4(z) = 767 \cdot 7001;$$

whence $g_1 = 0 \cdot 5374; \qquad g_2 = -0 \cdot 7750.$

(iii) Linear interpolation gives the quartiles

$$Q_1 = 19 \cdot 1411; \quad Q_2 = 21 \cdot 6033; \quad Q_3 = 26 \cdot 1545;$$

S.I.Q.R. $= 3 \cdot 5067$; 5 per cent point $= 17 \cdot 0578$; 95 per cent point $= 30 \cdot 4271$.

17   *Class-interval 2 units*

(i) Let x_i be the midpoint of the ith class-interval with frequency f_i. Make the transformation $x_i = 183 + 2z_i$. Then,

$$\sum f_i z_i = -37; \quad \sum f_i z_i^2 = 13\ 451; \quad \sum f_i z_i^3 = -2989; \quad \sum f_i z_i^4 = 834\ 203.$$

Hence, in terms of the original units of measurement,

Mean $= 182 \cdot 8767$ Standard deviation $= 9 \cdot 4767$.

(ii) The first four moments of z about the origin are:

$m_1'(z) = -0 \cdot 061\ 667; \ m_2'(z) = 22 \cdot 4183; \ m_3'(z) = -4 \cdot 9817; \ m_4'(z) = 1390 \cdot 3383.$

Hence the central moments

$$m_2(z) = 22 \cdot 4145; \quad m_3(z) = -0 \cdot 8348; \quad m_4(z) = 1389 \cdot 6210;$$

whence $g_1 = -0 \cdot 0079; \qquad g_2 = -0 \cdot 2341.$

(iii) Linear interpolation gives the quartiles

$$Q_1 = 176 \cdot 4211; \quad Q_2 = 183 \cdot 1045; \quad Q_3 = 189 \cdot 7838;$$

S.I.Q.R. $= 6 \cdot 6814$; 5 per cent point $= 166 \cdot 3750$; 95 per cent point $= 197 \cdot 5238$.

Class-interval 4 units

(i) Let x_i be the midpoint of the ith class-interval with frequency f_i. Make the transformation $x_i = 184 + 4z_i$. Then,

$$\sum f_i z_i = -176; \quad \sum f_i z_i^2 = 3496; \quad \sum f_i z_i^3 = -3302; \quad \sum f_i z_i^4 = 56\ 500.$$

Hence, in terms of the original units of measurement,

Mean $= 182 \cdot 8267$ Standard deviation $= 9 \cdot 5918$.

(ii) The first four moments of z about the origin are:

$m_1'(z) = -0 \cdot 293\ 333; \ m_2'(z) = 5 \cdot 8267; \ m_3(z) = -5 \cdot 5033; \ m_4'(z) = 94 \cdot 1667.$

Hence the central moments

$$m_2(z) = 5 \cdot 7407; \quad m_3(z) = -0 \cdot 4263; \quad m_4(z) = 90 \cdot 6954;$$

whence $g_1 = -0 \cdot 0310; \qquad g_2 = -0 \cdot 2480.$

(iii) Linear interpolation gives the quartiles

$$Q_1 = 176 \cdot 0000; \quad Q_2 = 183 \cdot 2650; \quad Q_3 = 189 \cdot 7838;$$

S.I.Q.R. $= 6 \cdot 8919$; 5 per cent point $= 166 \cdot 3529$; 95 per cent point $= 197 \cdot 5745$.

18 (i) Let x_i be the midpoint of the ith class-interval with frequency f_i. Make the transformation $x_i = 12 \cdot 5 + 0 \cdot 5z_i$. Then,

$$\sum f_i z_i = -25; \quad \sum f_i z_i^2 = 172\ 751; \quad \sum f_i z_i^3 = 436\ 175; \quad \sum f_i z_i^4 = 17\ 400\ 287.$$

Hence, in terms of the original units of measurement,

Mean $= 12 \cdot 4974$ Standard deviation $= 2 \cdot 9801$.

(ii) The first four moments of z about the origin are:

$m_1'(z) = -0 \cdot 005\ 140; \ m_2'(z) = 35 \cdot 5162; \ m_3'(z) = 89 \cdot 6741; \ m_4'(z) = 3577 \cdot 3616.$

Hence the central moments

$$m_2(z) = 35 \cdot 5162; \quad m_3(z) = 90 \cdot 2218; \quad m_4(z) = 3579 \cdot 2109;$$

whence $g_1 = 0.4263;$ $g_2 = -0.1625.$

(iii) Linear interpolation gives the quartiles

$$Q_1 = 10.1041; \quad Q_2 = 12.4380; \quad Q_3 = 14.4094;$$

S.I.Q.R. = 2.1526; 5 per cent point = 8.0359; 95 per cent point = 17.6596.

19   Bordighera: Mean = 26.7179   Standard deviation = 3.7750.

    Guernsey: Mean = 27.9558   Standard deviation = 4.8025.

    Surrey: Mean = 32.9380   Standard deviation = 5.7096.

    Dorset: Mean = 35.5386   Standard deviation = 5.9916.

Pooled data

(i) Let x_i be the midpoint of the ith class-interval with frequency f_i. Make the transformation $x_i = 27.5 + 2z_i$. Then,

$$\sum f_i z_i = 3262; \ \sum f_i z_i^2 = 25\ 824; \ \sum f_i z_i^3 = 145\ 186; \ \sum f_i z_i^4 = 1\ 273\ 872.$$

Hence, in terms of the units of the original measurements,

    Mean = 30.5358   Standard deviation = 6.2345.

(ii) The first four moments of z about the origin are:

$$m_1'(z) = 1.517\ 915; \ m_2'(z) = 12.0168; \ m_3'(z) = 67.5598; \ m_4'(z) = 592.7743.$$

Hence the central moments

$$m_2(z) = 9.7127; \quad m_3(z) = 19.8332; \quad m_4(z) = 332.7730;$$

whence $g_1 = 0.6552;$ $g_2 = 0.5275.$

20   (i) Let x_i be the midpoint of the ith class-interval with frequency f_i. Make the transformation $x_i = 48 + 3z_i$. Then,

$$\sum f_i z_i = 891; \ \sum f_i z_i^2 = 15\ 061; \ \sum f_i z_i^3 = 28\ 563; \ \sum f_i z_i^4 = 607\ 309.$$

Hence, in terms of the units of the original measurements,

    Mean = 50.6730   Standard deviation = 11.3372.

(ii) The first four moments of z about the origin are:

$$m_1'(z) = 0.891; \ m_2'(z) = 15.061; \ m_3'(z) = 28.563; \ m_4'(z) = 607.309.$$

Hence the central moments

$$m_2(z) = 14.2671; \quad m_3(z) = -10.2804; \quad m_4(z) = 575.3597;$$

whence $g_1 = -0.1908;$ $g_2 = -0.1734.$

(iii) Linear interpolation gives the quartiles

$$Q_1 = 43 \cdot 4032; \quad Q_2 = 51 \cdot 1875; \quad Q_3 = 58 \cdot 3404;$$

S.I.Q.R. $= 7 \cdot 4686$; 5 per cent point $= 30 \cdot 5625$; 95 per cent point $= 68 \cdot 8714$.

21 If x is the weight of grain in a sack, then the probability that the sack is underweight is

$$P(x < 112) = \Phi\left(\frac{112 - 112 \cdot 375}{0 \cdot 226}\right) = 1 - \Phi(1 \cdot 6593) = 0 \cdot 0486.$$

If μ is the new mean weight, then $P(x < 112) = \Phi\left(\dfrac{112 - \mu}{0 \cdot 226}\right) = 0 \cdot 05$,

which gives $\mu - 112 = 0 \cdot 226 \times 1 \cdot 645$ or $\mu = 112 \cdot 371\ 77$.

Therefore the average decrease in the weight of grain per sack is $0 \cdot 003\ 23$. Hence the saving to the farmer in 10 000 sacks is 32 lb.

To allow for evaporation, the mean weight per sack should be $112 \cdot 621\ 77$. This means an increase of $0 \cdot 246\ 77$ over the old mean. Hence the total increase in grain for 10 000 sacks is $2467 \cdot 7$ or 2468 lb.

22 Let x, y, and z be the times of the journey through the city centre, the village, and the ring-road, respectively. Then

$$P(x > 30) = 1 - \Phi\left(\frac{30 - 28}{7 \cdot 5}\right) = 1 - \Phi(0 \cdot 2667) = 0 \cdot 3948;$$

$$P(x > 35) = 1 - \Phi\left(\frac{35 - 28}{7 \cdot 5}\right) = 1 - \Phi(0 \cdot 9333) = 0 \cdot 1753;$$

$$P(y > 30) = 1 - \Phi\left(\frac{30 - 32}{4 \cdot 3}\right) = \Phi(0 \cdot 4651) = 0 \cdot 6791;$$

$$P(y > 35) = 1 - \Phi\left(\frac{35 - 32}{4 \cdot 3}\right) = 1 - \Phi(0 \cdot 6977) = 0 \cdot 2427;$$

$$P(z > 30) = 1 - \Phi\left(\frac{30 - 33}{1 \cdot 2}\right) = \Phi(2 \cdot 5) = 0 \cdot 9938;$$

$$P(z > 35) = 1 - \Phi\left(\frac{35 - 33}{1 \cdot 2}\right) = 1 - \Phi(1 \cdot 6667) = 0 \cdot 0478.$$

Hence for (i) the route through the city centre is best whilst for (ii) the route via the ring-road. This is based on the criterion of least probability of being delayed.

23 Let x and y be the lifetimes of lamps produced by A and B respectively. Then

$$P(x < 3000) = \Phi\left(\frac{3000 - 3675}{430}\right) = 1 - \Phi(1 \cdot 5698) = 0 \cdot 0582;$$

$$P(y < 3000) = \Phi\left(\frac{3000 - 4500}{980}\right) = 1 - \Phi(1 \cdot 5306) = 0 \cdot 0629;$$

$$P(x < 3500) = \Phi\left(\frac{3500 - 3675}{430}\right) = 1 - \Phi(0 \cdot 4070) = 0 \cdot 3420;$$

$$P(y < 3500) = \Phi\left(\frac{3500 - 4500}{980}\right) = 1 - \Phi(1 \cdot 0204) = 0 \cdot 1538.$$

Hence A is the better process when the customer's specification is 3000 hours, whilst B is the better process when the specification is 3500 hours.

$$E(x - y) = -825; \quad \text{var}(x - y) = 1\ 145\ 300; \quad \text{s.d.}(x - y) = 1070 \cdot 19.$$

Hence

$$P(x - y > 0) = 1 - \Phi(825/1070 \cdot 19) = 1 - \Phi(0 \cdot 7709) = 0 \cdot 2204.$$

24 Let x denote the diameter of a rod; $E(x) = \mu$ and s.d.$(x) = \sigma$. Then

$$P(x \leqslant 3 \cdot 995) = 1 - \Phi\left(\frac{\mu - 3 \cdot 995}{\sigma}\right) = 0 \cdot 05;$$

$$P(x \geqslant 4 \cdot 005) = 1 - \Phi\left(\frac{4 \cdot 005 - \mu}{\sigma}\right) = 0 \cdot 12.$$

Hence $\mu - 3 \cdot 995 = 1 \cdot 6450\,\sigma\,; \quad 4 \cdot 005 - \mu = 1 \cdot 1750\,\sigma$.

Therefore $\mu = 4 \cdot 000\ 833\ 3$ and $\sigma = 0 \cdot 003\ 546\ 1$.

If the new standard deviation is σ_i, then

$$P(x \geqslant 4 \cdot 005) = 1 - \Phi(0 \cdot 004\ 166\ 7/\sigma_i) = 0 \cdot 02,$$

whence $0 \cdot 004\ 166\ 7/\sigma_i = 2 \cdot 05375$ or $\sigma_i = 0 \cdot 002\ 028\ 8$.
The percentage reduction in s.d.(x) is $42 \cdot 79$.

25 Let x be the weight of the contents of a carton. Then the proportion of overweight cartons is

$$P(x > 20) = 1 - \Phi\left(\frac{20 - 20 \cdot 37}{0 \cdot 5634}\right) = \Phi(0 \cdot 6567) = 0 \cdot 7443.$$

If the new mean and standard deviation of x are μ and σ respectively, then

$$P(x > 20) = 1 - \Phi\left(\frac{20 - \mu}{\sigma}\right) = 0 \cdot 372\ 15;$$

$$P(x < 19\cdot9) = \Phi\left(\frac{19\cdot9 - \mu}{\sigma}\right) = 0\cdot01.$$

Hence $20 - \mu = 0\cdot3262\sigma; \quad \mu - 19\cdot9 = 2\cdot3263\sigma.$

Therefore $\mu = 19\cdot987\,702$ and $\sigma = 0\cdot0377.$

The average percentage saving in detergent is $1\cdot88.$

26 Suppose x is the resistance of a carbon rod. Then the proportion of rods with resistances less than 318 is

$$P(x < 318) = \Phi\left(\frac{318 - 319\cdot98}{1\cdot7645}\right) = 1 - \Phi(1\cdot1221) = 0\cdot1309;$$

and the proportion with resistances greater than 321 is

$$P(x > 321) = 1 - \Phi\left(\frac{321 - 319\cdot98}{1\cdot7645}\right) = 1 - \Phi(0\cdot5781) = 0\cdot2816.$$

If μ and σ are the new mean and standard deviation, then

$$P(x < 318) = 1 - \Phi\left(\frac{\mu - 318}{\sigma}\right) = 0\cdot065\,45;$$

$$P(x > 321) = 1 - \Phi\left(\frac{321 - \mu}{\sigma}\right) = 0\cdot140\,80.$$

Hence $\mu - 318 = 1\cdot5106\sigma; \quad 321 - \mu = 1\cdot0767.$

Therefore $\mu = 319\cdot7516$ and $\sigma = 1\cdot1595.$

With these new values of μ and σ,

$$P(x < 317) = \Phi\left(\frac{317 - 319\cdot7516}{1\cdot1595}\right) = 1 - \Phi(2\cdot3731) = 0\cdot008\,82;$$

and $P(x > 322) = 1 - \Phi\left(\dfrac{322 - 319\cdot7516}{1\cdot1595}\right) = 1 - \Phi(1\cdot9391) = 0\cdot0262.$

The new limits satisfy the upper but not the lower specification limit of the second customer.

27 Let x and y be the amounts of petrol delivered by the pumps of the two companies for each gallon metered. Then

(i) $P(x < 1) = \Phi\left(\dfrac{1 - 1\cdot004}{0\cdot007\,568}\right) = 1 - \Phi(0\cdot5285) = 0\cdot2986;$

and $P(y < 1) = \Phi\left(\dfrac{1 - 1\cdot0038}{0\cdot007\,249}\right) = 1 - \Phi(0\cdot5242) = 0\cdot3001.$

The first company's meters have a smaller probability than those of the second for giving short measure.

(ii) Since $\text{var}(x - y) = 0 \cdot 000\ 109\ 823$, s.d.$(x - y) = 0 \cdot 010\ 480$.

Therefore $P(x - y > 0) = 1 - \Phi\left(\dfrac{-0 \cdot 0002}{0 \cdot 010\ 480}\right) = \Phi(0 \cdot 0191) = 0 \cdot 5076$.

Thus the probability is $> \frac{1}{2}$ that the amount delivered by pumps of the first company will be greater than that delivered by those of the second.

(iii) $E(x) = 1 \cdot 004$ and $E(y) = 1 \cdot 0038$ so that $E(x) > E(y)$. Therefore, in the long run, customers will obtain a greater average amount from the first company's pumps than from those of the second.

28   Let x and y be random variables representing the weights of cartons of A and B respectively such that $E(x) = \theta_1$, s.d.$(x) = \sigma_1$, $E(y) = \theta_2$, s.d.$(y) = \sigma_2$. Then, for cartons of A,

$$P(x < 24 \cdot 5) = 1 - \Phi\left(\frac{\theta_1 - 24 \cdot 5}{\sigma_1}\right) = 0 \cdot 15;$$

$$P(x > 26) = 1 - \Phi\left(\frac{26 - \theta_1}{\sigma_1}\right) = 0 \cdot 06;$$

whence $\theta_1 - 24 \cdot 5 = 1 \cdot 0364\ \sigma_1;$ $26 - \theta_1 = 1 \cdot 5548\ \sigma_1$.
Therefore $\theta_1 = 25 \cdot 1000$ and $\sigma_1 = 0 \cdot 5789$.
Similarly, for cartons of B,

$$P(y < 24) = 1 - \Phi\left(\frac{\theta_2 - 24}{\sigma_2}\right) = 0 \cdot 05;$$

$$P(y > 26 \cdot 5) = 1 - \Phi\left(\frac{26 \cdot 5 - \theta_2}{\sigma_2}\right) = 0 \cdot 08;$$

whence $\theta_2 - 24 = 1 \cdot 6450\ \sigma_2;$ $26 \cdot 5 - \theta_2 = 1 \cdot 4051\ \sigma_2$.
Therefore $\theta_2 = 25 \cdot 3483$ and $\sigma_2 = 0 \cdot 8196$.

(i) Since $\theta_2 > \theta_1$, it follows that, on the average, detergent B is bette value than detergent A.

(ii) Since $\text{var}(y - x) = 1 \cdot 006\ 869\ 37$,   s.d.$(y - x) = 1 \cdot 0035$. Hence

$$P(y - x > 0) = 1 - \Phi(-0 \cdot 2483/1 \cdot 0035) = \Phi(0 \cdot 2474) = 0 \cdot 5977.$$

Therefore the probability is $> \frac{1}{2}$ that detergent B is better value than detergent A.

(iii)
$$P(x < 25) = \Phi\left(\frac{25 - 25 \cdot 1000}{0 \cdot 5789}\right) = 1 - \Phi(0 \cdot 1727) = 0 \cdot 4314;$$

$$P(y < 25) = \Phi\left(\frac{25 - 25 \cdot 3483}{0 \cdot 8196}\right) = 1 - \Phi(0 \cdot 4250) = 0 \cdot 3354.$$

Therefore the probability of obtaining short measure is less for detergent B than for detergent A.

Price of B 4 per cent more than that of A

(i) Now B will be a better buy on the average if $\theta_2 > 1\cdot04\theta_1$. But $1\cdot04\theta_1 = 26\cdot1040$ which is $> \theta_2$. Hence B is no longer a better buy than A.

(ii) To determine $P(y - 1\cdot04x > 0)$ observe that

$$\mathrm{var}(y - 1\cdot04x) = \mathrm{var}(y) + (1\cdot04)^2\,\mathrm{var}(x) = 1\cdot034\ 216,$$

so that s.d. $(y - 1\cdot04x) = 1\cdot0169$. Hence

$$P(y - 1\cdot04x > 0) = 1 - \Phi(0\cdot7557/1\cdot0169) = 1 - \Phi(0\cdot7431) = 0\cdot2287.$$

Since $P(y - 1\cdot04x > 0)$ is $< \frac{1}{2}$, detergent B is no longer a better buy than detergent A.

(iii) To be comparable, $P(x < 25)$ corresponds to $P(y < 26)$. But $P(x < 25) = 0\cdot4314$; and

$$P(y < 26) = \Phi\left(\frac{26 - 25\cdot3483}{0\cdot8196}\right) = \Phi(0\cdot7951) = 0\cdot7867.$$

The probability of obtaining short measure of A is now less than that of B.

29 For £9 the commercial traveller can buy 36 gallons of standard grade petrol, and to cover the journey he needs an average performance of at least $1150/36 = 31\cdot9444$ miles per gallon. If x is the random variable denoting the miles per gallon obtained, then

$$P(x \geqslant 31\cdot9444) = 1 - \Phi\left(\frac{31\cdot9444 - 28\cdot75}{3\cdot28}\right) = 1 - \Phi(0\cdot9739) = 0\cdot1650.$$

For £9 the traveller can buy $33\cdot9623$ gallons of superfine petrol. Therefore to cover the journey, he needs an average performance of at least $1150/33\cdot9623 = 33\cdot8611$ miles per gallon. If y is the random variable denoting the miles per gallon obtained, then

$$P(y \geqslant 34\cdot0741) = 1 - \Phi\left(\frac{33\cdot8611 - 29\cdot62}{4\cdot25}\right) = 1 - \Phi(0\cdot9979) = 0\cdot1592.$$

Hence the standard grade of petrol is the better buy since it gives the traveller a higher probability of finishing the journey within his petrol allowance.

30 If the undergraduates buy the mini, then £12 is available for petrol. This gives 48 gallons, and so, to cover the journey, the mini should average at least $2000/48 = 41\cdot6667$ miles per gallon. If x is the random variable representing the average mileage attained with the mini, then

$$P(x \geqslant 41\cdot6667) = 1 - \Phi\left(\frac{41\cdot6667 - 39\cdot65}{4\cdot58}\right) = 1 - \Phi(0\cdot4403) = 0\cdot3299.$$

If the undergraduates buy the saloon, then £20 is available for petrol. This gives 80 gallons, and to cover the journey the saloon should average at least $2000/80 = 25$ miles per gallon. If y is the random variable representing the average mileage attained with the saloon, then

$$P(y \geqslant 25) = 1 - \Phi\left(\frac{25 - 23 \cdot 95}{2 \cdot 89}\right) = 1 - \Phi(0 \cdot 3633) = 0 \cdot 3582.$$

The saloon is the better buy because it gives a higher probability of covering the journey within the petrol allowance.

31   Suppose x is the random variable representing the diameter of a bolt. If $E(x) = 0 \cdot 500\ 24$, then

$$P(x < 0 \cdot 485) = \Phi\left(\frac{0 \cdot 485 - 0 \cdot 500\ 24}{0 \cdot 009\ 45}\right) = 1 - \Phi(1 \cdot 6127) = 0 \cdot 0534;$$

$$P(x > 0 \cdot 515) = 1 - \Phi\left(\frac{0 \cdot 515 - 0 \cdot 500\ 24}{0 \cdot 009\ 45}\right) = 1 - \Phi(1 \cdot 5619) = 0 \cdot 0592.$$

If $E(x)$ is altered to $0 \cdot 499\ 76$, then

$$P(x < 0 \cdot 485) = \Phi\left(\frac{0 \cdot 485 - 0 \cdot 499\ 76}{0 \cdot 009\ 45}\right) = 0 \cdot 0592;$$

$$P(x > 0 \cdot 515) = 1 - \Phi\left(\frac{0 \cdot 515 - 0 \cdot 499\ 76}{0 \cdot 009\ 45}\right) = 0 \cdot 0534.$$

If the reduced s.d. $(x) = \sigma$ and $E(x) = 0 \cdot 500\ 24$, then

$$P(x > 0 \cdot 515) = 1 - \Phi(0 \cdot 014\ 76/\sigma) = 0 \cdot 025,$$

whence $0 \cdot 014\ 76/\sigma = 1 \cdot 96$ or $\sigma = 0 \cdot 007\ 531.$
With this new standard deviation and $E(x) = 0 \cdot 500\ 24$,

$$P(x < 0 \cdot 485) = \Phi\left(\frac{0 \cdot 485 - 0 \cdot 500\ 24}{0 \cdot 007\ 531}\right) = 1 - \Phi(2 \cdot 0236) = 0 \cdot 0215.$$

32   Let x be the random variable representing the weight (in oz) of a cake of soap. If $E(x) = \mu$, then

$$P(x < 6 \cdot 5) = \Phi\left(\frac{6 \cdot 5 - \mu}{0 \cdot 078}\right) = 0 \cdot 004,$$

whence $\mu - 6 \cdot 5 = 2 \cdot 6518 \times 0 \cdot 078$ or $\mu = 6 \cdot 706\ 840.$
Therefore the expected cost per cake is $9 \cdot 790\ 034$p.
If the new s.d.$(x) = 0 \cdot 039$, then the equation for μ is

$$\mu - 6 \cdot 5 = 2 \cdot 6518 \times 0 \cdot 039 \quad \text{or} \quad \mu = 6 \cdot 603\ 420.$$

Therefore the expected cost per cake is 9·771 067p. Hence the decrease in the expected cost per cake is 0·018 967p. Therefore the expected total saving per week is

$$£\frac{0·018\ 967 \times 450\ 000}{100} = £85·35 \quad \text{or} \quad £85.$$

33   Let x be the random variable denoting the weight (in oz) of a packet. If $E(x) = \mu$ and s.d.$(x) = \sigma$, then

$$P(x < 16) = \Phi\left(\frac{16 - \mu}{\sigma}\right) = 0·10;$$

$$P(x > 16·25) = 1 - \Phi\left(\frac{16·25 - \mu}{\sigma}\right) = 0·05.$$

Hence $\mu - 16 = 1·2816\,\sigma$; $16·25 - \mu = 1·6450\,\sigma$;

whence $\mu = 16·109\ 479$ and $\sigma = 0·085\ 424$.
Hence the expected cost per packet is

$$16·75 + 0·2689\mu - 0·000\ 54(\sigma^2 + \mu^2) = 20·941\ 697p.$$

Therefore the expected total daily cost is

$$£\ \frac{20·941\ 697 \times 375\ 000}{100} = £78\ 531·36.$$

After economies and the increase in the price of edible oils, the expected cost per packet is

$$16·734 + 0·2795\mu - 0·000\ 39(\sigma^2 + \mu^2) = 21·135\ 385p.$$

Therefore the increased expected total daily cost is

$$£\frac{21·135\ 385 \times 375\ 000}{100} = £79\ 257·69.$$

Net expected increase in daily cost is $= £726·33$ or £726.

34   Under the old system of bottling, let $E(y) = \mu$ and s.d.$(y) = \sigma$. Then

$$P(y < 1) = \Phi\left(\frac{1 - \mu}{\sigma}\right) = 0·08;$$

$$P(y > 1·05) = 1 - \Phi\left(\frac{1·05 - \mu}{\sigma}\right) = 0·30.$$

Hence $\mu - 1 = 1·4051\,\sigma$; $1·05 - \mu = 0·5244\,\sigma$.

Therefore $\mu = 1·036\ 411$ and $\sigma = 0·025\ 913$.
   Hence $E(x) = 8E(y) = 8·291\ 288$ and $E(x^2) = 64E(y^2) = 64(\sigma^2 + \mu^2)$
$= 68·788\ 416$.
   Therefore the expected cost of x gallons of milk in new pence is

$$22 \cdot 67 + 0 \cdot 1675 \, E(x) - 0 \cdot 000 \, 478 \, E(x^2) = 24 \cdot 025 \, 910.$$

The total quantity of milk processed in a day is

$$500 \, 000 \, x/8 \text{ gallons.}$$

Therefore the total expected daily cost is

$$\text{£} \frac{24 \cdot 025 \, 910 \times 500 \, 000}{8 \times 100} = \text{£}15 \, 016 \cdot 19.$$

Under the new bottling system, suppose $E(y) = \mu_1$ and s.d.$(y) = \sigma_1$.

Then
$$P(y < 1) = \Phi \left(\frac{1 - \mu_1}{\sigma_1} \right) = 0 \cdot 08;$$

$$P(y > 1 \cdot 05) = 1 - \Phi \left(\frac{1 \cdot 05 - \mu_1}{\sigma_1} \right) = 0 \cdot 15.$$

Hence $\quad \mu_1 - 1 = 1 \cdot 4051 \, \sigma_1; \quad 1 \cdot 05 - \mu_1 = 1 \cdot 0364 \, \sigma_1.$

Therefore $\quad \mu_1 = 1 \cdot 028 \, 775 \quad$ and $\quad \sigma_1 = 0 \cdot 020 \, 479.$
Hence $\quad E(x) = 8 \cdot 230 \, 200 \quad$ and $\quad E(x^2) = 67 \cdot 763 \, 008.$
Therefore the new expected cost of x gallons of milk in new pence is

$$22 \cdot 78 + 0 \cdot 1784 \, E(x) - 0 \cdot 000 \, 596 \, E(x^2) = 24 \cdot 207 \, 881.$$

Therefore the new total expected daily cost is

$$\text{£} \frac{500 \, 000 \times 24 \cdot 207 \, 881}{8 \times 100} = \text{£}15 \, 129 \cdot 93.$$

Net expected increase in daily cost is = £113·74 or £114.

35 (i) If x_i is the midpoint of the ith class-interval with frequency f_i, make the transformation $x_i = 134 \cdot 5 + 4z_i$. Then,

$$\sum f_i z_i = -25 \qquad \sum f_i z_i^2 = 1663.$$

Hence, in terms of the units of the original measurements,

Mean = 134·3559   Standard deviation = 6·1947.

(ii)

Upper limit of class-interval	Observed frequency	Expected frequency	Upper limit of class-interval	Observed frequency	Expected frequency
116·5	0	1·374	140·5	148	141·625
120·5	8	7·405	144·5	69	76·264
124·5	28	29·953	148·5	15	27·413
128·5	82	80·865	152·5	16	6·600
132·5	140	145·573	∞	0	1·180
136·5	188	175·749			
Total				694	694·001

Expected frequency below 116·5 is 1·374, above 152·5 is 1·180.

36   (i) Let x_i be the midpoint of the ith class-interval with frequency f_i. Make the transformation $x_i = 49 \cdot 5 + 4z_i$. Then,

$$\sum f_i z_i = -447; \quad \sum f_i z_i^2 = 2365; \quad \sum f_i z_i^3 = -2649; \quad \sum f_i z_i^4 = 15\ 601.$$

The first four moments of z about the origin are:

$$m_1'(z) = -0 \cdot 4470; \quad m_2'(z) = 2 \cdot 3650; \quad m_3'(z) = -2 \cdot 6490; \quad m_4'(z) = 15 \cdot 6010.$$

Hence the central moments

$$m_2(z) = 2 \cdot 165\ 191; \quad m_3(z) = 0 \cdot 343\ 836; \quad m_4(z) = 13 \cdot 580\ 107;$$

whence $\qquad\qquad\qquad g_1 = 0 \cdot 1079; \qquad g_2 = -0 \cdot 1033.$

In terms of the original units of measurement, the first four moments are:

$$m_1'(x) = 47 \cdot 7120; \quad m_2'(x) = 34 \cdot 6431; \quad m_3'(x) = 22 \cdot 0055; \quad m_4'(x) = 3476 \cdot 5074.$$

The variance of the frequency distribution is $34 \cdot 6778$ and the s.d. is $5 \cdot 8888$.

(ii)

Upper limit of class-interval	Observed frequency	Expected frequency	Upper limit of class-interval	Observed frequency	Expected frequency
27·5	0	0·30	51·5	263	252·24
31·5	1	2·65	55·5	156	167·20
35·5	14	16·09	59·5	67	70·28
39·5	56	62·47	63·5	23	18·97
43·5	172	155·80	67·5	3	3·28
47·5	245	248·33	∞	0	0·39
Total				1000	1000·00

(iii) Error in lower tail is $P(x < 27 \cdot 5) = 0 \cdot 0003$.

Error in upper tail is $P(x > 67 \cdot 5) = 0 \cdot 0004$.

(iv)

$$\text{(a) } P(x < 35) = \Phi\left(\frac{35 - 47 \cdot 712}{5 \cdot 8888}\right) = 1 - \Phi(2 \cdot 159) = 0 \cdot 015\ 43;$$

$$\text{(b) } P(x > 50) = 1 - \Phi\left(\frac{50 - 47 \cdot 712}{5 \cdot 8888}\right) = 1 - \Phi(0 \cdot 389) = 0 \cdot 348\ 64;$$

$$\text{(c) } P(x > 40) = 1 - \Phi\left(\frac{40 - 47 \cdot 712}{5 \cdot 8888}\right) = 1 - \Phi(-1 \cdot 310) = 0 \cdot 904\ 90;$$

$$P(x < 48) = \Phi\left(\frac{48 - 47 \cdot 712}{5 \cdot 8888}\right) = \Phi(0 \cdot 049) = 0 \cdot 519\ 54.$$

Then $\qquad P(40 < x < 48) = P(x < 48) + P(x > 40) - 1 = 0 \cdot 424\ 44.$

(v) $P(a) = [P(x < 35)]^2 = 0 \cdot 000\ 24;$

$P(b) = 2P(x < 35)P(x > 50) = 0 \cdot 010\ 76;$

$$P(c) = [P(40 < x < 48)]^2 = 0\cdot180\ 15;$$

$$P(d) = [P(x > 50)]^2 = 0\cdot121\ 55.$$

(e) If X is the event $x > 35$ and Y the event $x < 50$, then X and Y are not mutually exclusive. Define the mutually exclusive events

$$A : x < 35; \quad B : x > 50; \quad C : 35 < x < 50.$$

Then $X = B + C$ and $Y = B + C$. Therefore $XY = AB + AC + BC + C^2$. Hence $P(e) = 2P(A)P(B) + 2P(A)P(C) + 2P(B)P(C) + P^2(C)$

$$= 1 - P^2(A) - P^2(B), \text{ since } P(A) + P(B) + P(C) = 1$$

$$= 0\cdot878\ 21.$$

37   The probability that there are w defective fuses in a box is

$$B(w) = \binom{24}{w}(0\cdot03)^w(0\cdot97)^{24-w}, \quad \text{for } 0 \leqslant w \leqslant 24,$$

whence
$$B(0) = 0\cdot481\ 417; \quad B(1) = 0\cdot357\ 340;$$

$$B(2) = 0\cdot127\ 095; \quad B(3) = 10^{-1}\times 0\cdot288\ 257.$$

The probability that a box contains at least one defective fuse is

$$1 - B(0) = 0\cdot518\ 58.$$

$$P(i) = [1 - B(0)]^4 = 0\cdot072\ 32;$$

$$P(ii) = [B(1)]^4 = 0\cdot016\ 31;$$

$$P(iii) = \left[\sum_{w=0}^{3} B(w)\right]^4 = 0\cdot978\ 88;$$

$$P(iv) = \binom{4}{1}\left[\sum_{w=0}^{1} B(w)\right]^3\left[1 - \sum_{w=0}^{1} B(w)\right] = 0\cdot380\ 58.$$

(v) If X is the event $w \leqslant 3$ and Y the event $w \geqslant 3$, then X and Y are not mutually exclusive. If we define the mutually exclusive events

$$A : w \leqslant 2, \quad B : w = 3, \quad C : w \geqslant 4,$$

then
$$X = A + B \quad \text{and} \quad Y = B + C.$$

Also, $P(A) = 0\cdot965\ 852; \quad P(B) = 0\cdot028\ 826; \quad P(C) = 0\cdot005\ 322.$

Since the four boxes are statistically independent, the realisation of (v) is equivalent to the realisation of two X's and two Y's in any order. Therefore the distinct combinations of events constituting (v) are:

$$A^2B^2, \ AB^3, \ B^4, \ A^2BC, \ AB^2C, \ B^3C, \ A^2C^2, \ ABC^2, \ B^2C^2,$$

as can be seen from the algebraic product $(B + C)^2(A + B)^2$. Therefore the combinations of events complementary to (v) are:

$$A^4, \ C^4, \ A^3B, \ A^3C, \ AC^3, \ BC^3.$$

Hence, taking permutations into consideration,

$P(v) = 1 - P^4(A) - P^4(C) -$

$\qquad - \binom{4}{1} [P^3(A)P(B) + P^3(A)P(C) + P^3(C)P(A) + P^3(C)P(B)]$

$= 1 - P^3(A)[4 - 3P(A)] - P^3(C)[4 - 3P(C)]$, since $P(A) + P(B) + P(C) = 1$

$= 0 \cdot 006\ 68.$

    (vi) The expected numbers are:

        (a) $100B(2) = 13$;

        (b) $100[1 - B(0) - B(1)] = 16$;

        (c) $100[B(0) + B(1) + B(2)] = 97.$

38   The probability that exactly w boys are born is

$$B(w) = \binom{20}{w} (0 \cdot 51)^w (0 \cdot 49)^{20-w}, \quad \text{for } 0 \leqslant w \leqslant 20.$$

Hence $B(0) = 10^{-6} \times 0 \cdot 636\ 681$; $B(1) = 10^{-4} \times 0 \cdot 132\ 534$;

$\qquad\quad B(2) = 10^{-3} \times 0 \cdot 131\ 046$; $B(3) = 10^{-3} \times 0 \cdot 818\ 369$;

$\qquad\quad B(4) = 10^{-2} \times 0 \cdot 362\ 003$; $B(5) = 10^{-1} \times 0 \cdot 120\ 569$;

$\qquad\quad B(6) = 10^{-1} \times 0 \cdot 313\ 725$; $B(7) = 10^{-1} \times 0 \cdot 653\ 060$;

$\qquad\quad B(8) = 0 \cdot 110\ 454$; $B(9) = 0 \cdot 153\ 283$;

$\qquad\quad B(10) = 0 \cdot 175\ 493.$

Hence $P(i) = B(10) = 0 \cdot 1755$;

$\qquad\quad P(ii) = P(w \leqslant 9) = 0 \cdot 3771$;

$\qquad\quad P(iii) = P(w \geqslant 8) = 1 - P(w \leqslant 7) = 0 \cdot 8867$;

$\qquad\quad P(iv) = P(w \geqslant 8) = 0 \cdot 8867.$

$$P(v) = (0 \cdot 51)^5 \times \binom{15}{5} (0 \cdot 51)^5 (0 \cdot 49)^{10}$$

$$= 3003 (0 \cdot 2499)^{10} = 10^{-2} \times 0 \cdot 2852.$$

$$P(vi) = (0 \cdot 51)^5 \times \sum_{r=8}^{15} S(r)$$

$$= (0 \cdot 51)^5 \times \left[1 - \sum_{r=0}^{7} S(r) \right],$$

where $S(r) = \binom{15}{r} (0 \cdot 51)^r (0 \cdot 49)^{15-r}$, for $0 \leqslant r \leqslant 15$,

whence $S(0) = 10^{-4} \times 0 \cdot 225\ 393$; $S(1) = 10^{-3} \times 0 \cdot 351\ 889$;

$\qquad\quad\ \ S(2) = 10^{-2} \times 0 \cdot 256\ 376$; $S(3) = 10^{-1} \times 0 \cdot 115\ 631$;

$$S(4) = 10^{-1} \times 0 \cdot 361\ 052; \qquad S(5) = 10^{-1} \times 0 \cdot 826\ 735;$$
$$S(6) = 0 \cdot 143\ 413; \qquad\qquad\quad S(7) = 0 \cdot 191\ 914.$$

Therefore $\quad P(\text{vi}) = (0 \cdot 51)^5 \times 0 \cdot 531\ 392 = 10^{-1} \times 0 \cdot 1833.$

39   The probability that exactly w applications will receive financial assistance is

$$B(w) = \binom{15}{w}(0 \cdot 9)^{w}(0 \cdot 1)^{15-w}, \quad \text{for } 0 \leqslant w \leqslant 15,$$

whence $B(0) = 10^{-14} \times 0 \cdot 100\ 000; \qquad B(1) = 10^{-12} \times 0 \cdot 135\ 000;$
$$B(2) = 10^{-11} \times 0 \cdot 850\ 500; \qquad B(3) = 10^{-9} \times 0 \cdot 331\ 695;$$
$$B(4) = 10^{-8} \times 0 \cdot 895\ 576; \qquad B(5) = 10^{-6} \times 0 \cdot 177\ 324;$$
$$B(6) = 10^{-5} \times 0 \cdot 265\ 986; \qquad B(7) = 10^{-4} \times 0 \cdot 307\ 784;$$
and $\quad B(15) = (0 \cdot 9)^{15} = 0 \cdot 205\ 891.$

Therefore
$$P(\text{i}) = B(15) = 0 \cdot 2059;$$
$$P(\text{ii}) = P(w \leqslant 5) = 10^{-6} \times 0 \cdot 1866;$$
$$P(\text{iii}) = P(w \leqslant 7) = 10^{-4} \times 0 \cdot 3362.$$

(iv) Let x be the least number of applications rejected. Then

$$x = \tfrac{3}{2}(15 - x) \quad \text{or } x = 9.$$

Hence $\quad P(\text{iv}) = P(x \geqslant 9) = P(w \leqslant 6) = 10^{-5} \times 0 \cdot 2846.$

The expected amount of advance is

$$£0 \cdot 9 \times 15 \times 2750 = £37\ 125.$$

During the financial stringency, the probability that exactly w applications will receive financial assistance is

$$S(w) = \binom{15}{w}(0 \cdot 8)^{w}(0 \cdot 2)^{15-w}, \quad \text{for } 0 \leqslant w \leqslant 15,$$

whence $S(0) = 10^{-10} \times 0 \cdot 327\ 680; \qquad S(1) = 10^{-8} \times 0 \cdot 196\ 608;$
$$S(2) = 10^{-7} \times 0 \cdot 550\ 502; \qquad S(3) = 10^{-6} \times 0 \cdot 954\ 203;$$
$$S(4) = 10^{-4} \times 0 \cdot 114\ 504; \qquad S(5) = 10^{-3} \times 0 \cdot 100\ 764;$$
$$S(6) = 10^{-3} \times 0 \cdot 671\ 760; \qquad S(7) = 10^{-2} \times 0 \cdot 345\ 477;$$
and $\quad S(15) = (0 \cdot 8)^{15} = 10^{-1} \times 0 \cdot 351\ 844.$

Therefore
$$P(\text{i}) = S(15) = 10^{-1} \times 0 \cdot 3518;$$
$$P(\text{ii}) = P(w \leqslant 5) = 10^{-3} \times 0 \cdot 1132;$$
$$P(\text{iii}) = P(w \leqslant 7) = 10^{-2} \times 0 \cdot 4240;$$
$$P(\text{iv}) = P(w \leqslant 6) = 10^{-3} \times 0 \cdot 7850.$$

The reduced amount of expected advance is

$$£0 \cdot 8 \times 15 \times 0 \cdot 9 \times 2750 = £297 \ 00.$$

The percentage saving is $\dfrac{7425 \times 100}{37 \ 125} = 20.$

40 Leicester, 1970

The probability that a batch contains exactly w defective tubes is

$$B(w) = \binom{24}{w}(0 \cdot 06)^{w}(0 \cdot 94)^{24-w}, \quad \text{for } 0 \leqslant w \leqslant 24,$$

whence

$$B(0) = 0 \cdot 226 \ 500; \qquad B(1) = 0 \cdot 346 \ 979;$$
$$B(2) = 0 \cdot 254 \ 697; \qquad B(3) = 0 \cdot 119 \ 220.$$

Therefore

$$P(\text{i}) = B(3) = 0 \cdot 1192;$$
$$P(\text{ii}) = P(w \geqslant 4) = 1 - P(w \leqslant 3) = 0 \cdot 0526;$$
$$P(\text{iii}) = P(w \leqslant 3) = 0 \cdot 9474.$$

The expected number of defective tubes per batch is $0 \cdot 06 \times 24 = 1 \cdot 44$. Hence the expected number of defective tubes in 12 batches is $12 \times 1 \cdot 44 = 17 \cdot 28$. Therefore the expected refund on the defective tubes in one month's supply is $£0 \cdot 375 \times 17 \cdot 28 = £6 \cdot 48$.

Under the new arrangement, the probability of a refund being payable on a batch is $P(w \geqslant 2) = 1 - P(w \leqslant 1) = 0 \cdot 426 \ 521$.

Therefore the point-probabilities of the truncated distribution of w, the number of defective tubes in a batch, is

$$\frac{P(w = r)}{1 - P(w \leqslant 1)} = \frac{\binom{n}{r} p^{r} \ q^{n-r}}{1 - P(w \leqslant 1)}, \quad \text{for } 2 \leqslant w \leqslant 24,$$

where $n = 24$, $p = 1 - q = 0 \cdot 06$. The expectation of w for this distribution is

$$\sum_{r=2}^{n} r \binom{n}{r} p^{r} \ q^{n-r} / [1 - P(w \leqslant 1)]$$

$$= \frac{np(1 - q^{n-1})}{1 - P(w \leqslant 1)} = 2 \cdot 562 \ 65.$$

Thus the expected number of defective tubes in a batch for batches having at least two defective tubes is $2 \cdot 562 \ 65$. Therefore the expected refund payable on such a batch is $128 \cdot 132 \ 50$p.

Under the new system the expected refund payable on any batch is

$$0 \times 0 \cdot 573 \ 479 + 128 \cdot 1325 \times 0 \cdot 426 \ 521 = 54 \cdot 651 \ 20\text{p}.$$

Therefore the expected refund payable on 12 batches is

$$£\frac{12 \times 54 \cdot 651 \ 20}{100} = £6 \cdot 56.$$

41   The probability that the undergraduate makes exactly w journeys in a week by the city bus is

$$B(w) = \binom{6}{w}(0.84)^w(0.16)^{6-w}, \quad \text{for } 0 \leqslant w \leqslant 6,$$

whence

$$B(0) = 10^{-4} \times 0.167\ 772; \qquad B(1) = 10^{-3} \times 0.528\ 482;$$
$$B(2) = 10^{-2} \times 0.693\ 633; \qquad B(3) = 10^{-1} \times 0.485\ 543;$$
$$B(4) = 0.191\ 183; \qquad B(5) = 0.401\ 484;$$
$$B(6) = 0.351\ 298.$$

Therefore
$$P(i) \ \ = B(6) = 0.3513;$$
$$P(ii) \ = B(5) = 0.4015;$$
$$P(iii) = P(w \leqslant 4) = 0.2472;$$
$$P(iv) = P(w \leqslant 3) = 10^{-1} \times 0.5604;$$
$$P(v) \ = P(w \geqslant 4) = 0.9440.$$

(vi) The expected cost of one fare is $6 \times 0.84 + 9 \times 0.16 = 6.48\text{p}$.

Therefore the expected cost of 54 fares is $54 \times 6.48\text{p} = £3.50$.

If one journey per week is by the long-distance bus, then the weekly cost is 39p. Hence the total cost per term is $9 \times 39\text{p} = £3.51$.

(vii) Let X be the event $w \geqslant 4$ and Y the event $w \leqslant 2$. Then

$$P(X) = 0.9440 \quad \text{and} \quad P(Y) = 10^{-2} \times 0.7482.$$

Therefore
$$P(vii) = \binom{9}{5}[P(X)]^4[P(Y)]^5$$

$$= 126 \times 10^{-10} \times (0.9440)^4(0.7482)^5$$

$$= 10^{-8} \times 0.2346.$$

42   (i) The expected cost of a tin of tobacco is
$$65 \times 0.85 + 67.5 \times 0.15 = 65.375\text{p}.$$

Therefore the expected cost of a month's supply is $8 \times 65.375\text{p} = £5.23$.

The probability of buying w tins of A and the rest of B in a month is

$$B(w) = \binom{8}{w}(0.85)^w(0.15)^{8-w}, \quad \text{for } 0 \leqslant w \leqslant 8,$$

whence

$$B(0) = 10^{-6} \times 0.256\ 289; \qquad B(1) = 10^{-4} \times 0.116\ 184;$$
$$B(2) = 10^{-3} \times 0.230\ 432; \qquad B(3) = 10^{-2} \times 0.261\ 156;$$
$$B(4) = 10^{-1} \times 0.184\ 986; \qquad B(5) = 10^{-1} \times 0.838\ 603;$$
$$B(6) = 0.237\ 604.$$

Therefore
$$P(ii) \ = P(w \leqslant 6) = 0.3428;$$
$$P(iii) = P(w \geqslant 4) = 1 - P(w \leqslant 3) = 0.9971;$$

$$P(\text{iv}) = \binom{12}{6} [P(w \leqslant 6)]^6 [1 - P(w \leqslant 6)]^6$$

$$= 924(0 \cdot 3428)^6 (0 \cdot 6572)^6$$

$$= 0 \cdot 120\ 809 = 0 \cdot 1208 \text{ to 4 significant figures.}$$

$$P(\text{v}) = \sum_{r=6}^{12} S(r), \quad \text{where} \quad S(r) = \binom{12}{r} [P(w \leqslant 6)]^r [1 - P(w \leqslant 6)]^{12-r}.$$

Therefore

$$S(6) = 0 \cdot 120\ 809; \qquad\qquad S(7) = 10^{-1} \times 0 \cdot 540\ 127;$$
$$S(8) = 10^{-1} \times 0 \cdot 176\ 084; \qquad S(9) = 10^{-2} \times 0 \cdot 408\ 207;$$
$$S(10) = 10^{-3} \times 0 \cdot 638\ 771; \qquad S(11) = 10^{-4} \times 0 \cdot 605\ 795;$$
$$S(12) = 10^{-5} \times 0 \cdot 263\ 322.$$

Hence

$$P(\text{v}) = 0 \cdot 1972.$$

$$P(\text{vi}) = \binom{12}{8} [B(4)]^8 [1 - B(4)]^4$$

$$= 495(10^{-1} \times 0 \cdot 1850)^8 (0 \cdot 981\ 50)^4$$

$$= 10^{-11} \times 0 \cdot 6303.$$

$$P(\text{vii}) = \sum_{r=0}^{1} \binom{12}{r} [B(4)]^r [1 - B(4)]^{12-r}$$

$$= (0 \cdot 9815)^{11} \times 1 \cdot 2035$$

$$= 0 \cdot 9800.$$

43   The probability that the candidate will have a score of exactly w points is

$$B(w) = \binom{20}{w} (0 \cdot 46)^w (0 \cdot 54)^{20-w}, \quad \text{for } 0 \leqslant w \leqslant 20,$$

whence

$$B(0) = 10^{-5} \times 0 \cdot 444\ 504; \qquad B(1) = 10^{-4} \times 0 \cdot 757\ 303;$$
$$B(2) = 10^{-3} \times 0 \cdot 612\ 855; \qquad B(3) = 10^{-2} \times 0 \cdot 313\ 237;$$
$$B(4) = 10^{-1} \times 0 \cdot 113\ 403; \qquad B(5) = 10^{-1} \times 0 \cdot 309\ 128;$$
$$B(6) = 10^{-1} \times 0 \cdot 658\ 328; \qquad B(7) = 0 \cdot 112\ 160;$$
$$B(8) = 0 \cdot 155\ 259; \qquad\qquad\qquad B(9) = 0 \cdot 176\ 344;$$
$$B(10) = 0 \cdot 165\ 241; \qquad\qquad\quad B(11) = 0 \cdot 127\ 964;$$
$$B(12) = 10^{-1} \times 0 \cdot 817\ 548; \qquad B(13) = 10^{-1} \times 0 \cdot 428\ 572;$$
$$B(14) = 10^{-1} \times 0 \cdot 182\ 540; \qquad B(15) = 10^{-2} \times 0 \cdot 621\ 988;$$
$$B(16) = 10^{-2} \times 0 \cdot 165\ 576; \qquad B(17) = 10^{-3} \times 0 \cdot 331\ 874;$$
$$B(18) = 10^{-4} \times 0 \cdot 471\ 179; \qquad B(19) = 10^{-5} \times 0 \cdot 422\ 500;$$
$$B(20) = 10^{-6} \times 0 \cdot 179\ 954.$$

Therefore $P(\mathrm{i})$ $= P(w \geqslant 16) = 10^{-2} \times 0 \cdot 2039;$

             $P(\mathrm{ii})$ $= P(w \geqslant 12) = 0 \cdot 1511;$

             $P(\mathrm{iii})$ $= P(w \leqslant 5) = 10^{-1} \times 0 \cdot 4608;$

             $P(\mathrm{iv})$ $= P(w \geqslant 10) = 1 - P(w \leqslant 9) = 0 \cdot 4443;$

             $P(\mathrm{v})$ $= [P(w \geqslant 10)]^{6} = (0 \cdot 4443)^{6} = 10^{-2} \times 0 \cdot 7692;$

             $P(\mathrm{vi})$ $= [P(w \leqslant 5)]^{6} = 10^{-6} \times (0 \cdot 4608)^{6} = 10^{-8} \times 0 \cdot 9574;$

             $P(\mathrm{vii}) = [P(w \geqslant 16)]^{6} = 10^{-12} \times (0 \cdot 2039)^{6} = 10^{-16} \times 0 \cdot 7186.$

(viii) Since the events $X : w \geqslant 10$ and $Y : w \geqslant 16$, are not mutually exclusive, define the mutually exclusive events

$$A : w \geqslant 16 \quad \text{and} \quad B : w = 10, 11, 12, 13, 14, 15,$$

so that $X = A + B$ and $Y = A.$

Then $P(A) = 10^{-2} \times 0 \cdot 2039$ and $P(B) = 0 \cdot 4423.$

The event defined in (viii) means the realisation of two X's and four Y's in the six independent trials. But $X^{2}Y^{4} = (A + B)^{2}A^{4}$, so that the mutually exclusive events corresponding to (viii) are A^{6}, $A^{5}B$ and $A^{4}B^{2}$. Hence, taking permutations into consideration, we have

$$P(\mathrm{viii}) = P^{6}(A) + \binom{6}{1}P^{5}(A)P(B) + \binom{6}{2}P^{4}(A)P^{2}(B)$$

$$= P^{4}(A)[P^{2}(A) + 6P(A)P(B) + 15P^{2}(B)]$$

$$= 10^{-10} \times 0 \cdot 5082.$$

(ix)   Let X and Y be as defined above and Z be the event $w \leqslant 5$. Then X, Y and Z are not mutually exclusive. If C is the event $w \leqslant 5$, then A, B and C are mutually exclusive and

$$X = A + B, \quad Y = A, \quad Z = C; \quad P(C) = 10^{-1} \times 0 \cdot 4607.$$

The event defined in (ix) means the realisation of three X's, two Z's and one Y. But $X^{3}Z^{2}Y = (A + B)^{3}C^{2}A$, so that the mutually exclusive events corresponding to (ix) are $A^{4}C^{2}$, $A^{3}BC^{2}$, $A^{2}B^{2}C^{2}$ and $AB^{3}C^{2}$. Hence, taking permutations into consideration, we have

$$P(\mathrm{ix}) = \binom{6}{2}P^{4}(A)P^{2}(C) + \frac{6!}{3! \, 2! \, 1!} P^{3}(A)P^{2}(C)P(B) +$$

$$+ \frac{6!}{2! \, 2! \, 2!} P^{2}(A)P^{2}(B)P^{2}(C) + \frac{6!}{1! \, 3! \, 2!} P(A)P^{3}(B)P^{2}(C)$$

$$= 15P(A)P^{2}(C)[P^{2}(A)\{P(A) + 4P(B)\} + 2P^{2}(B)\{2P(B) + 3P(A)\}]$$

$$= 10^{-4} \times 0 \cdot 2262.$$

44   The probability that w potatoes in a bag are damaged is

$$B(w) = \binom{320}{w}(0 \cdot 0125)^{w}(1 - 0 \cdot 0125)^{320-w} \sim e^{-4} 4^{w}/w! \equiv Q(w), \quad \text{for } w \geqslant 0.$$

Therefore $Q(0) = e^{-4} = 10^{-1} \times 0.183\ 156$; $Q(1) = 10^{-1} \times 0.732\ 624$;

$\qquad Q(2) = 0.146\ 525$; $\qquad\qquad Q(3) = 0.195\ 367$;

$\qquad Q(4) = 0.195\ 367$; $\qquad\qquad Q(5) = 0.156\ 294$;

$\qquad Q(6) = 0.104\ 196$; $\qquad\qquad Q(7) = 10^{-1} \times 0.595\ 406$;

$\qquad Q(8) = 10^{-1} \times 0.297\ 703$.

The probability that the seed-store will have to pay compensation on a bag of potatoes supplied is $P(w \geqslant 9) = 1 - P(w \leqslant 8) = 0.021\ 362$.

(i) The average compensation per bag is 47½p. Therefore £2·85 of average compensation will arise from 6 bags each of which contains at least 9 damaged potatoes. The probability that compensation is payable on r of the 120 bags supplied is

$$C(r) = \binom{120}{r}(0.021\ 362)^r (1 - 0.021\ 362)^{120-r}$$

$$\sim e^{-2.5634}(2.5634)^r/r! \equiv S(r), \quad \text{for } r \geqslant 0.$$

Therefore $S(0) = e^{-2.5634} = 10^{-1} \times 0.770\ 424$; $S(1) = 0.197\ 490$;

$\qquad S(2) = 0.253\ 123$; $\qquad\qquad\qquad S(3) = 0.216\ 285$;

$\qquad S(4) = 0.138\ 606$; $\qquad\qquad\qquad S(5) = 10^{-1} \times 0.710\ 605$.

The compensation will be at least £2·85 on an average if 6 or more bags are damaged, with at least 9 or more damaged potatoes in each. The required probability is $P(r \geqslant 6) = 1 - P(r \leqslant 5) = 0.046\ 393$.

(ii) Out of 120 bags supplied, the expected number requiring compensation is $120 \times 0.021\ 362 = 2.563\ 440$; and on each bag the average compensation is 47½p. Hence the expected compensation for a year is £2·563 440 $\times$ 0·475 = £1·22.

45   The probability that a batch contains w defective bulbs is

$$B(w) = \binom{144}{w}(0.015)^w(1 - 0.015)^{144-w} \sim e^{-2.16}(2.16)^w/w! \equiv Q(w), \quad \text{for } w \geqslant 0.$$

Therefore $Q(0) = e^{-2.16} = 0.115\ 325$; $Q(1) = 0.249\ 102$;

$\qquad Q(2) = 0.269\ 030$; $\qquad\qquad Q(3) = 0.193\ 702$;

$\qquad Q(4) = 0.104\ 599$; $\qquad\qquad Q(5) = 10^{-1} \times 0.451\ 868$.

Hence

$P(w \geqslant 6) = 0.023\ 055$; $P(3 \leqslant w \leqslant 5) = 0.343\ 488$; $P(w \leqslant 2) = 0.633\ 457$.

Therefore $P(\text{i}) = \dfrac{3!}{(1!)^3} P(w \geqslant 6)P(3 \leqslant w \leqslant 5)P(w \leqslant 2) = 0.030\ 10$.

(ii) The expected refund on a random batch is

$$0 \times P(w \leqslant 2) + 15 \times P(3 \leqslant w \leqslant 5) + 25 \times P(w \geqslant 6) = 6\text{p}.$$

(iii) If X is a random variable denoting the amount of refund payable on a batch, then for the conditional distribution of X

$$P(X = 3 \,|\, w \geqslant 3) = \frac{P(3 \leqslant w \leqslant 5)}{1 - P(w \leqslant 2)} \quad \text{and} \quad P(X = 5 \,|\, w \geqslant 3) = \frac{P(w \geqslant 6)}{1 - P(w \leqslant 2)}.$$

Hence the required conditional expectation of X is

$$\frac{15P(3 \leqslant w \leqslant 5) + 25P(w \geqslant 6)}{1 - P(w \leqslant 2)} = \frac{5 \cdot 728\ 695}{0 \cdot 366\ 543} = 16\text{p}.$$

46   The expected amount of petrol sold in a week is

$$48 \times 7 \times 3 \cdot 85 \times 4 \cdot 25 = 5497 \cdot 8 = 5498 \text{ gallons, to nearest integer.}$$

(i) If x is the increased arrival rate of customers, then the week's supply will be just adequate if $48 \times 7 \times 4 \cdot 25 \times x = 5800$, or $x = 4 \cdot 0616$. Hence the percentage increase in the mean arrival rate of customers is

$$100(4 \cdot 0616 - 3 \cdot 85)/3 \cdot 85 = 5 \cdot 50.$$

(ii) A week has $48 \times 7 = 336$ half-hourly intervals. Suppose that $x_1, x_2, \ldots, x_n$ $(n = 336)$ are the actual numbers of arrivals of customers in a week. Then each x_i is an independent Poisson variable such that

$$E(x_i) = \text{var}(x_i) = 3 \cdot 85.$$

If $\bar{x}$ is the average of the x_i's, then

$$E(\bar{x}) = 3 \cdot 85; \quad \text{var}(\bar{x}) = \frac{3 \cdot 85}{336} = 0 \cdot 011\ 458\ 3; \quad \text{s.d.}(\bar{x}) = 0 \cdot 107\ 043.$$

Hence the probability that a week's supply will be exhausted is

$$P(\bar{x} \geqslant 4 \cdot 0616) = 1 - P(\bar{x} < 4 \cdot 0616)$$

$$\sim 1 - \Phi\!\left(\frac{4 \cdot 0616 - 3 \cdot 85}{0 \cdot 107\ 043}\right)$$

$$= 1 - \Phi(1 \cdot 977) = 0 \cdot 0240 \text{ to 4 decimal places.}$$

(iii) Let X be the event that supplies are exhausted in a week; then

$$P(X) = 0 \cdot 0240 \quad \text{and} \quad P(\bar{X}) = 0 \cdot 9760.$$

Therefore the probability that X occurs w times in 52 weeks is

$$B(w) = \binom{52}{w}(0 \cdot 0240)^{w}(0 \cdot 9760)^{52 - w} \sim e^{-1 \cdot 248}(1 \cdot 248)^{w}/w! = Q(w), \quad \text{for } w \geqslant 0.$$

Then $Q(0) = e^{-1 \cdot 248} = 0 \cdot 287\ 078;$ $Q(1) = 0 \cdot 358\ 273;$

$\qquad\quad Q(2) = 0 \cdot 223\ 562;$ $Q(3) = 10^{-1} \times 0 \cdot 930\ 018.$

Hence $P(\text{iii}) = P(w \geqslant 4) = 1 - P(w \leqslant 3) = 0 \cdot 038\ 08.$

47   Leicester, 1969.

$$P(X = r) = \frac{Q(r)}{1 - Q(0)}, \quad \text{for } X \geqslant 1,$$

where $Q(r) = e^{-1 \cdot 86} (1 \cdot 86)^r / r!$,   for $r \geqslant 0$.

$Q(0) = e^{-1 \cdot 86} = 0 \cdot 155 \ 673$; $1 - Q(0) = 0 \cdot 844 \ 327$;

$Q(1) = 0 \cdot 289 \ 552$; $Q(2) = 0 \cdot 269 \ 283$;

$Q(3) = 0 \cdot 166 \ 955$; $Q(4) = 10^{-1} \times 0 \cdot 776 \ 341$.

Hence $P(X = 1) = 0 \cdot 342 \ 938$; $P(X = 2) = 0 \cdot 318 \ 932$;

$P(X = 3) = 0 \cdot 197 \ 737$; $P(X = 4) = 10^{-1} \times 0 \cdot 919 \ 479$;

and $P(X \geqslant 5) = 0 \cdot 048 \ 445$.

Therefore $P(i) = 0 \cdot 3429$; $P(ii) = 0 \cdot 1947$; $P(iii) = 0 \cdot 048 \ 44$.

Suppose A is the event "failed first attempt" and B the event "passed second attempt". Then the required probability is

$$P(B \mid A) = \frac{P(BA)}{P(A)} = \frac{P(X = 2)}{1 - P(X = 1)} = 0 \cdot 4854.$$

The expected cost of passing the test is

$$16 \times P(X = 1) + 27 \times P(X = 2) + 43 \times P(X = 3) + \sum_{r=4}^{\infty} (70 + r)P(X = r)$$

$$= 16 \times P(X = 1) + 27 \times P(X = 2) + 43 \times P(X = 3) + 70 \times P(X \geqslant 4) +$$

$$+ 1 \cdot 86 \times P(X \geqslant 3),$$

since $\sum_{r=4}^{\infty} r \, e^{-\mu} \mu^r / r! = \mu \sum_{s=3}^{\infty} e^{-\mu} \mu^s / s!$,

$= 33 \cdot 057 \ 295 = $ £33 to the nearest £.

48   (i) Suppose the stock ordered is $175 \cdot 5y$ gross. Then, for $0 \leqslant x \leqslant y$, profit is

$$175 \cdot 5x \times 54 \times \frac{3}{2} + 175 \cdot 5(y - x) \times 54 \times \frac{85}{100} - 175 \cdot 5y \times 54$$

$$= 175 \cdot 5(35 \cdot 1x - 8 \cdot 1y).$$

Again, for $x > y$, profit is $175 \cdot 5y \times 54 \times \frac{3}{2} - 175 \cdot 5y \times 54 = 175 \cdot 5(27y)$.

Hence, for given y, the expected profit is

$$G(y) = 175 \cdot 5 \left[\sum_{x=0}^{y} (35 \cdot 1x - 8 \cdot 1y) f(x) + \sum_{x=y+1}^{\infty} 27y \, f(x) \right],$$

where $f(x) \equiv e^{-4 \cdot 96} (4 \cdot 96)^x / x!$

$$= 175 \cdot 5 \left[35 \cdot 1 \sum_{x=0}^{y} (x - y) f(x) + 27y \right].$$

Hence $\Delta G(y) = G(y + 1) - G(y)$

$$= 175 \cdot 5 \left[27 - 35 \cdot 1 \sum_{x=0}^{y} (x - y) f(x) + 35 \cdot 1 \sum_{x=0}^{y+1} (x - y - 1) f(x) \right]$$

$$= 175 \cdot 5 [27 - 35 \cdot 1 F(y)],$$

where $F(x)$ is the distribution function of the Poisson variable x. Hence $\Delta G(y)$ is positive so long as

$$F(y) < \frac{27}{35 \cdot 1} = 0 \cdot 769\ 231.$$

We have $f(0) = e^{-4 \cdot 96} = 10^{-2} \times 0 \cdot 701\ 293;$ $f(1) = 10^{-1} \times 0 \cdot 347\ 841;$

$f(2) = 10^{-1} \times 0 \cdot 862\ 646;$ $f(3) = 0 \cdot 142\ 624;$

$f(4) = 0 \cdot 176\ 854;$ $f(5) = 0 \cdot 175\ 439;$

$f(6) = 0 \cdot 145\ 030;$ $f(7) = 0 \cdot 102\ 764.$

so that $F(6) = 0 \cdot 768\ 009$ and $F(7) = 0 \cdot 870\ 773$. Hence the smallest integral value of y for which $\Delta G(y) < 0$ is 7, and so the optimum stock is 1228·5 gross.

(ii) For $y = 7$, $F(y) = 0 \cdot 870\ 773$ and $\sum_{x=0}^{y} x f(x) = 3 \cdot 809\ 325$. Therefore

$G(7) = 175 \cdot 5\,(27 \times 7 + 35 \cdot 1 \times 3 \cdot 809\ 325 - 35 \cdot 1 \times 7 \times 0 \cdot 870\ 773)$

$= 19\ 087 \cdot 10 = £19\ 087$ to the nearest integer.

(iii) The total outlay is £66 339. Therefore the maximum expected percentage profit is

$$\frac{19\ 087 \cdot 10 \times 100}{66\ 339} = 28 \cdot 77.$$

(iv) The probability of demand exceeding stock is

$$P(x \geqslant 8) = 1 - P(x \leqslant 7) = 0 \cdot 1292.$$

49

$$\sum_{r=\rho}^{\infty} r^{(k)} P(Z = r) = \mu^k \sum_{r=\rho}^{\infty} e^{-\mu} \mu^{r-k} /(r - k)!$$

$$= \mu^k \sum_{t=\rho-k}^{\infty} e^{-\mu} \mu^t /t!$$

$$= \mu^k P(Z \geqslant \rho - k). \qquad (1)$$

The probability of a batch containing w defective batteries is

$$B(w) = \binom{75}{w}(0 \cdot 0128)^w (1 - 0 \cdot 0128)^{75-w}$$

$$\sim e^{-0 \cdot 96} (0 \cdot 96)^w /w! = Q(w),\ \text{say, for } w \geqslant 0.$$

Then

$Q(0) = e^{-0 \cdot 96} = 0 \cdot 382\ 892\ 9;$ $Q(1) = 0 \cdot 367\ 577\ 2;$

$Q(2) = 0 \cdot 176\ 437\ 1;$ $P(w \geqslant 3) = 0 \cdot 073\ 092\ 8.$

If X is a random variable denoting the amount of refund (in new pence) payable per defective battery, then

$$E(X) = 0 \times Q(0) + 12 \times Q(1) + 15 \times Q(2) + \sum_{r=3}^{\infty} (12 + 2r)Q(r),$$

whence, using (1),

$$= 12Q(1) + 15Q(2) + 12P(w \geqslant 3) + 2 \times 0 \cdot 96 P(w \geqslant 2)$$
$$= 13 \cdot 92 + 12Q(1) + 15Q(2) - 12P(w \leqslant 2) - 1 \cdot 92P(w \leqslant 1). \qquad (2)$$

Similarly,
$$E(X^2) = 12^2 Q(1) + 15^2 Q(2) + \sum_{r=3}^{\infty} (12 + 2r)^2 Q(r)$$

$$= 144Q(1) + 225Q(2) + \sum_{r=3}^{\infty} (144 + 48r + 4r^2)Q(r)$$

$$= 144Q(1) + 225Q(2) + 144P(w \geqslant 3) + 48 \sum_{r=3}^{\infty} rQ(r) +$$

$$+ 4 \sum_{r=3}^{\infty} [r(r-1) + r]Q(r),$$

whence, using (1),

$$= 144Q(1) + 225Q(2) + 144P(w \geqslant 3) + 49 \cdot 92P(w \geqslant 2) +$$
$$+ 3 \cdot 6864P(w \geqslant 1)$$
$$= 197 \cdot 6064 + 144Q(1) + 225Q(2) - 144P(w \leqslant 2) -$$
$$- 49 \cdot 92P(w \leqslant 1) - 3 \cdot 6864P(w = 0). \qquad (3)$$

Hence, from (2) and (3), we have

$$E(X) = 8 \cdot 413\ 694; \qquad E(X^2) = 117 \cdot 886\ 264.$$

Therefore, correct to 4 decimal places,

$$E(X) = 8 \cdot 4137; \qquad \text{var}(X) = 47 \cdot 0960.$$

50   If the random variable X has the Poisson distribution with mean 4·6, then

$$P(X = w) = e^{-4 \cdot 6} (4 \cdot 6)^{w}/w! = Q(w), \text{ say, for } X \geqslant 0.$$

Therefore $Q(0) = e^{-4 \cdot 6} = 10^{-1} \times 0 \cdot 100\ 518$; $Q(1) = 10^{-1} \times 0 \cdot 462\ 383$;
$\qquad Q(2) = 0 \cdot 106\ 348$; $\qquad\qquad Q(3) = 0 \cdot 163\ 067$;
$\qquad Q(4) = 0 \cdot 187\ 527$; $\qquad\qquad Q(5) = 0 \cdot 172\ 525$;
$\qquad Q(6) = 0 \cdot 132\ 269$; $\qquad\qquad Q(7) = 10^{-1} \times 0 \cdot 869\ 196$;
$\qquad Q(8) = 10^{-1} \times 0 \cdot 499\ 788$; $\qquad Q(9) = 10^{-1} \times 0 \cdot 255\ 447$;
$\qquad Q(10) = 10^{-1} \times 0 \cdot 117\ 506$.

Now if Y is a random variable denoting the number of copies bought at a time by individual retailers, then

$$P(Y = r) = \frac{P(X = r)}{1 - P(X = 0)}, \quad \text{for } Y \geqslant 1.$$

Hence, if Z is the random variable denoting the discount per copy sold, then

$$P(Z = 0) = P(Y = 1)$$
$$P(Z = r) = P(Y = r), \quad \text{for } 2 \leqslant Y \leqslant 10,$$

and

$$P(Z = 13) = P(Y \geqslant 11).$$

Therefore

$$E[Z] = \left[\sum_{r=2}^{10} r\, P(X = r) + 13 \sum_{r=11}^{\infty} P(X = r) \right] / [1 - Q(0)]$$

$$= [4 \cdot 6 P(1 \leqslant X \leqslant 9) + 13\{1 - P(0 \leqslant X \leqslant 10)\}] / [1 - Q(0)]. \quad (1)$$

Similarly,

$$E[Z^{(2)}] = \left[\sum_{r=2}^{10} r^{(2)}\, P(X = r) + 156 \sum_{r=11}^{\infty} P(X = r) \right] / [1 - Q(0)]$$

$$= [(4 \cdot 6)^2\, P(0 \leqslant X \leqslant 8) + 156\{1 - P(0 \leqslant X \leqslant 10)\}] / [1 - Q(0)]. \quad (2)$$

Hence, from (1) and (2), we have

$$E(Z) = 4 \cdot 611\ 404; \qquad E[Z^{(2)}] = 21 \cdot 637\ 234.$$

Therefore, correct to 4 decimal places,

$$E(Z) = 4 \cdot 6114; \qquad \text{var}(Z) = 4 \cdot 9836.$$

The expected discount as percentage of the net price of the book is $461 \cdot 14/125 = 3 \cdot 69$.

The expected return on the 3700 copies sold is

$$\pounds \frac{120 \cdot 388\ 596 \times 3700}{100} = \pounds 4454 \cdot 38 = \pounds 4454 \text{ to nearest integer.}$$

51
$$\sum_{k=1}^{N} \frac{P(Z = a + k)}{a + k} = \sum_{k=1}^{N} \frac{e^{-\mu} \mu^{a+k}}{(a + k) \times (a + k)!}$$

$$> \frac{1}{\mu} \sum_{k=1}^{N} \frac{e^{-\mu} \mu^{a+k+1}}{(a + k + 1)!} = \frac{1}{\mu} \sum_{j=a+2}^{a+N+1} \frac{e^{-\mu} \mu^{j}}{j!}$$

or
$$\sum_{k=1}^{N} P(Z = a + k)/(a + k) > \frac{1}{\mu} P(a + 2 \leqslant Z \leqslant a + N + 1). \quad (1)$$

Again,

$$\sum_{k=1}^{N} \frac{e^{-\mu} \mu^{a+k}}{(a + k) \times (a + k)!} = \frac{a + 2}{a + 1} \times \sum_{k=1}^{N} \frac{a + 1}{(a + 2)(a + k)} \times \frac{e^{-\mu} \mu^{a+k}}{(a + k)!}.$$

But, for $k \geqslant 1$, $\dfrac{a + 1}{(a + 2)(a + k)} \leqslant \dfrac{1}{a + k + 1}$,

since the inequality implies $1 \leqslant k$. Therefore

$$\sum_{k=1}^{N} \frac{e^{-\mu} \mu^{a+k}}{(a+k).(a+k)!} < \frac{a + 2}{a + 1} \times \sum_{k=1}^{N} \frac{e^{-\mu} \mu^{a+k}}{(a + k + 1)!}$$

or

$$\sum_{k=1}^{N} P(Z = a + k)/(a + k) < \left(\frac{a + 2}{a + 1}\right)\frac{1}{\mu} P(a + 2 \leqslant Z \leqslant a + N + 1). \qquad (2)$$

Let X be a random variable denoting the number of calls in an hour emanating from the office telephone. Then

$$P(X = r) = e^{-15} \ 15^r/r!, \quad \text{for } X \geqslant 0.$$

Again, let Y be a random variable denoting the average cost (in new pence) of a call to the office. Then

(i) $Y = 0$ if $X = 0$;

(ii) $Y = 3$ if $1 \leqslant X \leqslant 10$.

If $X = 10 + s$, where $1 \leqslant s \leqslant 10$, then the average cost of the calls is $\dfrac{30 + 6s}{10 + s}$. Therefore

(iii) $Y = \dfrac{30 + 6s}{10 + s}$ if $X = 10 + s, \ 1 \leqslant s \leqslant 10$.

Finally, if $X = 20 + t$, where $1 \leqslant t \leqslant \infty$, then the average cost of the calls is $\dfrac{90 + 9t}{20 + t}$. Therefore

(iv) $Y = \dfrac{90 + 9t}{20 + t}$ if $X = 20 + t, \ t \geqslant 1$.

Hence, using (i) to (iv), we have

$$E(Y) = 0 \times P(X = 0) + 3 \times \sum_{r=1}^{10} P(X = r) + \sum_{s=1}^{10} \left(\frac{30 + 6s}{10 + s}\right) P(X = 10 + s) +$$

$$+ \sum_{t=1}^{\infty} \left(\frac{90 + 9t}{20 + t}\right) P(X = 20 + t)$$

$$= 3 \times P(1 \leqslant X \leqslant 10) + 6 \times P(11 \leqslant X \leqslant 20) + 9 \times P(21 \leqslant X < \infty) -$$

$$- 30 \sum_{s=1}^{10} \frac{P(X = 10 + s)}{10 + s} - 90 \sum_{t=1}^{\infty} \frac{P(X = 20 + t)}{20 + t}.$$

But by (1) with $N = 10$, $a = 10$, we have

$$\sum_{s=1}^{10} \frac{P(X = 10 + s)}{10 + s} > \frac{1}{15} P(12 \leqslant X \leqslant 21)$$

and for $N \to \infty$, $a = 20$,

$$\sum_{t=1}^{\infty} \frac{P(X = 20 + t)}{20 + t} > \frac{1}{15} P(X \geqslant 22).$$

Hence $E(Y) < 3P(1 \leqslant X \leqslant 10) + 6P(11 \leqslant X \leqslant 20) + 9P(X \geqslant 21) -$

$$- 2P(12 \leqslant X \leqslant 21) - 6P(X \geqslant 22). \qquad (3)$$

Similarly, by using (2), we have

$$\sum_{s=1}^{10} \frac{P(X = 10 + s)}{10 + s} < \frac{12}{11} \times \frac{1}{15} P(12 \leqslant X \leqslant 21)$$

and $$\sum_{t=1}^{\infty} \frac{P(X = 20 + t)}{20 + t} < \frac{22}{21} \times \frac{1}{15} P(X \geqslant 22).$$

Hence $E(Y) > 3P(1 \leqslant X \leqslant 10) + 6P(11 \leqslant X \leqslant 20) + 9P(X \geqslant 21) -$

$$- \frac{24}{11} P(12 \leqslant X \leqslant 21) - \frac{44}{7} P(X \geqslant 22). \qquad (4)$$

Now by using the normal approximation, we have

$$P(1 \leqslant X \leqslant 10) \sim \Phi\left(\frac{10 - 15 + 0 \cdot 5}{\sqrt{15}}\right) - \Phi\left(\frac{1 - 15 - 0 \cdot 5}{\sqrt{15}}\right)$$

$$= \Phi(3 \cdot 744) - \Phi(1 \cdot 162) = 0 \cdot 122\ 53;$$

$$P(11 \leqslant X \leqslant 20) \sim \Phi\left(\frac{20 - 15 + 0 \cdot 5}{\sqrt{15}}\right) - \Phi\left(\frac{11 - 15 - 0 \cdot 5}{\sqrt{15}}\right)$$

$$= \Phi(1 \cdot 420) - [1 - \Phi(1 \cdot 162)] = 0 \cdot 799\ 58;$$

$$P(12 \leqslant X \leqslant 21) \sim \Phi\left(\frac{21 - 15 + 0 \cdot 5}{\sqrt{15}}\right) - \Phi\left(\frac{12 - 15 - 0 \cdot 5}{\sqrt{15}}\right)$$

$$= \Phi(1 \cdot 678) - [1 - \Phi(0 \cdot 904)] = 0 \cdot 770\ 32;$$

$$P(X \geqslant 21) \sim 1 - \Phi\left(\frac{21 - 15 - 0 \cdot 5}{\sqrt{15}}\right)$$

$$= 1 - \Phi(1 \cdot 420) = 0 \cdot 077\ 80;$$

$$P(X \geqslant 22) \sim 1 - \Phi\left(\frac{22 - 15 - 0 \cdot 5}{\sqrt{15}}\right)$$

$$= 1 - \Phi(1 \cdot 678) = 0 \cdot 046\ 68.$$

Hence, from (3) and (4), we have

$$3 \cdot 891\ 15 < E(Y) < 4 \cdot 044\ 55$$

or $$1 \cdot 955\ 45 < 6 - E(Y) < 2 \cdot 108\ 85,$$

so that $$32 \cdot 59 < \text{percentage reduction} < 35 \cdot 15.$$

If the office makes 30 or less calls per hour, then the new system is not less favourable than the old system; the average cost per call for a rate of 30 calls per hour is the same for the two systems. Hence the probability of losing under the new system is

$$P(X \geqslant 31) = 1 - P(X \leqslant 30)$$

$$\sim 1 - \Phi\left(\frac{30 - 15 + 0 \cdot 5}{\sqrt{15}}\right) = 1 - \Phi(4 \cdot 002) \sim 0 \cdot 000\ 03.$$

Therefore it is practically certain that the new system will prove more economical than the standard rate.

Aliter

If T is a random variable denoting the *total* amount (in new pence) paid according to the new system for calls made in an hour, then

$$E(T) = 0 \times P(X = 0) + \sum_{r=1}^{10} 3rP(X = r) + \sum_{s=1}^{10} (30 + 6s)P(X = 10 + s) +$$

$$+ \sum_{t=1}^{\infty} (90 + 9t)P(X = 20 + t)$$

$$= 3 \sum_{r=1}^{10} rP(X = r) + 6 \sum_{u=11}^{20} uP(X = u) + 9 \sum_{v=21}^{\infty} vP(X = v) -$$

$$- 30P(11 \leqslant X \leqslant 20) - 90P(X \geqslant 21). \qquad (5)$$

But for any two integers α and β such that $1 \leqslant \alpha < \beta < \infty$, we have

$$\sum_{r=\alpha}^{\beta} rP(X = r) = 15P(\alpha - 1 \leqslant X \leqslant \beta - 1).$$

Therefore

$$E(T) = 45P(0 \leqslant X \leqslant 9) + 90P(10 \leqslant X \leqslant 19) + 135P(X \geqslant 20) -$$

$$- 30P(11 \leqslant X \leqslant 20) - 90P(X \geqslant 21).$$

Now using the normal approximation for the distribution of X, we have

$$P(0 \leqslant X \leqslant 9) \sim \Phi\left(\frac{9 - 15 + 0 \cdot 5}{\sqrt{15}}\right) - \Phi\left(\frac{0 - 15 - 0 \cdot 5}{\sqrt{15}}\right)$$

$$= 1 - \Phi(1 \cdot 420) - [1 - \Phi(4 \cdot 002)] = 0 \cdot 077\ 77;$$

$$P(10 \leqslant X \leqslant 19) \sim \Phi\left(\frac{19 - 15 + 0 \cdot 5}{\sqrt{15}}\right) - \Phi\left(\frac{10 - 15 - 0 \cdot 5}{\sqrt{15}}\right)$$

$$= \Phi(1 \cdot 162) - [1 - \Phi(1 \cdot 420)] = 0 \cdot 799\ 58;$$

$$P(X \geqslant 20) \sim 1 - \Phi\left(\frac{19 - 15 + 0 \cdot 5}{\sqrt{15}}\right) = 1 - \Phi(1 \cdot 162) = 0 \cdot 122\ 62;$$

and, as previously obtained,

$$P(11 \leqslant X \leqslant 20) = 0 \cdot 799\ 58; \quad P(X \geqslant 21) = 0 \cdot 077\ 80.$$

Hence, from (5), $E(T) = 61 \cdot 026\ 15.$

Therefore the average cost per call according to the new system is

$$\frac{61 \cdot 026\ 15}{15} = 4 \cdot 068\ 41 \text{p}.$$

Therefore the percentage reduction in cost is $193 \cdot 159/6 = 32 \cdot 19.$

52   Let X_i $(i = 0, 1, 2, 3)$ be the event that A has i spades and let Y be the event that A wins. Then

$$Y = \sum_{i=0}^{3} Y X_i.$$

Therefore $P(Y) = \sum_{i=0}^{3} P(Y X_i) = \sum_{i=0}^{3} P(Y \mid X_i) P(X_i),$

where $$P(X_i) = \binom{13}{i}\binom{39}{3-i} \Big/ \binom{52}{3}$$

and $$P(Y \mid X_i) = \binom{13-i}{i}\binom{36+i}{5-i} \Big/ \binom{49}{5}.$$

Hence
$$P(Y) = \sum_{i=0}^{3} \binom{13}{i}\binom{39}{3-i}\binom{13-i}{i}\binom{36+i}{5-i} \Big/ \binom{52}{3}\binom{49}{5}$$

$$= \frac{12\ 516\ 725\ 988}{42\ 142\ 136\ 400} = 0 \cdot 2970.$$

The probability of three consecutive wins for A is

$$(0 \cdot 2970)^3 = 0 \cdot 0262 \text{ which is } < 0 \cdot 05.$$

Hence it is reasonable to doubt the randomness of the card distribution.

53   Leicester, 1966.
   By the hypergeometric distribution

$$P(r) = \binom{13}{r}\binom{39}{13-r} \Big/ \binom{52}{13}, \quad \text{for } 0 \leqslant r \leqslant 13.$$

Therefore $P(0) = \binom{13}{0}\binom{39}{13} \Big/ \binom{52}{13} = \dfrac{17\ 063\ 919}{1\ 334\ 062\ 100} = 10^{-1} \times 0 \cdot 127\ 909.$

Also, $P(r+1) = \binom{13}{r+1}\binom{39}{12-r} \Big/ \binom{52}{13},$

so that $\dfrac{P(r)}{P(r+1)} = \dfrac{(r+1)(27+r)}{(13-r)^2},$

whence the recurrence relation. Hence

$$P(1) = 10^{-1} \times 0.800\ 616; \qquad P(2) = 0.205\ 873;$$
$$P(3) = 0.286\ 329; \qquad\qquad P(4) = 0.238\ 608;$$
$$P(5) = 0.124\ 692; \qquad\qquad P(6) = 10^{-1} \times 0.415\ 640;$$
$$P(7) = 10^{-2} \times 0.881\ 661; \qquad P(8) = 10^{-2} \times 0.116\ 690;$$
$$P(9) = 10^{-4} \times 0.926\ 111; \qquad P(r \geqslant 10) = 0.000\ 004.$$

Therefore $P(\text{i}) = (0.286\ 329)^3 = 0.0235;$

$$P(\text{ii}) = \binom{3}{2}(0.205\ 873)^2(0.124\ 691) = 0.0159.$$

54   Leicester, 1967.

By expansion of $G(\theta)$,

$$P(X = x) = \binom{x + r - 1}{r - 1} p^r q^x, \quad \text{for } X \geqslant 0$$

and

$$P(X = x - 1) = \binom{x + r - 2}{r - 1} p^r q^{x-1}, \quad \text{for } X \geqslant 1.$$

Therefore $\dfrac{P(X = x)}{P(X = x - 1)} = \dfrac{q(x + r - 1)}{x},$

whence the stated recurrence relation.

(i) For the given values of p, q and r,

$$P(X = 0) = (0.062\ 88)^{0.051\ 69},$$

so that $\log_{10} P(X = 0) = \bar{1}.937\ 895\ 1$

and $P(X = 0) = 0.866\ 752\ 4.$

$$P(X = 1) = 10^{-1} \times 0.419\ 852\ 5; \qquad P(X = 2) = 10^{-1} \times 0.206\ 894\ 9;$$
$$P(X = 3) = 10^{-1} \times 0.132\ 597\ 5; \qquad P(X = 4) = 10^{-2} \times 0.948\ 005\ 7;$$
$$P(X = 5) = 10^{-2} \times 0.719\ 900\ 3; \qquad P(X \geqslant 6) = 0.040\ 634\ 0.$$

Therefore, correct to 5 decimal places,

$$P(X = 0) = 0.866\ 75; \quad P(X = 1) = 0.041\ 99; \quad P(X = 2) = 0.020\ 69;$$
$$P(X = 3) = 0.013\ 26; \quad P(X = 4) = 0.009\ 48; \quad P(X = 5) = 0.007\ 20;$$
$$P(X \geqslant 6) = 0.040\ 63.$$

(ii) Let A, B and C be the events

$$A : X \geqslant 3; \quad B : X = 4; \quad C : X \geqslant 5.$$

Since A, B and C are not mutually exclusive events, define the mutually exclusive events

$$E_1 : X = 0, 1, 2; \quad E_2 : X = 3; \quad E_3 : X = 4; \quad E_4 : X = 5; \quad E_5 : X \geqslant 6.$$

and $P(E_1) = 0{\cdot}929\ 43;$ $P(E_2) = 0{\cdot}013\ 26;$ $P(E_3) = 0{\cdot}009\ 48;$
$P(E_4) = 0{\cdot}007\ 20;$ $P(E_5) = 0{\cdot}040\ 63.$

Then in terms of the E_i,

$$A : E_2, E_3, E_4, E_5; \quad B : E_3; \quad C : E_1, E_2, E_3, E_4.$$

Therefore the possible realisations of A, B and C are obtained from the algebraic product $E_3(E_1 + E_2 + E_3 + E_4)(E_2 + E_3 + E_4 + E_5)$, which gives the distinct combinations

$$E_3[(E_1E_2 + E_1E_4 + E_1E_5 + E_2E_4 + E_2E_5 + E_4E_5) +$$
$$+ (E_1E_3 + E_2E_3 + E_3E_4 + E_3E_5 + E_2^2 + E_4^2) + E_3^2].$$

Hence, taking account of the permutations and remembering that the E_i are mutually exclusive, we have

$$P(\text{ii}) = P(E_3)[6\{P(E_1)P(E_2) + P(E_1)P(E_4) + P(E_2)P(E_4) + P(E_1)P(E_5) +$$
$$+ P(E_2)P(E_5) + P(E_4)P(E_5)\} + 3\{P(E_1)P(E_3) + P(E_2)P(E_3) +$$
$$+ P(E_3)P(E_4) + P(E_3)P(E_5) + P^2(E_2) + P^2(E_4)\} + P^2(E_3)]$$
$$= 0{\cdot}009\ 48(0{\cdot}346\ 234 + 0{\cdot}028\ 943) = 0{\cdot}0036.$$

55 Based on past experience, the expected amount paid out per claim is

$$£(0{\cdot}60 \times 85 + 0{\cdot}40 \times 35) = £65.$$

The expected number of insurers making claims in a year is

$$0{\cdot}2 \times 178\ 540 = 35\ 708.$$

Hence the expected amount paid out in claims in a year is

$$£65 \times 35\ 708 = £2\ 321\ 020.$$

In the future, the expected amount paid out per claim is

$$£(0{\cdot}65 \times 108 + 0{\cdot}35 \times 52) = £88{\cdot}40.$$

The expected number of policy-holders is

$$178\ 540 + 17\ 854 = 196\ 394.$$

The expected number of claims per year is

$$0{\cdot}25 \times 196\ 394 = 49\ 098{\cdot}5.$$

Therefore the expected amount paid out in claims is

$$£88{\cdot}40 \times 49\ 098{\cdot}5 = £4\ 340\ 307{\cdot}40.$$

Hence total expected increase in claim costs is

$$£4\ 340\ 307{\cdot}40 - £2\ 321\ 020{\cdot}00 = £2\ 019\ 287{\cdot}40.$$

The percentage increase in the expected cost per claim is

$$\frac{23{\cdot}4 \times 100}{65} = 36{\cdot}00.$$

At current rates, the total amount received in premiums is

$$£15 \times 178\ 540 = £2\ 678\ 100.$$

Therefore the difference between current income and expected claim costs is £2 678 100 − £2 321 020 = £357 080.

The new expected income should be

$$£4\ 340\ 307 \cdot 40 + £357\ 080 = £4\ 697\ 387 \cdot 40.$$

Hence the average premium per policy should be

$$£4\ 697\ 387 \cdot 40/196\ 394 = £23 \cdot 9182.$$

Therefore the percentage increase in the average premium is

$$(8 \cdot 9182 \times 100)/15 = 59 \cdot 45.$$

56   The probability that there are w defective bulbs in the sample of 20 is

$$B(w) = \binom{20}{w}(0 \cdot 05)^{w}(0 \cdot 95)^{20-w}, \quad \text{for } 0 \leqslant w \leqslant 20.$$

Therefore $B(0) = (0 \cdot 95)^{20} = 0 \cdot 358\ 486;$ $B(1) = 0 \cdot 377\ 354;$

$$B(2) = 0 \cdot 188\ 677.$$

Hence $P(\text{i}) = 1 - P(w \leqslant 2) = 10^{-1} \times 0 \cdot 7548\,;$

$$P(\text{ii}) = \binom{30}{10}[1 - P(w \leqslant 2)]^{10}[P(w \leqslant 2)]^{20}$$

$$= 30\ 045\ 015 \times 10^{-10} \times (0 \cdot 7548)^{10}(0 \cdot 924\ 52)^{20} = 10^{-4} \times 0 \cdot 3753.$$

$P(\text{iii}) = 1 - $ Probability of rejecting not more than four batches

$$= 1 - \sum_{r=0}^{4} S(r),$$

where $S(r) = \binom{30}{r}(0 \cdot 075\ 48)^{r}(0 \cdot 924\ 52)^{30-r}, \quad \text{for } 0 \leqslant r \leqslant 30.$

Therefore $S(0) = 10^{-1} \times 0 \cdot 949\ 488;$ $S(1) = 0 \cdot 232\ 555;$

$$S(2) = 0 \cdot 275\ 302;   S(3) = 0 \cdot 209\ 779\,;$$

$$S(4) = 0 \cdot 115\ 606.$$

Hence $P(\text{iii}) = 10^{-1} \times 0 \cdot 7181.$

If a batch contains 8 per cent defectives, then the probability that a sample of 20 from such a batch will contain w defectives is

$$C(w) = \binom{20}{w}(0 \cdot 08)^{w}(0 \cdot 92)^{20-w}, \quad \text{for } 0 \leqslant w \leqslant 20.$$

Therefore $C(0) = (0.92)^{20} = 0.188\ 693;$ $C(1) = 0.328\ 162;$

$C(2) = 0.271\ 090.$

Hence the probability of rejecting a batch is now

$1 - \sum\limits_{w=0}^{2} C(w) = 0.212\ 055 = 0.2121$ to 4 significant figures.

For a population consisting of the two kinds of batches, the probability of rejecting a random batch is

$0.75 \times 0.075\ 483 + 0.25 \times 0.212\ 055 = 0.1096$ to 4 significant figures.

Hence, in this case, we have

$$P(\text{ii}) = \binom{30}{10}(0.1096)^{10}(0.8904)^{20}$$

$$= 30\ 045\ 015 \times 10^{-10} \times (1.096)^{10}(0.8904)^{20}$$

$$= 10^{-3} \times 0.7372.$$

Again, the probability of rejecting r out of 30 batches from the mixed population is

$$T(r) = \binom{30}{r}(0.1096)^{r}(0.8904)^{30-r}, \quad \text{for } 0 \leqslant r \leqslant 30.$$

Therefore $T(0) = 10^{-1} \times 0.307\ 294;$ $T(1) = 0.113\ 475;$

$T(2) = 0.202\ 532;$ $T(3) = 0.232\ 679;$

$T(4) = 0.193\ 325.$

Hence we now have

$P(\text{iii}) = 0.227\ 260 = 0.2273$ to 4 significant figures.

The percentage increase in the proportion of batches rejected is

$$\frac{34\ 143 \times 100}{75\ 483} = 45.2.$$

57   The probability that a packet contains w defective blades is

$$B(w) = \binom{5}{w}(0.32)^{w}(0.68)^{5-w}, \quad \text{for } 0 \leqslant w \leqslant 5.$$

Therefore $B(0) = (0.68)^{5} = 0.145\ 393;$ $B(1) = 0.342\ 101;$

$B(2) = 0.321\ 977;$ $B(3) = 0.151\ 519;$

$B(4) = 10^{-1} \times 0.356\ 515;$ $B(5) = 10^{-2} \times 0.335\ 544.$

Hence $P(\text{i}) = P(w \leqslant 1) = 0.4875;$

$P(\text{ii}) = P(w \geqslant 3) = 0.1905;$

$P(\text{iii}) = P(w \leqslant 3) = 0.9610;$

$P(\text{iv}) = [1 - P(w \leqslant 1)]^{12} = (0.5125)^{12} = 10^{-3} \times 0.3283.$

(v) If r out of 12 packets contain at least one defective blade each, then the probability of this event is

$$S(r) = \binom{12}{r} [P(w \geqslant 1)]^r [1 - P(w \geqslant 1)]^{12-r}$$

$$= \binom{12}{r} (0 \cdot 8546)^r (0 \cdot 1454)^{12-r}, \quad \text{for } 0 \leqslant r \leqslant 12.$$

Therefore
$$P(v) = 1 - \sum_{r=0}^{6} S(r),$$

where $S(0) = (0 \cdot 1454)^{12} = 10^{-10} \times 0 \cdot 892\ 838;$ $S(1) = 10^{-8} \times 0 \cdot 629\ 727;$

$S(2) = 10^{-6} \times 0 \cdot 203\ 570;$ $S(3) = 10^{-5} \times 0 \cdot 398\ 833;$

$S(4) = 10^{-4} \times 0 \cdot 527\ 439;$ $S(5) = 10^{-3} \times 0 \cdot 496\ 010;$

$S(6) = 10^{-2} \times 0 \cdot 340\ 123.$

Hence
$$P(v) = 0 \cdot 9960.$$

(vi) The probability that a packet contains not more than two defective blades is $P(w \leqslant 2) = 0 \cdot 8095$. If r packets out of 12 contain not more than two defective blades each, then the probability of this event is

$$T(r) = \binom{12}{r} [P(w \leqslant 2)]^r [1 - P(w \leqslant 2)]^{12-r}, \quad \text{for } 0 \leqslant r \leqslant 12,$$

and
$$P(vi) = \sum_{r=0}^{4} T(r),$$

where

$T(0) = (0 \cdot 1905)^{12} = 10^{-8} \times 0 \cdot 228\ 423;$ $T(1) = 10^{-6} \times 0 \cdot 116\ 478;$

$T(2) = 10^{-5} \times 0 \cdot 272\ 225;$ $T(3) = 10^{-4} \times 0 \cdot 385\ 593;$

$T(4) = 10^{-3} \times 0 \cdot 368\ 666.$

Hence
$$P(vi) = 10^{-3} \times 0 \cdot 4101.$$

58   The probability that the lady will obtain exactly w deep red plants in her selection of 10 is

$$B(w) = \binom{10}{w} (0 \cdot 25)^w (0 \cdot 75)^{10-w}, \quad \text{for } 0 \leqslant w \leqslant 10.$$

Therefore

$B(0) = (0 \cdot 75)^{10} = 10^{-1} \times 0 \cdot 563\ 135;$ $B(1) = 0 \cdot 187\ 712;$

$B(2) = 0 \cdot 281\ 568;$ $B(3) = 0 \cdot 250\ 283;$

$B(4) = 0 \cdot 145\ 998;$ $B(5) = 10^{-1} \times 0 \cdot 583\ 992;$

$B(6) = 10^{-1} \times 0 \cdot 162\ 220.$

Hence $P(i) = B(6) = 10^{-1} \times 0.1622$;

$\qquad P(ii) = 1 - P(w \leqslant 5) = 10^{-1} \times 0.1973$;

$\qquad P(iii) = B(1) = 0.1877$.

(a)   If one specified plant is definitely deep red, then the probability of getting r deep red plants amongst the other 9 is

$$S(r) = \binom{9}{r}(0.25)^r(0.75)^{9-r}, \quad \text{for } 0 \leqslant r \leqslant 9.$$

Therefore $S(0) = (0.75)^9 = 10^{-1} \times 0.750\ 847$; $S(1) = 0.225\ 254$;

$\qquad\qquad S(2) = 0.300\ 339$; $\qquad\qquad\qquad\qquad S(3) = 0.233\ 597$;

$\qquad\qquad S(4) = 0.116\ 798$; $\qquad\qquad\qquad\qquad S(5) = 10^{-1} \times 0.389\ 327$.

Hence we now have

$$P(i) = P(r = 5) = 10^{-1} \times 0.3893;$$
$$P(ii) = 1 - P(r \leqslant 4) = 10^{-1} \times 0.4893;$$
$$P(iii) = P(r = 0) = 10^{-1} \times 0.7508.$$

(b)   Define the events

X :   at least one of the 10 plants is deep red;

Y_1:   there are exactly 6 deep red plants in the 10;

Y_2:   there are at least 6 deep red plants in the 10;

Y_3:   there is exactly one deep red plant in the 10.

Then

$$P(i) = P(Y_1 \mid X) = P(XY_1)/P(X) = P(Y_1)/P(X)$$

$$= \frac{B(6)}{1 - B(0)} = 10^{-1} \times 0.1719;$$

$$P(ii) = P(Y_2 \mid X) = P(XY_2)/P(X) = P(Y_2)/P(X)$$

$$= \frac{P(w \geqslant 6)}{1 - B(0)} = 10^{-1} \times 0.2090;$$

$$P(iii) = P(Y_3 \mid X) = P(XY_3)/P(X) = P(Y_3)/P(X)$$

$$= \frac{B(1)}{1 - B(0)} = 0.1989.$$

With 40 plants, the probability that the lady will get w deep red plants is

$$T(w) = \binom{40}{w}(0.25)^w(0.75)^{40-w}, \quad \text{for } 0 \leqslant w \leqslant 40.$$

Then

$$T(0) = (0\cdot75)^{40} = 10^{-4} \times 0\cdot100\ 566; \qquad T(1) = 10^{-3} \times 0\cdot134\ 088;$$
$$T(2) = 10^{-3} \times 0\cdot871\ 572; \qquad\qquad T(3) = 10^{-2} \times 0\cdot367\ 997;$$
$$T(4) = 10^{-1} \times 0\cdot113\ 466; \qquad\qquad T(5) = 10^{-1} \times 0\cdot272\ 318.$$

Therefore the probability that the lady will get less than six deep red plants is

$$P(w \leqslant 5) = 10^{-1} \times 0\cdot4328.$$

59   The probability that there are w wet days in a week is

$$B(w) = \binom{7}{w}(0\cdot42)^{w}(0\cdot58)^{7-w}, \quad \text{for } 0 \leqslant w \leqslant 7.$$

Therefore

$$B(0) = (0\cdot58)^{7} = 10^{-1} \times 0\cdot220\ 798; \qquad B(1) = 0\cdot111\ 922;$$
$$B(2) = 0\cdot243\ 141; \qquad\qquad\qquad\qquad B(3) = 0\cdot293\ 446;$$
$$B(4) = 0\cdot212\ 495.$$

Hence
$$P(\text{i}) = (0\cdot42)^{4}(0\cdot58)^{3} = 10^{-2} \times 0\cdot6071;$$
$$P(\text{ii}) = B(4) = 0\cdot2125;$$
$$P(\text{iii}) = 1 - P(w \leqslant 3) = 0\cdot3294;$$
$$P(\text{iv}) = P(w \leqslant 4) = 0\cdot8831;$$
$$P(\text{v}) = \sum_{r=2}^{5} \binom{5}{r}[P(w \geqslant 4)]^{r}[1 - P(w \geqslant 4)]^{5-r}$$
$$= 1 - \sum_{r=0}^{1} \binom{5}{r}(0\cdot3294)^{r}(0\cdot6706)^{5-r} = 0\cdot5313;$$
$$P(\text{vi}) = [P(w \leqslant 4)]^{5} = (0\cdot8831)^{5} = 0\cdot5371;$$
$$P(\text{vii}) = \sum_{r=2}^{5} \binom{5}{r}[P(w \leqslant 4)]^{r}[1 - P(w \leqslant 4)]^{5-r}$$
$$= 1 - \sum_{r=0}^{1} \binom{5}{r}(0\cdot8831)^{r}(0\cdot1169)^{5-r} = 0\cdot9992.$$

60   The probability that the house-agent can provide suitable accommodation to w of the 18 clients is

$$B(w) = \binom{18}{w}(0\cdot75)^{w}(0\cdot25)^{18-w}, \quad \text{for } 0 \leqslant w \leqslant 18.$$

Therefore $B(0) = (0\cdot25)^{18} = 10^{-10} \times 0\cdot145\ 519; \ B(1) = 10^{-9} \times 0\cdot785\ 803;$
$$B(2) = 10^{-7} \times 0\cdot200\ 380; \qquad\qquad B(3) = 10^{-6} \times 0\cdot320\ 608;$$

$B(4) = 10^{-5} \times 0 \cdot 360\ 684$; $B(5) = 10^{-4} \times 0 \cdot 302\ 975$;

$B(6) = 10^{-3} \times 0 \cdot 196\ 934$; $B(7) = 10^{-2} \times 0 \cdot 101\ 280$;

$B(8) = 10^{-2} \times 0 \cdot 417\ 780.$

Hence $P(i) = P(w \leqslant 5) = 10^{-4} \times 0 \cdot 3425$;

$P(ii) = B(8) = 10^{-2} \times 0 \cdot 4178$;

$P(iii) = 1 - P(w \leqslant 6) = 0 \cdot 9998$;

$P(iv) = (0 \cdot 75)^4\ S(6),$

where $S(r) = \dbinom{14}{r}(0 \cdot 75)^r (0 \cdot 25)^{14-r}$,   for $0 \leqslant r \leqslant 14.$

We have $S(0) = (0 \cdot 25)^{14} = 10^{-8} \times 0 \cdot 372\ 529$; $S(1) = 10^{-6} \times 0 \cdot 156\ 462$;

$S(2) = 10^{-5} \times 0 \cdot 305\ 101$; $S(3) = 10^{-4} \times 0 \cdot 366\ 121$;

$S(4) = 10^{-3} \times 0 \cdot 302\ 050$; $S(5) = 10^{-2} \times 0 \cdot 181\ 230$;

$S(6) = 10^{-2} \times 0 \cdot 815\ 535$; $S(7) = 10^{-1} \times 0 \cdot 279\ 612$;

$S(8) = 10^{-1} \times 0 \cdot 733\ 982$; $S(9) = 0 \cdot 146\ 796$;

$S(10) = 0 \cdot 220\ 194.$

Therefore $P(iv) = (0 \cdot 75)^4 \times 10^{-2} \times 0 \cdot 815\ 535 = 10^{-2} \times 0 \cdot 2580$;

$$P(v) = (0 \cdot 75)^4 \left[1 - \sum_{r=0}^{3} S(r)\right] = 0 \cdot 3164;$$

$$P(vi) = (0 \cdot 75)^4 \sum_{r=0}^{1} S(r) = 10^{-7} \times 0 \cdot 5068;$$

$$P(vii) = (0 \cdot 75)^4 \sum_{r=0}^{10} S(r) = 0 \cdot 1515.$$

61 The probability that a room is occupied for w nights in a week is

$$B(w) = \dbinom{7}{w}(0 \cdot 85)^w (0 \cdot 15)^{7-w}, \quad \text{for } 0 \leqslant w \leqslant 7,$$

whence

$B(0) = (0 \cdot 15)^7 = 10^{-5} \times 0 \cdot 170\ 859$; $B(1) = 10^{-4} \times 0 \cdot 677\ 740$;

$B(2) = 10^{-2} \times 0 \cdot 115\ 216$; $B(3) = 10^{-1} \times 0 \cdot 108\ 815$;

$B(4) = 10^{-1} \times 0 \cdot 616\ 618.$

Hence $P(i) = B(4) = 10^{-1} \times 0 \cdot 6166$;

$P(ii) = 1 - P(w \leqslant 4) = 0 \cdot 9262$;

$P(iii) = P(w \leqslant 3) = 10^{-1} \times 0 \cdot 1210$;

$P(iv) = B(7) = (0 \cdot 85)^7 = 0 \cdot 3206.$

The expected numbers are

$$10P(\text{i}) = 1; \quad 10P(\text{ii}) = 9; \quad 10P(\text{iii}) = 0; \quad 10P(\text{iv}) = 3.$$

$$P(\text{v}) = \binom{10}{5}[B(4)]^5[1 - B(4)]^5$$

$$= 252(10^{-1} \times 0.6166)^5(0.938\ 34)^5 = 10^{-3} \times 0.1634;$$

$$P(\text{vi}) = \binom{10}{5}[P(w \leqslant 3)]^5[1 - P(w \leqslant 3)]^5$$

$$= 252(10^{-1} \times 0.1210)^5(0.987\ 90)^5 = 10^{-7} \times 0.6150;$$

$$P(\text{vii}) = [B(7)]^{10} = (0.3206)^{10} = 10^{-4} \times 0.1147;$$

$$P(\text{viii}) = 1 - \sum_{r=0}^{2} S(r),$$

where

$$S(r) = \binom{10}{r}[B(7)]^r[1 - B(7)]^{10-r}$$

$$= \binom{10}{r}(0.3206)^r(0.6794)^{10-r}, \quad \text{for } 0 \leqslant r \leqslant 10.$$

Therefore $S(0) = (0.6794)^{10} = 10^{-1} \times 0.209\ 534; \quad S(1) = 10^{-1} \times 0.988\ 764;$
$S(2) = 0.209\ 963.$

Hence $P(\text{viii}) = 0.6702.$

$$P(\text{ix}) = \frac{10!}{1!\,6!\,3!} B(4)[P(w \geqslant 5)]^6[P(w \leqslant 3)]^3$$

$$= 840 \times 10^{-1} \times 0.6166(0.926\ 24)^6(10^{-1} \times 0.1210)^3$$

$$= 10^{-4} \times 0.5794.$$

62 The probability that in a round the golfer will obtain a par or better score for w holes is

$$B(w) = \binom{18}{w}(0.43)^w(0.57)^{18-w}, \quad \text{for } 0 \leqslant w \leqslant 18.$$

Therefore

$B(0) = (0.57)^{18} = 10^{-4} \times 0.403\ 411; \quad B(1) = 10^{-3} \times 0.547\ 790;$
$B(2) = 10^{-2} \times 0.351\ 258; \qquad\qquad B(3) = 10^{-1} \times 0.141\ 325;$
$B(4) = 10^{-1} \times 0.399\ 801; \qquad\qquad B(5) = 10^{-1} \times 0.844\ 492;$
$B(6) = 0.138\ 032; \qquad\qquad\qquad\quad B(7) = 0.178\ 508;$
$B(8) = 0.185\ 163; \qquad\qquad\qquad\quad B(9) = 0.155\ 205.$

Hence
$$P(i) = B(9) = 0\cdot1552;$$
$$P(ii) = 1 - P(w \leqslant 3) = 0\cdot9818;$$
$$P(iii) = P(w \leqslant 5) = 0\cdot1427;$$
$$P(iv) = P(w \leqslant 8) = 0\cdot6444;$$
$$P(v) = 1 - P(w \leqslant 8) = 0\cdot3556;$$

$$P(vi) = \binom{3}{2}[P(w \leqslant 5)]^2[P(w \geqslant 9)]$$

$$= 3(0\cdot1427)^2(0\cdot3556) = 10^{-1} \times 0\cdot2172;$$

$$P(vii) = \binom{3}{2}[B(9)]^2[P(w \leqslant 8)]$$

$$= 3(0\cdot1552)^2(0\cdot6444) = 10^{-1} \times 0\cdot4657.$$

(viii)   The events $X : w \geqslant 4$, $Y : w \geqslant 6$, $Z : w \geqslant 9$ are not mutually exclusive. Define the mutually exclusive events

$$A : w = 4, 5; \quad B : w = 6, 7, 8; \quad C : w \geqslant 9.$$

Then
$$X = A + B + C, \quad Y = B + C, \quad Z = C,$$
and
$$P(A) = 0\cdot1244, \quad P(B) = 0\cdot5017, \quad P(C) = 0\cdot3556.$$

The realisations of (viii) are now obtained in terms of the distinct combinations of the product

$$XYZ = (A + B + C)(B + C)C,$$

which are seen to be ABC, AC^2, B^2C, BC^2, C^3. Hence, taking permutations into consideration and remembering that A, B, C are mutually exclusive, we have

$$P(viii) = 6P(A)P(B)P(C) + 3P(A)P^2C + 3P^2(B)P(C) + 3P(B)P^2(C) + P^3(C)$$

$$= 3P(B)P(C)[2P(A) + P(B)] + P^2(C)[3P(A) + 3P(B) + P(C)]$$

$$= 0\cdot6842.$$

63

$$P(i) = \binom{10}{5}(0\cdot48)^5(0\cdot52)^5 \times 2(0\cdot48)(0\cdot52) \times (0\cdot48)^2$$

$$= 504(0\cdot48)^8(0\cdot52)^6 = 10^{-1} \times 0\cdot2808;$$

$$P(ii) = \binom{10}{5}(0\cdot48)^5(0\cdot52)^5 \times (0\cdot48)^2$$

$$= 252(0\cdot48)^7(0\cdot52)^5 = 10^{-1} \times 0\cdot5625;$$

$$P(\text{iii}) = \binom{8}{3}(0\cdot48)^5(0\cdot52)^3 \times (0\cdot48)$$

$$= 56(0\cdot48)^6(0\cdot52)^3 = 10^{-1} \times 0\cdot9630;$$

$$P(\text{iv}) = \binom{7}{2}(0\cdot48)^5(0\cdot52)^2 \times (0\cdot48)$$

$$= 21(0\cdot48)^6(0\cdot52)^2 = 10^{-1} \times 0\cdot6945;$$

$$P(\text{v}) = P(\text{i})P(\text{ii})P(\text{iii}) = 10^{-3} \times 0\cdot1878;$$

$$P(\text{vi}) = P(\text{ii})P^2(\text{iii}) = 10^{-3} \times 0\cdot5216.$$

If the sets in (v) and (vi) are not ordered, then the required probabilities are

$$6P(\text{v}) = 10^{-2} \times 0\cdot1127 \quad \text{and} \quad 3P(\text{vi}) = 10^{-2} \times 0\cdot1565.$$

64   The probability that a housewife obtains w A coupons from the 25 packets purchased is

$$C(w) = \binom{25}{w}(0\cdot6)^w(0\cdot4)^{25-w}, \quad \text{for } 0 \leqslant w \leqslant 25.$$

Therefore

$C(0) = (0\cdot4)^{25} = 10^{-9} \times 0\cdot112\ 590;$ $C(1) = 10^{-8} \times 0\cdot422\ 212;$

$C(2) = 10^{-7} \times 0\cdot759\ 982;$ $C(3) = 10^{-6} \times 0\cdot873\ 979;$

$C(4) = 10^{-5} \times 0\cdot721\ 033;$ $C(5) = 10^{-4} \times 0\cdot454\ 251;$

$C(6) = 10^{-3} \times 0\cdot227\ 126;$ $C(7) = 10^{-3} \times 0\cdot924\ 727;$

$C(8) = 10^{-2} \times 0\cdot312\ 095;$ $C(9) = 10^{-2} \times 0\cdot884\ 269;$

$C(10) = 10^{-1} \times 0\cdot212\ 225;$ $C(11) = 10^{-1} \times 0\cdot434\ 097;$

$C(12) = 10^{-1} \times 0\cdot759\ 670;$ $C(13) = 0\cdot113\ 950;$

$C(14) = 0\cdot146\ 507;$ $C(15) = 0\cdot161\ 158;$

$C(16) = 0\cdot151\ 086;$ $C(17) = 0\cdot119\ 980;$

$C(18) = 10^{-1} \times 0\cdot799\ 867;$ $C(19) = 10^{-1} \times 0\cdot442\ 032;$

$C(20) = 10^{-1} \times 0\cdot198\ 914;$ $C(21) = 10^{-2} \times 0\cdot710\ 407;$

$C(22) = 10^{-2} \times 0\cdot193\ 747;$ $C(23) = 10^{-3} \times 0\cdot379\ 070;$

$C(24) = 10^{-4} \times 0\cdot473\ 838;$ $C(25) = 10^{-5} \times 0\cdot284\ 303.$

Hence $P(\text{i}) = 1 - P(w \leqslant 9) = 0\cdot9868;$

$$P(\text{ii}) = P(w \leqslant 10) = 10^{-1} \times 0\cdot3439;$$

$$P(\text{iii}) = P(w \leqslant 5) = 10^{-4} \times 0\cdot5359;$$

$$P(\text{iv}) = C(5) = 10^{-4} \times 0\cdot4543;$$

$$P(\text{v}) = [1 - C(10)]^2 + 2C(10)[1 - C(10)] = 0\cdot9995;$$

$$P(\text{vi}) = \left[\sum_{w=11}^{19} C(w) \right]^2 = 0\cdot 8765;$$

$$P(\text{vii}) = \binom{2}{1} \left[\sum_{w=0}^{9} C(w) \right] \left[\sum_{w=11}^{19} C(w) \right] = 10^{-1} \times 0\cdot 2466;$$

$$P(\text{viii}) = \binom{2}{1} \left[\sum_{w=0}^{9} C(w) \right] \left[\sum_{w=20}^{25} C(w) \right] = 10^{-2} \times 0\cdot 7733.$$

65   The probability of a tie is

$$\binom{16}{8}(0\cdot 5)^{16} = \frac{12\,870}{65\,536} = 0\cdot 1964;$$

$$P(\text{i}) = \binom{14}{7}(0\cdot 5)^{14} = \frac{3432}{16\,384} = 0\cdot 2095;$$

$$P(\text{ii}) = P(\text{i}) = 0\cdot 2095.$$

If one of the directors abstains, then the probability that the proposal will be carried by the narrowest majority is

$$\binom{15}{8}(0\cdot 5)^{15} = \frac{6435}{32\,768} = 0\cdot 1964.$$

$$P(\text{iii}) = \binom{6}{4}(0\cdot 1964)^4(0\cdot 8036)^2 = 10^{-1} \times 0\cdot 1441;$$

$$P(\text{iv}) = \sum_{r=4}^{6} S(r), \quad \text{where} \quad S(r) = \binom{6}{r}(0\cdot 1964)^r(0\cdot 8036)^{6-r},$$

so that $S(4) = 10^{-1} \times 0\cdot 144\,124; \qquad S(5) = 10^{-2} \times 0\cdot 140\,896;$
$S(6) = 10^{-4} \times 0\cdot 573\,916.$

Hence $P(\text{iv}) = 10^{-1} \times 0\cdot 1588;$
$$P(\text{v}) = 1 - P(\text{iv}) = 0\cdot 9841;$$
$$P(\text{vi}) = S(6) = 10^{-4} \times 0\cdot 5739.$$

66   The probability that exactly w miners will be killed in a year is

$$B(w) = \binom{10\,000}{w}\left(\frac{1}{2500}\right)^w\left(1 - \frac{1}{2500}\right)^{10\,000-w}, \quad \text{for } 0 \leqslant w \leqslant 10\,000,$$

$$\sim e^{-4}\, 4^w/w! \equiv Q(w), \quad \text{for } w \geqslant 0.$$

Hence $Q(0) = e^{-4} = 10^{-1} \times 0\cdot 183\,156; \quad Q(1) = 10^{-1} \times 0\cdot 732\,624;$

$$Q(2) = 0.146\ 525; \qquad\qquad Q(3) = 0.195\ 367;$$
$$Q(4) = 0.195\ 367.$$

Therefore
$$P(\text{i}) = 1 - Q(0) = 0.9817;$$
$$P(\text{ii}) = Q(2) = 0.1465;$$
$$P(\text{iii}) = P(w \leqslant 3) = 0.4335;$$
$$P(\text{iv}) = 1 - P(w \leqslant 3) = 0.5665.$$

On an average, there are four fatal accidents in a year, and so the expected amount of compensation paid in a year is $4 \times £3250 = £13\ 000$.

The probability that the compensation will exceed the expected amount is $P(w \geqslant 5) = 1 - P(w \leqslant 4) = 0.3712$.

67 The probability that there is demand for w sets in a week is

$$Q(w) = e^{-3.56}(3.56)^{w}/w!, \quad \text{for } w \geqslant 0.$$

Therefore

$$Q(0) = e^{-3.56} = 10^{-1} \times 0.284\ 388; \quad Q(1) = 0.101\ 242;$$
$$Q(2) = 0.180\ 211; \qquad\qquad\qquad Q(3) = 0.213\ 850;$$
$$Q(4) = 0.190\ 326; \qquad\qquad\qquad Q(5) = 0.135\ 512;$$
$$Q(6) = 10^{-1} \times 0.804\ 038; \qquad\quad Q(7) = 10^{-1} \times 0.408\ 911;$$
$$P(w \geqslant 8) = 10^{-1} \times 0.291\ 25.$$

Hence
$$P(\text{i}) = P(w \leqslant 4) = 0.7141;$$
$$P(\text{ii}) = 1 - P(w \leqslant 6) = 10^{-1} \times 0.7002;$$
$$P(\text{iii}) = 1 - P(w \leqslant 2) = 0.6901;$$

$$P(\text{iv}) = \binom{4}{2}[Q(5)]^{2}[1 - Q(5)]^{2} = 10^{-1} \times 0.8234.$$

(v) With six sets, the expected demand met is

$$\sum_{w=0}^{5} wQ(w) + 6P(w \geqslant 6) = 3.444\ 598.$$

Therefore the percentage of expected demand not met is

$$\frac{(3.56 - 3.444\ 598)100}{3.56} = 3.24.$$

Also, the expected percentage of unused stock is

$$\frac{(6 - 3.444\ 598)100}{6} = 42.59.$$

(vi) $P(\text{a}) = P(w \geqslant 6) = 0.1504.$

(b) With five sets, the expected demand satisfied is

$$\sum_{w=0}^{4} wQ(w) + 5P(w \geqslant 5) = 3.294\ 178.$$

Therefore the expected percentage of unsatisfied demand is

$$\frac{(3\cdot56 - 3\cdot294\ 178)100}{3\cdot56} = 7\cdot47.$$

(c) The percentage of available stock in excess of expected demand is

$$\frac{(5 - 3\cdot294\ 178)100}{5} = 34\cdot12.$$

68 If X is a random variable denoting the length in lines of the cinema's advertisements, then $P(X = w) = e^{-12}\ 12^{w}/w!$,   for $X \geqslant 0$.

At the reduced rates, we have

$$E(X) = 60 \times P(1\leqslant X\leqslant 8) + 100 \times P(9 \leqslant X \leqslant 12) + 125 \times P(13 \leqslant X \leqslant 16)+$$

$$+ \sum_{w=17}^{\infty} P(X = w) \times 9w$$

$$= 60P(1\leqslant X\leqslant 8) + 100P(9 \leqslant X \leqslant 12) + 125P(13 \leqslant X \leqslant 16) +$$

$$+ 108P(X \geqslant 16). \tag{1}$$

If we now use the normal approximation for the distribution of X, we obtain

$$P(1 \leqslant X\leqslant 8) \sim \Phi\left(\frac{8 - 12 + 0\cdot5}{\sqrt{12}}\right) - \Phi\left(\frac{1 - 12 + 0\cdot5}{\sqrt{12}}\right)$$

$$= 1 - \Phi(1\cdot010) - [1 - \Phi(3\cdot031)] = 0\cdot155\ 03;$$

$$P(9 \leqslant X \leqslant 12) \sim \Phi\left(\frac{12 - 12 + 0\cdot5}{\sqrt{12}}\right) - \Phi\left(\frac{9 - 12 - 0\cdot5}{\sqrt{12}}\right)$$

$$= \Phi(0\cdot144) - [1 - \Phi(1\cdot010)] = 0\cdot401\ 00;$$

$$P(13 \leqslant X \leqslant 16) \sim \Phi\left(\frac{16 - 12 + 0\cdot5}{\sqrt{12}}\right) - \Phi\left(\frac{12 - 12 + 0\cdot5}{\sqrt{12}}\right)$$

$$= \Phi(1\cdot299) - \Phi(0\cdot144) = 0\cdot345\ 78;$$

$$P(X \geqslant 16) = 1 - P(X \leqslant 15)$$

$$\sim 1 - \Phi\left(\frac{15 - 12 + 0\cdot5}{\sqrt{12}}\right) = 1 - \Phi(1\cdot010) = 0\cdot156\ 25.$$

Hence, from (1), we have $E(X) = 109\cdot499\ 30$p, and so the percentage reduction in expected cost is

$$\frac{(120 - 109\cdot499\ 30)100}{120} = 8\cdot75.$$

69 Leicester, 1971.

Let X be a random variable denoting the number of defects in a blanket. Then $P(X = w) = e^{-2\cdot48}\ (2\cdot48)^{w}/w!$,   for $X \geqslant 0$.

Therefore

$$P(X = 0) = e^{-2 \cdot 48} = 10^{-1} \times 0 \cdot 837\ 432; \quad P(X = 1) = 0 \cdot 207\ 683;$$
$$P(X = 2) = 0 \cdot 257\ 527; \quad\quad\quad\quad\quad\quad P(X = 3) = 0 \cdot 212\ 889;$$
$$P(X \geqslant 4) = 0 \cdot 238\ 158.$$

Hence the probabilities of a blanket being of A, B, or C grade are

$$P(A) = 10^{-1} \times 0 \cdot 8374; \quad P(B) = 0 \cdot 6781; \quad P(C) = 0 \cdot 2382.$$

Let Y be the event that the blanket has at least one defect and Z be the event that the blanket is of grade B. Then

$$P(Y) = 1 - P(A) = 0 \cdot 916\ 257; \quad\quad P(Z) = P(B) = 0 \cdot 678\ 099.$$

Therefore the required conditional probability is

$$P(Z \mid Y) = \frac{P(ZY)}{P(Y)} = \frac{P(Z)}{P(Y)} = 0 \cdot 7401.$$

(i)   The expected price of a randomly selected blanket is

$$4 \cdot 50 \times P(A) + 4 \cdot 00 \times P(B) + 3 \cdot 75 \times P(C) = £3 \cdot 98.$$

(ii)   The expected price of a blanket known to have at least one defect is

$$\frac{4 \cdot 00 \times P(B) + 3 \cdot 75 \times P(C)}{1 - P(A)} = £3 \cdot 94.$$

The expected price of a randomly selected blanket is £3·98 and so for ungraded blankets for the hospital, the manufacturer could tender for £3·95.

70   If X is a random variable denoting the number of buses breaking down in a day, then

$$P(X = w) = \binom{1000}{w}(0 \cdot 0024)^{w}(1 - 0 \cdot 0024)^{1000 - w}$$

$$\sim e^{-2 \cdot 4}(2 \cdot 4)^{w}/w! \equiv Q(w), \quad \text{for } X \geqslant 0.$$

Then $Q(0) = e^{-2 \cdot 4} = 10^{-1} \times 0 \cdot 907\ 180; \quad Q(1) = 0 \cdot 217\ 723;$
$Q(2) = 0 \cdot 261\ 268; \quad\quad\quad\quad\quad\quad\ Q(3) = 0 \cdot 209\ 014;$
$Q(4) = 0 \cdot 125\ 408; \quad\quad\quad\quad\quad\quad\ Q(5) = 10^{-1} \times 0 \cdot 601\ 958;$
$Q(6) = 10^{-1} \times 0 \cdot 240\ 783; \quad\quad\ Q(7) = 10^{-2} \times 0 \cdot 825\ 542;$
$Q(8) = 10^{-2} \times 0 \cdot 247\ 663.$

Hence $P(\mathrm{i}) = 1 - P(X \leqslant 6) = 10^{-1} \times 0 \cdot 115\ 95.$

(ii)   The probability that there is inadequate staff on r days of the year is

$$B(r) = \binom{365}{r}(0 \cdot 011\ 595)^{r}(1 - 0 \cdot 011\ 595)^{365 - r}$$

$$\sim e^{-4 \cdot 2322}(4 \cdot 2322)^{r}/r!, \quad \text{for } r \geqslant 0.$$

If Y is a random variable denoting the number of days in a year when there is inadequate staff, then $P(Y = r) = B(r)$. Therefore

$$P(Y = 0) = e^{-4 \cdot 2322} = 10^{-1} \times 0 \cdot 145 \ 204; \quad P(Y = 1) = 10^{-1} \times 0 \cdot 614 \ 532;$$
$$P(Y = 2) = 0 \cdot 130 \ 041; \qquad\qquad\qquad P(Y = 3) = 0 \cdot 183 \ 453.$$

Hence $P(\text{ii}) = P(Y \geqslant 4) = 1 - P(Y \leqslant 3) = 0 \cdot 6105.$

(iii)   Since $P(X \geqslant 9) = 0 \cdot 000 \ 863$, the maintenance staff should be increased to cope with 8 breakdowns in a day.

(iv)   With this increased strength, we now have

$$B(r) = \binom{365}{r} (0 \cdot 000 \ 863)^r (1 - 0 \cdot 000 \ 863)^{365 - r}$$

$$\sim e^{-0 \cdot 3150} (0 \cdot 3150)^r / r!, \quad \text{for } r \geqslant 0.$$

Therefore

$$P(Y = 0) = e^{-0 \cdot 3150} = 0 \cdot 729 \ 789; \quad P(Y = 1) = 0 \cdot 229 \ 884;$$
$$P(Y = 2) = 10^{-1} \times 0 \cdot 362 \ 067; \qquad P(Y = 3) = 10^{-2} \times 0 \cdot 380 \ 170.$$

Hence $P(\text{ii}) = 1 - P(Y \leqslant 3) = 0 \cdot 000 \ 318.$

71   The probability that there are w errors on a page-proof is

$$\binom{400}{w} \left(\frac{1}{1000} \right)^w \left(1 - \frac{1}{1000} \right)^{400 - w}$$

$$\sim e^{-0 \cdot 4} (0 \cdot 4)^w / w! \equiv Q(w), \quad \text{for } w \geqslant 0.$$

Therefore $Q(0) = e^{-0 \cdot 4} = 0 \cdot 670 \ 320; \qquad Q(1) = 0 \cdot 268 \ 128;$

and $P(w \geqslant 2) = 0 \cdot 061 \ 552.$

The expected payment to the proof-reader for a page is (in new pence)

$$25 \times Q(0) + 100 \times Q(1) + 125 \times P(w \geqslant 2) = 51 \cdot 2648 \text{p}.$$

Therefore the expected payment for a book of 250 pages is

$$\text{\pounds} \frac{250 \times 51 \cdot 2648}{100} = \text{\pounds} 128 \cdot 16.$$

If x is the new rate for a page on which the proof-reader detects no errors, then the equation for x is

$$x \times Q(0) + 100 \times Q(1) + 125 \times P(w \geqslant 2) = 75,$$

whence

$$x = 60 \cdot 4088 = 60 \text{p}.$$

The percentage increase in the expected payment per page to the proof-reader is

$$\frac{(75 - 51 \cdot 2648)100}{51 \cdot 2648} = 46 \cdot 30.$$

72 The probability that there is a demand for w batteries in a month is

$$Q(w) = e^{-5 \cdot 28}(5 \cdot 28)^{w}/w!, \quad \text{for } w \geqslant 0.$$

Therefore

$Q(0) = e^{-5 \cdot 28} = 10^{-2} \times 0 \cdot 509\ 243;$ $Q(1) = 10^{-1} \times 0 \cdot 268\ 880;$

$Q(2) = 10^{-1} \times 0 \cdot 709\ 843;$ $Q(3) = 0 \cdot 124\ 932;$

$Q(4) = 0 \cdot 164\ 910;$ $Q(5) = 0 \cdot 174\ 145;$

$Q(6) = 0 \cdot 153\ 248;$ $Q(7) = 0 \cdot 115\ 593;$

$Q(8) = 10^{-1} \times 0 \cdot 762\ 914;$ $Q(9) = 10^{-1} \times 0 \cdot 447\ 576;$

$Q(10) = 10^{-1} \times 0 \cdot 236\ 320$ $Q(11) = 10^{-1} \times 0 \cdot 113\ 434;$

$Q(12) = 10^{-2} \times 0 \cdot 499\ 110.$

Since $P(w \leqslant 10) = 0 \cdot 980\ 473$ and $P(w \leqslant 11) = 0 \cdot 991\ 816$, the monthly stock should be 10 batteries. This gives a probability of not meeting the demand in full as

$$1 - 0 \cdot 980\ 473 = 0 \cdot 019\ 527, \text{ nearest to } 0 \cdot 025.$$

With 6 batteries, the probability of meeting the demand in full is $P(w \leqslant 6) = 0 \cdot 720\ 199$. Hence the probability that in six months of short supplies there will be no loss of custom is $(0 \cdot 720\ 199)^{6} = 0 \cdot 1395$.

With 6 batteries, the expected demand met is

$$\sum_{w=0}^{5} wQ(w) + 6P(w \geqslant 6) = 4 \cdot 672\ 311.$$

Hence the percentage expected loss of custom is

$$(5 \cdot 28 - 4 \cdot 672\ 311)100/5 \cdot 28 = 11 \cdot 51.$$

73 The probability that w workers are absent on one day is

$$P(X = w) = \binom{1500}{w}(0 \cdot 0025)^{w}(1 - 0 \cdot 0025)^{1500 - w}$$

$$\sim e^{-3 \cdot 75}(3 \cdot 75)^{w}/w!, \quad \text{for } w \geqslant 0$$

$$\equiv Q(w), \text{ say.}$$

The average daily rate of absenteeism is $3 \cdot 75$.

$Q(0) = e^{-3 \cdot 75} = 10^{-1} \times 0 \cdot 235\ 177;$ $Q(1) = 10^{-1} \times 0 \cdot 881\ 914;$

$Q(2) = 0 \cdot 165\ 359;$ $Q(3) = 0 \cdot 206\ 699;$

$Q(4) = 0 \cdot 193\ 780;$ $Q(5) = 0 \cdot 145\ 335;$

$Q(6) = 10^{-1} \times 0 \cdot 908\ 344;$ $Q(7) = 10^{-1} \times 0 \cdot 486\ 613;$

$Q(8) = 10^{-1} \times 0 \cdot 228\ 100;$ $Q(9) = 10^{-2} \times 0 \cdot 950\ 417;$

$Q(10) = 10^{-2} \times 0 \cdot 356\ 406.$

Therefore $P(X \geqslant 11) = 1 - P(X \leqslant 10) = 0.001\ 745$.

The probability that there are not more than ten absentees on a day is

$$P(X \leqslant 10) = 0.998\ 255 = 0.9983 \text{ to 4 decimal places.}$$

The expected number of hours lost in a working year is

$$3.75 \times 8 \times 300 = 9000.$$

If it is known that not more than ten workers are absent on a day, then the average daily rate of absenteeism is

$$\sum_{w=0}^{10} wP(X = w)/P(X \leqslant 10) = \frac{3.730\ 088}{0.998\ 255} = 3.736\ 608.$$

Therefore the expected number of man-hours lost in a year is now

$$3.736\ 608 \times 8 \times 300 = 8968.$$

Let the number of absentees on the 300 days of the working year be $x_1, x_2, \dots, x_n$ $(n = 300)$. Then each x_i is an independent Poisson variable and the mean $\bar{x}$ is such that

$$E(\bar{x}) = 3.75; \quad \text{var}(\bar{x}) = \frac{3.75}{300} = 0.0125; \quad \text{s.d.}(\bar{x}) = 0.111\ 803.$$

Also, 5 per cent more than the average daily rate of absenteeism is

$$1.05 \times 3.75 = 3.9375.$$

Hence the required probability is

$$P(\bar{x} \geqslant 3.9375) = 1 - P(\bar{x} < 3.9375)$$

$$\sim 1 - \Phi\left(\frac{3.9375 - 3.75}{0.111\ 803}\right)$$

$$= 1 - \Phi(1.677) = 0.046\ 77.$$

74   The probability of making w errors on a stencil is

$$\binom{420}{w}\left(\frac{1}{1400}\right)^w \left(1 - \frac{1}{1400}\right)^{420-w}$$

$$\sim e^{-0.3}(0.3)^w/w! \equiv Q(w), \text{ say,}   \text{for } w \geqslant 0.$$

Hence $P(\text{i}) = Q(0) = e^{-0.3} = 0.740\ 818.$

(ii)   If μ is the average rate of errors on a stencil, then the expected cost of correcting the errors made on a stencil is

$$\sum_{r=0}^{\infty} e^{-\mu}\frac{\mu^r}{r!} \times \frac{2r(3r+2)}{r+1} = 2e^{-\mu}\sum_{r=0}^{\infty}\frac{\mu^r[3(r+1)r - (r+1) + 1]}{(r+1)!}$$

$$= 2e^{-\mu}\left[3\sum_{r=1}^{\infty}\frac{\mu^r}{(r-1)!} - \sum_{r=0}^{\infty}\frac{\mu^r}{r!} + \sum_{r=0}^{\infty}\frac{\mu^r}{(r+1)!}\right]$$

$$= 2\left[\left(3\mu - 1 + \frac{1}{\mu}\right) - \frac{1}{\mu}e^{-\mu}\right].$$

Hence, for $\mu = 0\cdot3$, the expected cost of correcting the errors on a stencil is $1\cdot527\ 88$p. Therefore the net expected profit on a stencil is $4 - 1\cdot527\ 88 = 2\cdot472\ 12$p; and the net expected profit on 500 stencils is

$$\frac{2\cdot472\ 12 \times 500}{100} = £12\cdot36.$$

(iii) $Q(0) = 0\cdot740\ 818;$ $Q(1) = 0\cdot222\ 245;$

$Q(2) = 10^{-1} \times 0\cdot333\ 368;$ $P(w \geqslant 3) = 0\cdot003\ 600.$

Therefore the probability that none of the 500 stencils has more than two errors is $(1 - 0\cdot003\ 600)^{500} \sim e^{-1\cdot8} = 0\cdot1653.$

(iv)   The expected number of errors per stencil is $0\cdot3$ and so the expectation increased by ten per cent is $0\cdot33$. Also, if $x_1, x_2, \ldots, x_n$ ($n = 500$) are the observed number of errors on the 500 stencils, then, for their average $\bar{x}$,

$$E(\bar{x}) = 0\cdot3; \quad \text{var}(\bar{x}) = \frac{0\cdot3}{500} = 0\cdot0006; \quad \text{s.d.}(\bar{x}) = 0\cdot024\ 494\ 9.$$

Hence the required probability is

$$P(\bar{x} > 0\cdot33) = 1 - P(\bar{x} \leqslant 0\cdot33)$$

$$\sim 1 - \Phi\left(\frac{0\cdot33 - 0\cdot3}{0\cdot024\ 494\ 9}\right)$$

$$= 1 - \Phi(1\cdot225) = 0\cdot1103.$$

75   On an average, the daily number of bills received for amounts not exceeding £5 is

$$485 \times 0\cdot4 = 194.$$

Hence the probability that, on an average, w incorrect bills are amongst the 194 received on any day is

$$\binom{194}{w}(0\cdot025)^w(1 - 0\cdot025)^{194-w} \sim e^{-4\cdot85}(4\cdot85)^w/w!, \quad \text{for } w \geqslant 0.$$

Thus the average daily rate of such incorrect bills is $4\cdot85$. Therefore the expected amount paid in excess on such bills during a year is

$$£4\cdot85 \times 300 \times 3\cdot39 = £4932 \text{ to nearest £.}$$

On the other hand, by the old system the expected amount spent on scrutiny in a year is

$$£194 \times 300 \times \frac{1}{8} = £7275.$$

Therefore the expected annual saving by the new system is £2343.

If the new daily rate of incorrect bills under £5 is x, then the equation for x is

$$300x \times 3 \cdot 39 = 7275,$$

whence $x = 7 \cdot 1534$.

Suppose $x_1, x_2, \ldots, x_n$ $(n = 300)$ are the daily rates of incorrect bills under £5 during the year. Then each x_i is an independent Poisson variable with mean 4·85, and $\bar{x}$, the average of the x_i, is such that

$$E(\bar{x}) = 4 \cdot 85; \quad \text{var}(\bar{x}) = \frac{4 \cdot 85}{300} = 0 \cdot 016\ 17; \quad \text{s.d.}(\bar{x}) = 0 \cdot 1272.$$

Hence the required probability for no saving by the new system is

$$P(\bar{x} \geqslant 7 \cdot 1534) = 1 - P(\bar{x} < 7 \cdot 1534)$$

$$\sim 1 - \Phi\left(\frac{7 \cdot 1534 - 4 \cdot 85}{0 \cdot 1272}\right) = 1 - \Phi(18 \cdot 11) \sim 0.$$

Hence it is practically impossible that no saving will be made by using the new system during a year.

3 Statistical inference

1 A national air transport authority proposes to locate a new airport in a certain county in order to cope with the increasing pressure on existing airports. However, there is considerable division of opinion regarding the project among the residents of the county and of adjoining districts which are likely to be affected. Those who favour the scheme point to the economic benefit to be expected from the development, whereas the objectors claim that the new airport would increase traffic, noise and pollution, with adverse effects on the amenities of the region. Electorally, the county and its neighbourhood are a marginal area for the government, which is therefore unwilling to support the scheme unless there is strong assurance that at least 60 per cent of the adult population of the area are in favour of it. Accordingly, to assess public opinion, the air transport authority decides to carry out an opinion survey in the county and its neighbourhood.

(i) If the true proportion of the population in favour of the airport is 64 per cent, how large a sample should be taken in the survey so that the chance of the sample proportion being not more than 62 per cent is 0·2 per cent?

(ii) Also, if the true proportion of the population in favour of the airport is 59 per cent, how large a sample should be taken in the survey so that the chance of the sample proportion being at least 60 per cent is 0·1 per cent?

It may be assumed that the population sampled is sufficiently large for the effects of finite sampling to be negligible.

2 Electric-light bulbs made by a standard process have a mean lifetime of 2000 hours with a standard deviation of 250 hours. A research engineer suggests a new production process which involves substantial capital outlay, and it is considered worth while to adopt the new process only if the mean lifetime can be increased by at least 15 per cent. In order to determine the average performance of bulbs produced by the new process, the engineer plans to test a certain number of bulbs. Calculate the size of the sample which should be examined if the engineer wishes the probability to be about 0·01 that he will fail to adopt the new process if, in fact, it produces bulbs with a mean lifetime of 2375 hours. It may

129

be assumed that the distributions of the lifetimes of bulbs produced by both processes are normal and have the same standard deviation.

Also, determine the size of the sample if the standard deviation of the new process is 320 hours.

3   A city has an adult population of 750 000 persons, and in a public opinion survey interest centres on finding out the proportion of adults who are in favour of the city council giving financial support to a newly established local theatre. In a random sample of 8500 adults questioned, 6893 were found to be in favour of subsidising the theatre. Estimate, correct to four decimal places, the proportion of the city's adult population which is in favour of supporting the theatre and determine, correct to four significant figures, the standard error of the estimate.

Also, estimate the total number of adults in the city who are in favour of the subsidy, and evaluate the standard error of the estimate.

How is the standard error of the estimated proportion of adults in favour of supporting the theatre affected if the adult population of the city is assumed to be "effectively" infinite?

4   In a certain country, the Ministry of Transport conducted a survey to assess the attitude of the people to the introduction of breathalyser tests as a measure for reducing road accidents due to drunken drivers. The survey was planned in two parts based on separate independent samples. In the first sample, 11 850 motorists were questioned and of these 6092 were in favour of the tests and the rest were against them. In the second sample of 17 532 non-motorists, 14 968 declared themselves in favour of the tests and the rest were against them. If the true proportions of motorists and non-motorists who are in favour of the tests are p_1 and p_2 respectively, determine, correct to four decimal places, their maximum-likelihood estimates and evaluate, correct to four significant figures, the standard errors of the estimates. It may be assumed that the populations of motorists and non-motorists are both sufficiently large for the effects of sampling from finite populations to be ignored.

Also, evaluate the best estimates of the parametric functions

$$u = \frac{n_1 p_1 + n_2 p_2}{n_1 + n_2} \quad \text{and} \quad v = \tfrac{1}{2}(p_1 + p_2),$$

where n_1 and n_2 are the two sample sizes, and determine the standard errors of the two estimates.

5   In a breeding experiment with a variety of *Primula sinensis*, a total of 4164 individual plants were observed in four distinct classes with the following frequencies:

Class	F, Ch	F, ch	f, Ch	f, ch
Frequency	2972	171	190	831

Source   D. de Winton and J.B.S. Haldane (1933), *Journal of Genetics*, Vol. 31, p.67.

It is known from genetical considerations that the expected proportions in the four classes are $\frac{1}{4}(2 + \theta)$, $\frac{1}{4}(1 - \theta)$, $\frac{1}{4}(1 - \theta)$, and $\frac{1}{4}\theta$ respectively. Calculate, correct to six decimal places, the maximum-likelihood estimate of the parameter θ. Determine the large-sample variance of the estimate and hence evaluate, correct to four significant figures, the standard error of the estimate. Also, calculate the expected frequencies in the four classes.

6 A biologist wishes to determine the effectiveness of a new insecticide when used in varying strengths. It is known from empirical considerations that $p(x)$, the probability of killing an insect at strength x of the insecticide in a fixed interval of time, is given by the relation

$$\log_e\left[\frac{p(x)}{1 - p(x)}\right] = a + x,$$

where a is an unknown parameter.

To estimate a, the insecticide is used for a fixed period on three random groups of n insects each, and the groups are subjected to different strengths of insecticide, which, on an appropriately chosen scale, correspond to the values $x = -1, 0, 1$. If the total number of insects killed in the three groups is r, show that $\hat{a}$, the maximum-likelihood estimate of a, is obtained from the cubic

$$y^3 + (1 - \phi)(e + 1 + e^{-1})y^2 + (1 - 2\phi)(e + 1 + e^{-1})y + (1 - 3\phi) = 0,$$

where $\hat{a} \equiv -\log_e y$, $\phi \equiv \dfrac{n}{r}$, and e is the Napierian constant.

Hence obtain the numerical value of $\hat{a}$, correct to four decimal places, in the particular case when $\phi = 11/6$.

7 In a genetical study of some characteristics of maize, M.T. Jenkins (*Genetics*, 1927, Vol. 12, p. 498) gives the results of three independent but comparable experiments. In each experiment progenies in four mutually distinct classes were obtained. The corresponding expected probabilities are known to be dependent upon an unknown positive parameter p. The observed frequencies and the expected probabilities are given below:

Class		AB	Ab	aB	ab	Total
Experiment I	Observed frequency	397	297	289	412	1395
	Expected probability	$\frac{1}{2}(1 - p)$	$\frac{1}{2}p$	$\frac{1}{2}p$	$\frac{1}{2}(1 - p)$	1
Experiment II	Observed frequency	78	136	120	80	414
	Expected probability	$\frac{1}{2}p$	$\frac{1}{2}(1 - p)$	$\frac{1}{2}(1 - p)$	$\frac{1}{2}p$	1
Experiment III	Observed frequency	461	161	515	130	1267
	Expected probability	$\frac{1}{4}(1 + p)$	$\frac{1}{4}(1 - p)$	$\frac{1}{4}(2 - p)$	$\frac{1}{4}p$	1

Derive the equation for $\hat{p}$, the maximum-likelihood estimate of p, and show that this equation can be put into the form

$$\hat{p} = \frac{1748 - 4357\hat{p}^2 + 3076\hat{p}^3}{2919}.$$

If it is known that $\hat{p}$ is approximately equal to $0\cdot5$, use an iterative procedure to determine the root correct to four decimal places.

[Hint   If p_r and p_{r+1} are the rth and $(r+1)$th approximations for $\hat{p}$, use $\frac{1}{2}(p_r + p_{r+1})$ instead of p_{r+1} to obtain p_{r+2}.]

8   In a breeding experiment with certain varieties of *Antirrhinum majus*, yellow plants were crossed with the same ivory individual, and in the resulting progeny there were 98 yellow and 49 ivory plants. In a second experiment two independent families were obtained by the self-pollination of two yellow plants with the following frequencies of yellow and ivory plants:

Colour	Yellow	Ivory
1st family	27	2
2nd family	15	12

Source   K. Mather (1937), *Annals of Eugenics*, Vol. 8, p. 96 (adapted).

According to the simplest Mendelian hypothesis, the plants in the first experiment should be in the ratio $1:1$ and those in the two families of the second experiment separately in the ratio $3:1$. Since these simple proportions do not agree with the observations, it is postulated that the deviations from the Mendelian proportions could be explained by a recessive lethal gene linked with the gene determining the yellow and ivory colours in the plants. It can then be shown that the expected frequencies of yellow and ivory plants in the first experiment are proportional to $2 - \theta$ and $1 + \theta$ respectively, where θ is an unknown parameter such that $0 < \theta < 1$. Similarly, the corresponding expected frequencies in the two families of the second experiment are proportional to $2 + (1 - \theta)^2$, $1 - (1 - \theta)^2$; and $2 + \theta^2$, $1 - \theta^2$.

Prove that the equation for $\hat{\theta}$, the maximum-likelihood estimate of θ, is

$$24 - 256\hat{\theta} - 654\hat{\theta}^2 + 1608\hat{\theta}^3 - 1905\hat{\theta}^4 + 1611\hat{\theta}^5 - 831\hat{\theta}^6 + 259\hat{\theta}^7 = 0.$$

Hence, by using a suitable iterative procedure, determine the value of $\hat{\theta}$ correct to six decimal places, and then test for the goodness of fit.

9   In the process of creosoting telephone poles, one of the important factors affecting the depth of penetration of the creosote is the depth of alburnum in the poles. The following data obtained from W.A. Shewhart's *Economic Control of Quality of Manufactured Product* (Van Nostrand, New York: 1931; p. 77) give the frequency distribution of the depth of alburnum (in inches) in a random sample of 1370 telephone poles that were examined.

Depth of alburnum	Frequency	Depth of alburnum	Frequency
0·85 –	2	3·25 –	151
1·15 –	29	3·55 –	123
1·45 –	62	3·85 –	82
1·75 –	106	4·15 –	48
2·05 –	153	4·45 –	27
2·35 –	186	4·75 –	14
2·65 –	193	5·05 –	5
2·95 –	188	5·35 –	1
Total			1370

Reproduced from Shewhart: *Economic Control of Quality of Manufactured Product*, copyright 1931 by the Van Nostrand Co., New York.

Calculate, correct to six decimal places, the mean and variance of the frequency distribution. Hence evaluate, correct to four significant figures, the standard error of the best estimate of the mean depth of the alburnum in the poles.

If the depth of alburnum is assumed to be a normally distributed random variable, evaluate, correct to four decimal places, the probability that the observed mean depth of alburnum exceeds the true mean depth by at least 0·045 inch.

10 During an experiment on some factors which might influence the weight of table ducklings some data were collected on the weights of eggs which hatched, and the weights of eggs which failed to do so. The eggs used in the experiment were of Aylesbury and Allport ducks, and were all laid between 26th February and 7th May 1939. After the elimination of infertile eggs there were 619 fertile eggs of Aylesbury ducks which hatched and 341 fertile ones which did not hatch. The corresponding figures for the eggs of the Allports were 583 and 347 respectively.

The table overleaf gives the frequency distributions of the weights (in grams) of the eggs which hatched and of those which did not hatch, obtained by combining the data from the Aylesbury and Allport ducks.

(i) If the populations of the weights of fertile eggs which hatch, and which do not hatch are normal with means θ_1 and θ_2 and variances σ_1^2 and σ_2^2 respectively, evaluate, correct to four significant figures, the best estimate of $\theta_1 - \theta_2$ and the standard error of this estimate. How is this standard error affected if it may be assumed that $\sigma_1 = \sigma_2$?

(ii) Estimate the true proportion of fertile eggs which hatch, and evaluate, correct to four significant figures, the standard error of the estimate.

Weight (Midpoint)	Frequency	
	Eggs which hatched	Eggs which did not hatch
51		1
54		2
57	7	4
60	13	16
63	68	43
66	144	74
69	197	103
72	204	112
75	208	98
78	160	83
81	101	63
84	54	38
87	25	19
90	13	16
93	4	9
96	2	6
99	1	1
102	0	
105	1	
Total	1202	688

Source J.M. Rendel (1943), *Biometrika*, Vol. 33, p. 48 (adapted).

11 A lady declares that by tasting a cup of tea made with milk she can discriminate whether the milk or the tea infusion was first added to the cup. In an experiment designed to test her claim, eight cups of tea are made, four by each method, and are presented to her for judgement in a random order. Calculate, correct to four significant figures, the probability of the cups being correctly divided into two groups of four, if such a division is entirely random. If the experiment is repeated ten times, and a "success" in an individual experiment is scored if the lady correctly divides the cups, show that a score of two or more successes in ten trials provides evidence, significant at the one per cent level, for her claim.

The lady considers this definition of success too stringent, so an alternative experiment is devised. In this experiment six cups of each kind are presented to her in random order, and she is regarded as scoring a success if she correctly distinguishes five or more of each kind. Calculate the probability of such a success, and also the probability of three or more successes in ten independent experiments, assuming her claim to be false. Explain clearly the difference, if any, between the stringency of these two procedures for testing the lady's claim.

12 A consumer research organisation conducts tests on the safety features of a new model of a standard make of electric steam iron. In an initial investigation, a random sample of twelve irons was tested, and it was found that in eight of them there was an improvement in safety as compared with older irons. Show that this experimental result provides reasonable evidence that, in fact, there has been no improvement in the safety performance of the new model irons.

It was felt that the sample examined was too small to reach a firm conclusion. Consequently, tests are carried out on a total of 120 new model irons, and it is observed that 71 of them had improved safety performance. Does this evidence again support the belief that there has been no improvement in the safety performance of the new irons?

How are the above conclusions altered if it is believed that there is an improvement in the safety performance of the new irons?

13 In a survey of student opinion in a large university, it was observed that in a random sample of 472 students questioned, 384 declared themselves in favour of the proposal that a part of their maintenance grants should be allocated specifically for the purchase of books and equipment. Estimate, correct to four significant figures, the true proportion of the students in the university who are in favour of the proposal, and evaluate the standard error of the estimate.

(i) It is believed that less than 75 per cent of the students in the university are in favour of the proposal. Are the sample results consistent with this hypothesis?

(ii) Also, test whether it is reasonable to accept the hypothesis that exactly 75 per cent of the students in the university support specific allocation for books and equipment.

(iii) Modify the above analysis if it is known that there are 4872 students in the university.

14 It is proposed to establish a local radio station in a large provincial city. An opinion survey conducted prior to the establishment of the station showed that in a random sample of 15 892 persons, a total of 7642 persons were in favour of the station and the rest against it. It is believed that more than 50 per cent of the persons in the city are, in fact, in favour of local radio. Do the sample data provide reasonable evidence for this belief?

Another independent survey is conducted in the city after the local station has been in operation for several months. It is now found that of 18 682 persons contacted, a total of 8935 are in favour of the station and the rest against it. Is there reasonable evidence to infer that the functioning of the radio station has not affected public opinion?

Also, test whether the data are in agreement with the hypothesis that the difference between the proportions of the city's population in favour of the radio station before and after its establishment is

(i) less than five per cent;

(ii) greater than seven per cent.

15 The following figures, based on samples, relate to hair-colour of girls in Edinburgh and Glasgow:

City	Of medium hair-colour	Sample size
Edinburgh	4 008	9 743
Glasgow	17 529	39 764

'Source   J. Gray (1907), *Journal of the Royal Anthropological Institute*, Vol. 37.

Estimate, correct to six decimal places, the difference between the true proportions of girls with medium hair-colour in the two cities, and evaluate, correct to six significant figures, the standard error of the estimate.

Is it reasonable to accept the hypothesis that there is no difference in the true proportions of girls with medium hair-colour in the two cities?

By pooling the data, determine the best estimate of the true proportion of girls with medium hair-colour in the two cities as a whole. Hence test the significance of the following hypotheses. The true proportion of girls with medium hair-colour in the combined population of the two cities is

(i) 45 per cent;

(ii) less than 45 per cent;

(iii) greater than 45 per cent.

16 During the years 1900–10, the figures for antitoxin and ordinary treatments in cases of diphtheria at the City of Toronto hospital were as follows:

Treatment	Cases	Deaths
Antitoxin	228	37
Ordinary	337	28

Estimate, correct to six decimal places, the difference in the death-rates for the two treatments. Also, evaluate, correct to six significant figures, the standard error of the estimated difference.

Is there a significant difference between the true death-rates?

Also, test whether the data are in agreement with the hypothesis that the difference between the death-rates amongst the antitoxin and ordinary treatments is

(i) greater than ten per cent;

(ii) less than seven per cent.

17 Over a period of ten years, the number of patients with cancer of the lung entering a certain hospital was 532, of whom 283 had been heavy smokers. Of 2791 other patients 728 had been heavy smokers. Estimate, correct to six decimal places, the difference in the true proportion of heavy smokers as between lung cancer and other patients. Also, evaluate the standard error of the estimated difference. Are these data in reasonable agreement with the hypothesis that

(i) there is no real difference in the proportion of heavy smokers amongst lung cancer and other patients;

(ii) there are 25 per cent more heavy smokers amongst lung cancer patients than amongst other patients?

State any reservations you might have about an association between lung cancer and heavy smoking.

18 An estimate of p, the true proportion of defective components produced in a factory, is found by choosing a sample of n items at random and calculating p^*, the proportion defective in the sample. Prove that the standard deviation of p^* cannot be greater than $1/(2\sqrt{n})$.

Samples of 450 components from one factory and of 680 from another are examined and the difference between the observed proportions defective is found to be 0·086. Show that the true proportions defective in the components produced by the two factories must be significantly different at the one per cent level, regardless of their actual values.

19 The accompanying table gives 100 four-figure random numbers.

3843	8801	1325	9051	7361	2281	7506	5237	3422	9745
7581	4023	7251	2388	0153	3866	9059	9682	6335	5803
5292	3040	3871	9229	0665	3950	8238	8608	3168	8216
5204	0831	0392	3583	0722	5499	3859	6992	0310	1949
3488	5433	5283	0652	7638	8238	0067	5419	9641	5446
2118	7059	7814	5052	5929	0303	4665	5197	9100	0705
5749	8163	0132	9731	9899	2113	6460	9682	2335	9943
0872	6206	7662	9607	2238	6116	6106	6710	9411	7753
3969	8406	5007	5688	9263	0643	4746	4412	3046	1072
0109	3363	2718	7636	4849	0598	9861	3655	2630	1914

Use these numbers to obtain from the age distribution of the female population of England and Wales in 1941 given below, a random sample of size 100.

Age-group	Proportion	Age-group	Proportion
0 –	0·065	45 –	0·131
5 –	0·131	55 –	0·109
15 –	0·148	65 –	0·071
25 –	0·161	75 –	0·028
35 –	0·152	85 –	0·004
Total			1·000

Calculate, correct to four decimal places, the sample mean and sample standard deviation. Also, evaluate the population mean and population standard deviation.

Assuming that the sample mean is normally distributed, use appropriate tests of significance to show that the difference between the population and sample means may be attributed to chance, when it is assumed that the population variance is (i) known; (ii) unknown.

Note   The class-marks used in the age distribution are, in fact, class-limits.

20   The accompanying table gives 690 two-digit random numbers, and by regarding each number as a proper fraction, these numbers may be considered as a random sample of 690 independent observations of a continuous random variable x having a uniform distribution in the (0, 1) interval.

By considering these individual x observations column-wise from top to bottom, obtain the 345 derived values of another random variable y by summing consecutively every two x observations. Use these 345 y observations to form a frequency distribution having class-intervals with class-marks $0.00-$; $0.11-$; $0.21-$; ... ; $1.91-$.

Calculate, correct to four significant figures, the mean and standard deviation of the frequency distribution, and compare these values with the corresponding ones obtained from the theoretical distribution of y.

It may be assumed that the theoretical distribution of y has the probability density function

$$f(y) = y, \qquad \text{for } 0 \leqslant y \leqslant 1,$$
$$= 2 - y, \quad \text{for } 1 \leqslant y \leqslant 2.$$

21   The table on page 140 gives 500 two-digit random numbers, and by regarding each number as a proper fraction, these numbers may be treated as a random sample of 500 independent observations of a continuous random variable x which is uniformly distributed in the (0, 1) interval.

By using the transformation $y = -\log_e x$ on the x observations obtain the corresponding independent observations of y. Make a frequency distribution of the y values, using class-intervals with class-marks $0.00-$; $0.26-$; $0.51-$; ... ; $4.76-$.

Calculate, correct to four significant figures, the mean and standard deviation of the frequency distribution, and compare these values with the corresponding ones obtained from the theoretical probability distribution of y.

Note   Evaluate the y values correct to two decimal places.

Table for Exercise 20

Col. 1	Col. 2	Col. 3	Col. 4	Col. 5	Col. 6	Col. 7	Col. 8	Col. 9	Col. 10	Col. 11	Col. 12	Col. 13	Col. 14	Col. 15
25	63	79	91	20	38	19	26	22	66	10	83	86	32	40
05	47	01	13	10	94	87	55	72	01	08	64	78	19	05
54	93	31	68	07	18	85	58	44	30	34	58	94	99	30
97	19	31	06	37	79	07	39	45	94	75	14	30	48	18
78	29	47	38	41	11	56	16	53	70	18	62	19	75	33
70	40	97	20	21	08	82	92	74	89	32	62	55	36	15
29	79	88	61	67	03	69	05	28	75	74	15	75	71	15
33	98	22	93	42	34	42	07	47	01	06	32	53	09	14
98	77	35	62	12	97	34	10	52	42	66	48	85	89	85
27	30	72	08	96	18	31	67	64	46	26	62	35	45	79
22	18	57	33	83	05	17	77	20	77	45	35	72	57	36
54	87	37	55	43	64	28	01	38	39	45	95	31	17	69
54	57	83	48	53	41	25	50	77	47	18	68	03	90	54
49	61	67	48	52	95	35	83	83	82	79	03	99	12	18
06	89	08	11	81	32	93	47	65	17	82	28	70	21	02
83	46	15	93	52	57	81	09	23	10	92	06	45	09	04
90	58	85	59	40	84	88	69	11	22	98	12	99	08	96
51	84	49	95	77	45	55	84	43	26	95	52	35	01	90
31	99	49	48	10	83	46	44	05	47	24	16	05	35	61
46	61	51	13	18	15	49	15	96	84	19	79	53	65	18
48	27	68	53	73	63	20	81	57	14	18	25	23	65	17
98	34	19	46	63	63	85	47	70	42	13	06	40	62	76
23	14	27	99	49	71	36	95	16	43	56	86	05	71	68
93	10	76	59	62	35	34	89	79	84	64	31	91	68	77
77	61	33	49	05	47	82	49	56	77	77	53	71	42	71
52	62	03	75	35	82	80	07	39	03	98	91	39	40	53
04	83	80	43	62	67	24	80	38	29	67	20	61	94	35
85	51	92	58	64	76	46	38	51	86	59	34	17	39	23
20	27	15	76	08	54	60	38	35	91	95	36	97	26	34
12	67	95	87	55	31	67	48	80	56	96	47	54	15	67
64	31	73	36	07	19	72	52	84	05	19	51	02	08	12
46	24	96	83	31	81	09	07	28	42	77	05	23	65	98
75	18	65	98	16	51	29	45	91	57	59	97	40	82	39
31	51	99	61	79	98	42	76	06	33	41	11	70	49	61
83	62	95	83	81	90	82	28	42	63	05	35	07	72	10
12	89	91	82	58	77	88	82	08	15	17	19	68	30	71
99	51	33	62	38	16	42	35	82	11	64	88	52	30	01
96	61	30	88	28	57	01	68	03	64	07	22	79	70	79
40	55	46	56	47	88	29	50	10	78	22	58	91	40	37
31	15	88	69	58	71	92	23	72	66	65	40	16	08	73
56	54	23	75	47	46	86	45	09	73	20	35	42	41	12
55	79	80	28	68	20	46	29	42	42	35	26	19	84	78
84	01	05	79	80	79	60	08	37	58	30	08	81	37	41
95	99	09	56	01	19	29	94	63	59	44	78	09	45	57
80	14	38	79	72	09	84	09	78	96	53	22	06	85	77
69	35	38	39	54	68	21	62	72	29	59	98	24	37	32

Table for Exercise 21

87	83	40	39	99	21	45	54	36	03	88	78	40	24	77	31	84	26	50	95	22	09	37	36	55
83	13	46	77	11	48	56	16	21	91	18	59	74	31	77	93	40	12	88	26	85	19	84	59	16
50	72	39	40	57	09	20	86	13	98	91	01	67	11	32	39	21	29	65	51	33	63	95	13	12
12	91	41	79	44	90	65	78	16	97	57	79	79	16	43	05	93	80	86	26	27	34	46	65	56
92	63	69	81	70	17	43	86	99	34	08	36	82	26	76	85	82	65	83	06	59	77	83	81	93
48	88	37	74	93	05	86	25	76	36	26	54	84	02	77	13	99	30	71	91	99	55	95	90	49
56	09	66	32	50	76	27	45	29	23	54	62	22	39	58	73	59	26	76	99	83	86	89	68	29
49	46	56	41	52	28	17	81	24	55	42	73	16	47	42	52	68	32	40	89	11	18	27	46	85
64	09	68	86	60	87	77	31	12	21	80	13	80	25	70	76	03	57	87	56	28	57	54	80	82
25	74	89	25	57	23	49	17	88	63	02	71	21	29	88	04	27	10	13	99	64	90	69	10	16
21	95	89	65	12	94	35	58	92	74	52	13	36	39	82	46	04	34	36	46	90	88	61	22	90
01	22	26	95	75	76	49	25	81	19	33	60	27	05	51	06	76	54	54	15	49	65	93	54	95
47	60	45	67	70	06	84	88	33	77	71	94	64	18	76	95	33	35	30	53	38	36	74	61	56
83	15	63	50	88	09	50	23	05	68	06	17	76	95	03	82	08	33	29	18	98	89	50	13	60
14	02	72	85	09	05	05	90	47	85	62	63	20	60	90	09	59	60	21	38	15	56	29	90	07
80	92	88	01	05	23	03	20	96	01	79	58	89	74	01	96	81	01	18	41	70	94	48	01	04
12	98	82	49	71	83	55	38	70	75	70	63	94	54	47	63	79	31	38	54	30	70	70	73	49
23	59	80	61	45	36	96	60	60	52	53	39	41	24	88	17	36	90	48	99	64	84	79	53	64
73	27	44	07	27	42	97	84	18	59	92	47	82	76	54	40	71	66	83	17	05	28	50	47	36
80	43	36	97	29	51	04	06	31	69	30	20	14	70	56	23	85	96	37	52	65	21	84	99	02

Table for Exercise 22

1·88	0·75	0·94	0·04	1·15	-0·01	1·40	-1·87	-0·30	-1·70	0·69	0·71	-0·10	1·07
0·13	-0·17	1·48	0·31	0·96	-1·35	0·90	-0·01	0·14	0·38	2·59	0·74	0·38	0·74
-1·08	-2·29	0·66	-0·52	0·92	0·81	0·11	0·69	-0·39	-1·51	-1·18	1·05	-1·86	-0·37
2·20	0·15	1·76	0·32	0·45	-1·14	-0·86	-0·27	0·13	0·69	0·36	0·47	-0·10	-0·55
-0·86	1·01	-0·15	-0·16	0·63	0·49	0·24	2·78	-1·30	0·39	0·54	0·52	-0·27	0·48
0·88	-0·94	-0·70	-2·12	-0·80	0·49	-0·70	2·91	-1·41	0·89	-1·32	0·49	0·68	0·47
-0·21	-0·27	0·27	1·28	-1·52	1·91	-0·05	1·18	0·54	-0·12	1·18	-0·07	-0·72	0·17
0·73	0·91	1·33	-0·63	2·60	0·49	0·47	-2·99	-0·89	1·06	-1·48	2·06	-0·28	-1·22
-0·66	1·84	-0·23	0·78	0·04	-0·89	1·24	-0·58	0·84	0·37	-1·21	-0·66	-1·19	-0·28
-0·15	1·34	0·78	0·95	-0·53	0·50	0·69	-0·97	-0·85	0·34	-1·93	1·42	0·58	1·05
-0·80	-0·44	-0·65	1·50	-0·80	-1·53	0·14	0·82	-1·05	-0·05	0·13	-1·32	-1·95	0·18
1·33	-0·51	-0·57	-0·36	-0·37	-0·50	-1·06	-0·31	0·38	-1·19	0·10	0·79	-0·19	-1·29
-1·18	-0·20	-0·16	-0·01	2·00	0·07	0·69	0·08	1·46	-0·53	-1·02	-1·63	-0·32	-0·27
1·63	-0·39	0·52	-0·80	-0·01	0·11	-0·16	-0·32	-0·69	2·22	-0·37	1·13	0·87	0·24
0·61	0·20	-0·32	-0·58	-1·35	1·46	2·18	0·75	1·22	-0·43	2·31	-0·41	-0·39	-1·57
-0·86	1·34	-0·73	0·17	0·84	0·30	0·49	-1·20	0·65	0·78	-0·79	-1·03	-1·17	0·80
1·22	-0·58	-0·77	-0·74	0·68	0·29	0·39	-1·69	0·02	0·26	-0·99	-0·69	0·23	0·42
1·12	-0·84	-0·62	1·18	-2·41	0·50	0·81	-0·81	0·82	1·88	2·44	0·94	1·66	0·34
-1·10	0·98	-0·57	-0·27	-0·13	-0·13	0·73	0·02	1·18	0·04	-0·57	0·25	0·81	-0·18
1·04	-0·04	0·88	0·15	-1·87	1·03	2·35	-1·07	-0·89	-0·62	-0·45	-0·56	1·24	0·87
1·16	1·93	-0·42	-0·26	0·76	1·25	-0·05	-0·14	1·77	0·86	0·56	0·12	-0·47	-0·06
-0·51	0·84	-2·43	0·24	1·11	0·29	0·90	0·44	0·03	-2·58	-0·18	0·45	-0·47	-0·75
0·62	-1·75	-1·13	-1·00	1·10	-0·25	0·70	-1·01	-0·11	-2·26	1·00	-2·02	0·47	-1·48
1·69	-1·65	0·52	0·39	-0·63	-1·02	1·08	-1·43	0·63	-2·02	-1·71	0·97	0·12	0·69
0·00	-0·59	-0·21	0·03	0·27	0·18	-1·10	-0·46	-2·36	1·42	1·89	-1·13	-1·42	-1·59
-0·44	0·55	-0·09	-0·77	1·37	0·33	2·10	-1·04	0·63	-1·95	0·26	-0·32	-2·54	-0·67
0·25	-0·41	0·52	0·97	-0·21	-1·25	-0·72	0·29	-0·06	0·27	0·03	0·29	-1·81	-0·32
-0·04	0·56	-0·20	1·39	0·84	-0·48	0·15	0·98	-0·28	-0·34	0·14	-0·29	1·37	0·36
1·17	0·83	-0·66	0·01	0·93	-0·63	1·20	0·05	0·53	-2·22	-1·58	0·49	1·96	-1·07
0·16	2·64	0·20	0·58	-0·60	-0·05	0·53	-1·52	-0·13	-0·56	-0·96	-0·13	1·51	0·34

Table for Exercise 23

92, 03	97, 11	99, 03	83, 17	87, 24	75, 03	94, 04	92, 14	97, 00	99, 00	84, 18	82, 04	87, 05	95, 11
87, 04	98, 05	79, 18	92, 07	77, 05	91, 08	97, 18	95, 02	97, 17	91, 22	97, 07	99, 15	98, 07	94, 05
94, 12	80, 22	81, 01	84, 03	97, 27	87, 27	96, 04	98, 14	98, 02	92, 18	96, 04	88, 04	96, 07	99, 08
98, 01	95, 11	90, 01	94, 09	73, 12	92, 00	88, 03	86, 01	89, 01	92, 08	86, 02	80, 23	78, 25	97, 16
98, 00	93, 00	99, 15	91, 16	79, 23	64, 02	86, 03	79, 05	93, 10	98, 02	86, 08	92, 00	90, 01	99, 05
97, 17	91, 39	93, 01	97, 00	90, 01	80, 07	76, 05	91, 03	98, 07	85, 03	95, 03	98, 03	93, 01	98, 04
68, 09	86, 13	78, 00	95, 03	95, 10	81, 13	91, 16	95, 02	97, 02	86, 22	77, 00	90, 11	99, 04	95, 16
93, 00	95, 01	89, 32	81, 01	98, 02	84, 01	95, 01	98, 14	67, 12	94, 03	87, 03	87, 10	97, 34	70, 02
99, 15	87, 21	89, 20	98, 02	86, 00	95, 08	79, 00	86, 01	99, 00	89, 04	91, 17	93, 06	85, 10	86, 16
85, 01	98, 01	87, 34	76, 01	94, 17	80, 12	96, 18	79, 05	69, 04	97, 01	99, 10	85, 00	96, 02	94, 05
91, 00	96, 10	91, 01	86, 04	99, 07	95, 01	92, 01	91, 03	96, 02	78, 05	90, 00	89, 00	95, 00	91, 02
72, 01	94, 08	99, 06	95, 11	93, 01	96, 05	90, 13	95, 28	90, 02	91, 00	96, 12	85, 03	81, 06	81, 14
89, 33	99, 00	89, 27	94, 17	99, 08	99, 06	98, 24	97, 06	75, 03	93, 28	83, 01	71, 03	94, 00	97, 23
95, 05	99, 01	96, 30	96, 00	66, 01	69, 04	90, 00	94, 12	74, 19	97, 07	88, 11	97, 00	98, 04	78, 02
93, 04	85, 02	94, 05	84, 14	87, 13	86, 06	98, 05	75, 06	99, 03	99, 21	99, 03	99, 14	92, 12	94, 17
93, 14	91, 06	99, 00	94, 12	81, 02	92, 00	83, 31	98, 01	96, 02	95, 22	99, 08	82, 01	96, 00	99, 11
93, 21	90, 11	98, 03	99, 27	97, 20	89, 03	95, 13	90, 12	90, 03	91, 02	94, 16	86, 13	99, 02	99, 09
94, 02	91, 21	98, 18	88, 02	77, 01	97, 38	99, 08	79, 00	94, 00	73, 02	98, 02	90, 03	85, 01	96, 09

Table for Exercise 23 (concluded)

90, 08	87, 05	96, 00	97, 10	98, 25	84, 11	99, 25	99, 05	96, 00	91, 12	74, 25	96, 05	91, 00	99, 08
93, 13	95, 17	82, 04	86, 08	96, 08	91, 30	95, 00	85, 07	97, 08	91, 00	83, 02	83, 09	88, 00	89, 05
97, 01	86, 12	85, 07	68, 00	89, 30	81, 01	99, 08	81, 16	98, 00	99, 12	57, 01	99, 02	98, 09	76, 06
84, 00	84, 01	99, 12	79, 00	90, 22	92, 09	73, 17	99, 15	84, 38	96, 42	95, 07	88, 13	96, 10	91, 14
91, 02	93, 04	82, 00	99, 12	91, 02	88, 04	91, 10	76, 19	87, 28	93, 04	96, 02	92, 27	99, 14	94, 02
92, 03	98, 01	84, 01	99, 01	98, 06	91, 11	98, 10	93, 01	97, 08	70, 00	92, 11	98, 03	98, 22	87, 06
89, 16	86, 03	92, 27	95, 20	95, 05	99, 14	99, 34	77, 12	60, 07	88, 06	94, 03	78, 05	98, 11	67, 06
96, 00	95, 08	97, 03	91, 11	98, 12	94, 15	99, 04	95, 11	99, 02	96, 02	98, 03	99, 07	92, 17	93, 01
90, 07	98, 02	97, 20	94, 08	96, 26	99, 01	92, 12	91, 01	96, 10	92, 15	92, 21	66, 02	98, 00	90, 27
98, 32	93, 19	98, 17	93, 06	89, 06	93, 07	99, 06	88, 02	95, 03	94, 05	99, 01	94, 30	98, 08	92, 00
97, 10	99, 04	98, 01	80, 01	86, 08	88, 04	78, 02	99, 07	85, 13	96, 06	89, 01	99, 20	92, 01	91, 01
97, 02	96, 02	95, 23	93, 08	99, 03	90, 03	91, 23	87, 12	84, 05	99, 19	92, 01	93, 12	88, 21	95, 08
95, 21	72, 01	82, 13	81, 18	96, 10	74, 10	84, 01	90, 13	64, 14	95, 00	96, 07	97, 19	95, 04	82, 01
79, 04	79, 07	97, 11	97, 08	89, 09	92, 03	87, 05	80, 02	87, 00	79, 14	96, 15	93, 04	88, 05	90, 06
97, 15	94, 13	91, 02	98, 06	90, 06	98, 06	75, 03	95, 03	91, 00	95, 19	93, 02	93, 25	95, 30	97, 07
96, 18	94, 07	74, 15	94, 07	78, 06	87, 13	99, 19	96, 07	86, 01	98, 09	92, 09	97, 05	75, 01	97, 01
80, 03	93, 15	85, 07	93, 02	98, 18	99, 07	91, 12	95, 04	96, 06	90, 05	97, 19	68, 05	91, 00	94, 03
95, 15	99, 01	92, 04	89, 16	95, 23	96, 08	98, 19	86, 11	66, 10	98, 12	68, 09	96, 16	95, 02	91, 13

22   If x is a unit normal variable and another random variable y is defined by the "probability transformation"

$$y = \frac{1}{\sqrt{2\pi}} \int_{-\infty}^{x} e^{-t^2/2}\, dt,$$

prove that y has a rectangular distribution in the $(0, 1)$ interval.

Determine the expectation and variance of y.

The table on page 141 gives 420 independent observations of x. Use these numbers to obtain, correct to two decimal places, a sample of 420 observations of y. Form a frequency distribution of these y observations by using the class-marks $0\cdot00-$; $0\cdot06-$; $0\cdot11-$; ... ; $0\cdot96-$.

Evaluate, correct to four decimal places, the mean and variance of the frequency distribution. Also, by using a suitable test of significance, show that the difference between the sample mean and its expectation can reasonably be attributed to chance errors of sampling.

23   The table on pages 142–3 gives the largest and smallest two-figure numbers, where each pair is obtained from one of 504 independent sets of ten two-figure random numbers.

Use these paired numbers to obtain a sample of 504 independent observations of the sample range determined from a random sample of size ten from a uniform distribution in the $(0, 1)$ interval. Form a frequency distribution of these observations and use the class-marks $0\cdot44-$; $0\cdot48-$; $0\cdot52-$; ... ; $0\cdot96-$.

Calculate, correct to four decimal places, the mean and standard deviation of the frequency distribution, and compare them with their theoretical values. Also, evaluate, correct to four significant figures, the standard deviation of the sample mean $\bar{x}$.

It may be assumed that if x is a random variable denoting the range of n independent observations from a uniform distribution in the interval $(0, 1)$, then the probability density function of the distribution of x is

$$n(n - 1)\, x^{n-2}(1 - x), \quad \text{for } 0 \leqslant x \leqslant 1.$$

24   The following table, compiled from some data of Mendel, gives the number of round and angular peas which were obtained from ten plants.

Plant no.	Round	Angular
1	45	12
2	27	8
3	24	7
4	19	10
5	32	11
6	26	6
7	88	24
8	22	10
9	28	6
10	25	7

If p is the true proportion of round peas, then by combining the data from the ten plants, calculate, correct to four decimal places, the maximum-likelihood estimate of p and evaluate the standard error of the estimate. Hence determine a symmetrical large-sample 95 per cent confidence interval for p.

According to Mendel's theory, the true value of p should be 0·75. Use the confidence interval obtained to test that, on the average, the observed proportion of round peas is consistent with the Mendelian expectation.

Also, use another approximate procedure to test the significance of the Mendelian hypothesis about p which does not use the standard error of the estimate.

25 An experiment consists of tossing a penny three times and observing the outcomes — heads (H) and tails (T). If p is the probability of obtaining a head at a trial, determine the probability distribution of the four distinct combinations possible — three heads, two heads and a tail, a head and two tails, and three tails.

The experiment is repeated 337 times and the observed frequencies in the four distinct classes are 32, 103, 122, and 80 respectively. Evaluate, correct to six decimal places, the maximum-likelihood estimate of p and also determine the standard error of the estimate. Hence calculate an approximate large-sample 98 per cent confidence interval for p. Is it reasonable to conclude that the penny is, in fact, unbiased in the light of the experimental evidence obtained?

26 In a breeding experiment with certain wheat varieties, the numbers of "fired" and normal plants raised were 161 and 276 respectively. The leaf-blades of "fired" plants change their colour soon after flowering through a greenish grey to brownish red and finally to the ordinary straw colour. On a genetical hypothesis, the true proportions of the two types of plant are $\frac{1}{64}(27 - \theta)$ and $\frac{1}{64}(37 + \theta)$ respectively, where θ is a positive parameter.

It is known that certain fired plants may sometimes look normal, and so in another experiment 60 normal plants were tested by growing progenies from them. It was thus determined that 5 of the plants were, in fact, fired and the rest were normal. The true proportions of fired and normal plants amongst the progenies of normal-looking plants are $\theta/(37 + \theta)$ and $37/(37 + \theta)$ respectively.

Source H.F. Smith (1937), *Annals of Eugenics*, Vol. 8, p.94.

In general, if the numbers of fired and normal plants in the two experiments are n_1, $N_1 - n_1$; n_2, $N_2 - n_2$, prove that the equation for $\hat{\theta}$, the maximum-likelihood estimate of θ, is

$$999n_2 + [27(N_1 - N_2) - (64n_1 + 10n_2)]\hat{\theta} - (N_1 - N_2 + n_2)\hat{\theta}^2 = 0,$$

and that for large samples

$$\text{var}\,(\hat{\theta}) = \frac{\theta(27 - \theta)(37 + \theta)^2}{N_1\,\theta(37 + \theta) + 37N_2\,(27 - \theta)}.$$

Hence determine, correct to six decimal places, $\hat{\theta}$ and its standard error. Also, evaluate a large-sample 95 per cent confidence interval for θ and use it to test the hypothesis that the true value of θ is 3·5.

27   The table below gives the frequency distribution of the lengths (measured in centimetres) of a sample of 1110 eggs of the common tern (*Sterna fluviatilis*). The measurements are based upon a census of the eggs made during the period from 3rd to 20th July, 1914, and contained in W. Rowan's *Fifth MS Report on the Faunistics of Blakeney Point*, a field station on the Norfolk coast.

Length	Frequency	Length	Frequency
3·25 –	1	4·10 –	146
3·30 –	1	4·15 –	118
3·35 –	1	4·20 –	155
3·40 –	0	4·25 –	75
3·45 –	0	4·30 –	118
3·50 –	1	4·35 –	59
3·55 –	1	4·40 –	50
3·60 –	0	4·45 –	50
3·65 –	4	4·50 –	30
3·70 –	4	4·55 –	17
3·75 –	6	4·60 –	11
3·80 –	6	4·65 –	3
3·85 –	18	4·70 –	0
3·90 –	33	4·75 –	2
3·95 –	33	4·80 –	1
4·00 –	67	4·85 –	2
4·05 –	97		
Total			1110

Source   "A Co-operative Study" (1919), *Biometrika*, Vol. 12, p. 308.

Evaluate, correct to four decimal places, the mean and variance of the frequency distribution. Hence evaluate the 99 per cent confidence interval for the true mean length of the eggs.

28   In an experimental study of the distribution of yeast cells in a solution, the yeast cells were killed by adding a little mercuric chloride to the water in which they had first been well mixed. A small quantity of this mixture was mixed with a ten per cent solution of gelatine, and after being well stirred a small amount was put on a haemocytometer, the thickness of the solution being 0·01 mm. This was then put on a plate of glass kept at a temperature just above the setting point of gelatine and allowed to cool slowly till the gelatine had set. The grid of the haemocytometer consisted of 400 squares, each of area 1/400 sq. mm. The table below gives the distribution of the number of yeast cells counted in each of the 400 squares.

2	2	4	4	4	5	2	4	7	7	4	7	5	2	8	6	7	4	3	4
3	3	2	4	2	5	4	2	8	6	3	6	6	10	8	3	5	6	4	4
7	9	5	2	7	4	4	2	4	4	4	3	5	6	5	4	1	4	2	6
4	1	4	7	3	2	3	5	8	2	9	5	3	9	5	5	2	4	3	4
4	1	5	9	3	4	4	6	6	5	4	6	5	5	4	3	5	9	6	4
4	4	5	10	4	4	3	8	3	2	1	4	1	5	6	4	2	3	3	3
3	7	4	5	1	8	5	7	9	5	8	9	5	6	6	4	3	7	4	4
7	5	6	3	6	7	4	5	8	6	3	3	4	3	7	4	4	4	5	3
8	10	6	3	3	6	5	2	5	3	11	3	7	4	7	3	5	5	3	4
1	3	7	2	5	5	5	3	3	4	6	5	6	1	6	4	4	4	6	4
4	2	5	4	8	6	3	4	6	5	2	6	6	1	2	2	2	5	2	2
5	9	3	5	6	4	6	5	7	1	3	6	5	4	2	8	9	5	4	3
2	2	11	4	6	6	4	6	2	5	3	5	7	2	6	5	5	1	2	7
5	12	5	8	2	4	2	1	6	4	5	1	2	9	1	3	4	7	3	6
5	6	5	4	4	5	2	7	6	2	7	3	5	4	4	5	4	7	5	4
8	4	6	6	5	3	3	5	7	4	5	5	5	6	10	2	3	8	3	5
6	6	4	2	6	6	7	5	4	5	8	6	7	6	4	2	6	1	1	4
7	2	5	7	4	6	4	5	1	5	10	8	7	5	4	6	4	4	7	5
4	3	1	6	2	5	3	3	3	7	4	3	7	8	4	7	3	1	4	4
7	6	7	2	4	5	1	3	12	4	2	2	8	7	6	7	6	3	5	4

Source "Student" (1907), *Biometrika*, Vol. 5, p. 351.

Form a frequency distribution of the data using a unit class-interval. Hence evaluate, correct to four decimal places, the mean and variance of the observed distribution.

By considering two separate models for the distribution of the yeast cells, calculate, correct to four decimal places, the 95 per cent confidence intervals for the true mean number of cells per square.

29 The ages at death of all persons dying over 70 years of age were extracted for a period of three complete years from the obituary notices in *The Times* appearing daily during the years 1910–12. The obituary notices are those of persons in a fairly limited class, which may be considered stable for the three years considered. Ignoring age differences, the table overleaf gives, separately for men and women, the frequency distributions of the number of deaths reported daily of persons who had reached the age of 70 years at the time of death.

Evaluate, correct to four decimal places, the means and variances of the two frequency distributions. If it may be assumed that the daily averages of deaths for men and women in the over 70 age-group are θ_1 and θ_2 respectively, determine the best estimates of θ_1 and θ_2, and the standard errors of the estimates. Also, calculate the 96 per cent confidence interval for the parametric difference $\theta_1 - \theta_2$. Is it reasonable to assume that the data are consistent with the hypothesis that $\theta_1 - \theta_2 = 0$?

Table for Exercise 29

No. of deaths per day	Frequency	
	Men	Women
0	33	46
1	110	140
2	170	207
3	246	221
4	187	169
5	142	119
6	84	87
7	69	44
8	31	35
9	19	18
10	4	4
11	1	4
12	0	1
13	0	1
Total	1096	1096

Source   L. Whitaker (1914), *Biometrika*, Vol. 10, p. 36.

Table for Exercise 30

Pulse per minute	Frequency	
	Normal convicts	Weak-minded convicts
45 –	2	
49 –	5	
53 –	16	1
57 –	47	10
61 –	74	16
65 –	131	19
69 –	99	21
73 –	103	29
77 –	84	25
81 –	66	20
85 –	46	17
89 –	30	12
93 –	10	5
97 –	10	8
101 –	6	3
105 –	2	3
109 –	2	1
113 –	2	1
Total	735	191

Source   M.H. Whiting (1915), *Biometrika*, Vol. 11, p. 1.

30   In an anthropometric study of English convicts, Dr Charles Goring made a variety of measurements on 500 inmates of Parkhurst prison. Of these convicts, 400 were of normal intelligence and 100 were classed as weak-minded. The table opposite gives separately the frequency distributions of pulse rates for the two groups. In the bulk of the cases, the pulse rates were observed twice at an interval of 14 days or more. However, due to the time-interval between duplication of measurements, it may be assumed that the frequency distributions are based on samples of independent observations.

Evaluate, correct to four decimal places, the means and variances of the two observed distributions. Hence, if θ_1 and θ_2 denote the true pulse rates of the normal and weak-minded convicts, determine, correct to four decimal places, the 95 per cent confidence intervals for θ_1, θ_2, and $\theta_2 - \theta_1$. Is it reasonable to conclude from the data that normal convicts have the same pulse rate as weak-minded convicts?

31   In 1905–6 an inquiry was held in the schools under the control of the School Board for Glasgow on the height and weight of schoolchildren in relation to parental and socio-economic conditions. In all, about 70 000 children between the ages from 5 to 18 years were investigated. The schools from which these children came were grouped into four classes according to the economic status of the areas in which the schools were located. The following table gives the frequency distributions of weights (measured in lb) of boys and girls aged between 10·5 and 11·5 years from the best districts included in the inquiry.

Weight	Frequency		Weight	Frequency	
	Boys	Girls		Boys	Girls
43·5 –	0	1	73·5 –	31	14
45·5 –	1	2	75·5 –	18	18
47·5 –	1	9	77·5 –	12	13
49·5 –	11	10	79·5 –	12	4
51·5 –	9	15	81·5 –	10	5
53·5 –	13	24	83·5 –	7	3
55·5 –	26	44	85·5 –	2	2
57·5 –	49	43	87·5 –	3	1
59·5 –	53	44	89·5 –	1	2
61·5 –	64	54	91·5 –	2	0
63·5 –	65	47	93·5 –	0	0
65·5 –	75	43	95·5 –	1	0
67·5 –	49	34	· · ·	· · ·	· · ·
69·5 –	48	35	111·5 –	0	1
71·5 –	31	30			
Total				594	498

Source   E.M. Elderton (1914), *Biometrika*, Vol. 10, p. 287.

Evaluate, correct to four decimal places, the means and variances of the two distributions. Hence calculate a 95 per cent confidence interval for the true difference between the weights of boys and girls in the specified age-group.

In the light of the sample information, test whether it is reasonable to accept the hypothesis that there is

(i) no difference between the true mean weights of boys and girls in the specified age-group;

(ii) on the average a difference of more than 2 lb in the true mean weights of boys and girls in the specified age-group.

Note   The class-marks used in the two frequency distributions are, in fact, class-limits.

32   R. Pearl (*Biometrika*, 1905, Vol. 4, p. 13) quotes the following frequency distributions of brain-weights (in grams) of Swedish males and females. The data were collected by A. Retzius (1900) mainly from the autopsies at the Sabbatsberg Krankenhaus in Stockholm, although some figures were obtained from autopsies at the Maria Krankenhaus. The original data of Retzius have been modified to the extent that incompletely recorded individuals and those falling outside the age-limits 20 to 80 years have been excluded from the frequency distributions.

Males		Females	
Weight	Frequency	Weight	Frequency
1100 –	1	900 –	1
1150 –	10	950 –	0
1200 –	21	1000 –	1
1250 –	44	1050 –	9
1300 –	53	1100 –	28
1350 –	86	1150 –	30
1400 –	72	1200 –	53
1450 –	60	1250 –	40
1500 –	28	1300 –	30
1550 –	25	1350 –	23
1600 –	12	1400 –	10
1650 –	3	1450 –	6
1700 –	1	1500 –	1
		1550 –	1
Total	416		233

(i) Calculate, correct to four decimal places, the mean and variance of each of the two frequency distributions.

(ii) Assuming that the populations from which the samples are postulated to be drawn have the same variance, obtain the pooled estimate of this common population variance.

(iii) By using the pooled estimate of variance, evaluate the 90 per cent confidence interval of the difference between the mean brain-weights of the male and female populations. Use this confidence interval to test the null hypothesis that there is a difference of 150 grams between the mean male and female brain-weights. Explain, very briefly, the population assumptions which underlie this test of significance.

33 In the Province of Verona, Italy, in the five years between 1875 and 1879 there were 16 203 conscripts, and from these, 3810 recruits were selected for the Italian army, the remaining conscripts being rejected as unfit for the army. Among many conditions which the conscripts had to satisfy for recruitment was a minimum requirement of stature. The accompanying table gives the frequency distributions of stature (measured in centimetres) of the recruits and the rejected conscripts.

(i) Analyse these data to determine, correct to four decimal places, the mean and variance of each of the two frequency distributions.

(ii) Assuming that the conceptual populations of the statures of recruits and rejected conscripts have the same variance, evaluate the best estimate of this common parameter. Hence, using the two per cent level of significance, test the null hypotheses that, on the average, the difference between the statures of recruits and rejected conscripts is (a) more than two centimetres; (b) less than three centimetres.

(iii) How are the conclusions of the tests of significance in (ii) affected if it cannot be assumed that the two populations of the statures of recruits and rejected conscripts have the same variance?

(iv) Obtain a 95 per cent confidence interval for the true difference in the mean statures of recruits and rejected conscripts on the assumption that the stature populations do not have equal variance.

34 In a quantitative study of certain morphological characteristics of the fruit of the bloodroot (*Sanguinaria canadensis*), a sample of 1000 fruiting stalks was collected in the spring of 1906 at Meramec Highlands, near St Louis, Missouri, U.S.A. In the spring of 1907 another lot of 400 fruits was obtained from the same habitat.

The fruit of the bloodroot is one-celled with two parietal placentae and, amongst other characteristics, a count was made of the number of seeds developing on each placenta. The table on page 153 gives the frequency distributions of the number of seeds per placenta, separately for the two years.

(i) Calculate, correct to four decimal places, the mean and variance of each of the two frequency distributions.

(ii) Assuming that the conceptual populations of the number of seeds per placenta for the two years have the same variance, evaluate the best estimate of the population variance. Hence, using the one per cent level

Table for Exercise 33

Stature	Frequency	
	Recruits	Rejected conscripts
124 –		4
126 –		0
128 –		2
130 –		1
132 –		1
134 –		1
136 –		0
138 –		2
140 –		7
142 –		7
144 –		22
146 –		33
148 –		55
150 –		117
152 –		225
154 –	16	610
156 –	148	577
158 –	239	788
160 –	354	1 149
162 –	471	1 398
164 –	542	1 523
166 –	538	1 587
168 –	425	1 278
170 –	414	1 111
172 –	256	802
174 –	170	529
176 –	133	251
178 –	55	150
180 –	35	96
182 –	9	36
184 –	3	21
186 –	0	7
188 –	2	2
190 –		1
Total	3 810	12 393

Source   K. Pearson (1906), *Biometrika*, Vol. 4, p. 505.

of significance, test the null hypotheses that, on an average, the difference between the number of seeds per placenta in 1906 and 1907 is (a) more than 1·5; (b) less than 2; (c) exactly 2.

(iii) Also, test the significance of the null hypotheses in (ii) if it cannot be assumed that the 1906 and 1907 seed populations have the same variance.

(iv) Obtain a 99·9 per cent confidence interval for the true difference in the mean number of seeds per placenta in 1906 and 1907 on the assumption that the two seed populations have equal variance.

Table for Exercise 34

No. of seeds per placenta	Frequency		No. of seeds per placenta	Frequency	
	1906	1907		1906	1907
0	1	4	15	105	26
1	4	5	16	71	16
2	6	7	17	62	12
3	9	23	18	55	13
4	16	23	19	42	8
5	31	38	20	23	5
6	76	50	21	12	0
7	115	78	22	15	2
8	150	83	23	5	2
9	236	88	24	10	2
10	246	94	25	1	2
11	221	77	26	2	1
12	196	69	27	2	
13	162	40	28	1	
14	125	32			
Total				2000	800

Source J.A. Harris (1910), *Biometrika*, Vol. 7, p. 305.

35 In a project on a species of hermit-crab (*Eupagurus prideauxi*, Heller) at the Naples Biological Station, about 2000 specimens were collected during the winter of 1901–2. These were separated into two main groups:

(a) shallow water forms, from a depth of 35 metres or less; and

(b) deep water forms, from a depth of over 35 metres.

The animals in each of these groups were further subdivided into males and females, thus giving in all four distinct groups with about 500 specimens in each. One of the measurements made on the specimens pertained to the length of the carapace along the median line. The table overleaf gives the frequency distribution of the length (in millimetres and measured to the nearest tenth of a mm) of the carapaces in each of the four groups.

Length of carapace	Frequency			
	Deep water males	Shallow water males	Deep water females	Shallow water females
4·1–		3	2	
4·6–	3	3	7	8
5·1–	10	12	5	13
5·6–	23	20	22	42
6·1–	23	24	29	73
6·6–	40	39	56	115
7·1–	49	59	93	119
7·6–	37	57	107	108
8·1–	51	77	98	57
8·6–	49	63	36	16
9·1–	51	58	13	3
9·6–	52	52	2	
10·1–	27	43		
10·6–	33	28		
11·1–	18	9		
11·6–	15	3		
12·1–	2			
Total	483	550	470	554

Source   E.H.J. Schuster (1903), *Biometrika*, Vol. 2, p. 191.

(i) Calculate, correct to four decimal places, the mean and variance of each of the above four frequency distributions.

(ii) If it may be assumed that the conceptual populations of carapace lengths of shallow water and deep water males have the same variance, determine the best estimate of the common variance. Similarly, obtain the best estimate of the common variance of the populations of shallow water and deep water females.

(iii) Use the best estimates evaluated in (ii) to calculate the 95 per cent confidence intervals for the true difference between the mean carapace lengths of

(a) shallow water and deep water males;
(b) shallow water and deep water females;
(c) shallow water males and females;
(d) deep water males and females; and
(e) males and females ignoring water depth.

(iv) How are the confidence intervals in the first four comparisons of (iii) affected if the assumptions of the equality of variances in (ii) are not tenable?

36   In the early nineteen-twenties some field experiments conducted by
Sir Oliver Lodge gave results which were interpreted as demonstrating
that increased plant growth followed electric excitation of a certain
intensity and duration supplied from a charged network suspended above
the plants. Soon after the announcement of Lodge's results, similar
electroculture experiments were carried out in the Biophysical Laboratory
of the Bureau of Plant Industry at Washington, D.C., during the years
from 1924 to 1928.

In 1926, ten experiments were tried with maize seedlings in the
Washington laboratory. In each of these experiments seedlings were
grown in groups of 100 and spaced one inch apart in each direction in
wooden boxes of sandy loam. Four such boxes were used in each experiment. The seedlings in two of the four boxes used in an experiment were
treated with a current of 10^{-9} amperes during the night, the other two
boxes being used as controls. An experiment extended over a period of
about two weeks from the planting of the seeds, and the plants were
treated for six to ten days.

The mean elongation (mean final height—mean initial height) of the
seedlings was used as an index of plant response to treatment. The table
below gives the difference (measured in millimetres) between the mean
elongation of the treated and control seedlings in parallel pairs of boxes
used in the experiments, a positive difference showing that the electrical
treatment increased the rate of growth.

Experiment no.	Difference in mean elongation
1	6·0
4	1·3
5	10·2
6	23·9
8	3·1
9	6·8
11	− 1·5
13	− 14·7
20	− 3·3
21	11·1

Source   G.N. Collins, L.H. Flint, and J.W. McLane (1929), *Journal of Agricultural Research*, Vol. 38, p. 585.

Analyse these data to determine the 95 per cent confidence interval
for the true difference in the mean elongation of the treated and control
plants. Hence test whether the electrical treatment produced a different
rate of growth of the maize seedlings.

37   The following data are from an experimental study made of the effect
of storage at different temperatures upon the changes in the constituents
of Conference pears. The figures refer to measurements made on ten
pears of the 1926 season which were stored for 152 days at 1°C at the
Low Temperature Research Station, Cambridge.

Pear no.	Percentage loss of weight (x)	Acidity (y)	Specific gravity (z)
1	25·9	0·104	1·061
2	25·0	0·101	1·062
3	23·8	0·042	1·058
4	26·7	0·098	1·067
5	28·9	0·048	1·064
6	29·1	0·069	1·062
7	30·9	0·115	1·059
8	27·6	0·062	1·057
9	43·1	0·080	1·065
10	30·3	0·094	1·061

Source   A.M. Emmet (1929), *Annals of Botany*, Vol. 43, p. 269.

Analyse these data to determine the 98 per cent confidence intervals for the true means of the three characteristics of the pears considered after storage for 152 days — percentage loss of weight, acidity and specific gravity.

38   To examine the growth of cork, borings of constant diameter were taken from the northern and eastern directions of the trunk for 28 trees in a block of plantations. The weights of these borings (measured in centigrams) are given in the table below.

Tree no.	Northern (x)	Eastern (y)	Tree no.	Northern (x)	Eastern (y)
1	72	66	15	91	79
2	60	53	16	56	68
3	56	57	17	79	65
4	41	29	18	81	80
5	32	32	19	78	55
6	30	35	20	46	38
7	39	39	21	39	35
8	42	43	22	32	30
9	37	40	23	60	50
10	33	29	24	35	37
11	32	30	25	39	36
12	63	45	26	50	34
13	54	46	27	43	37
14	47	51	28	48	54

Source   C.R. Rao (1948), *Biometrika*, Vol. 35, p. 58.

Analyse these data to test

(i) whether there is a significant difference between the mean weights of the cork borings on the northern (x) and eastern (y) sides of the trees; and

(ii) whether the mean of the differences in weight between the northern and eastern sides $(x - y)$ differs significantly from zero.

Discuss briefly which of these tests seems preferable in this situation, explaining any apparent contradiction between the tests.

39 The following table gives the values of the cephalic index found in two random samples of skulls, one consisting of 15 and the other of 13 individuals.

Sample I	74·1	77·7	74·4	74·0	73·8	79·3	75·8	82·8
	72·2	75·2	78·2	77·1	78·4	76·3	76·8	
Sample II	70·8	74·9	74·2	70·4	69·2	72·2	76·8	
	72·4	77·4	78·1	72·8	74·3	74·7		

If it is known that the distribution of cephalic indices for a homogeneous population is normal, analyse the above data to answer the following questions:

(i) Is the observed variation in the first sample consistent with the hypothesis that the standard deviation of the population from which the sample was drawn is 3·0?

(ii) Is it possible that the second sample was obtained from a population in which the mean cephalic index is 72·0?

(iii) Use a more sensitive test for (ii) if it is known that the two samples were obtained from populations having the same but unknown variance.

(iv) Obtain the 90 per cent confidence interval for the ratio of the variances of the populations from which the two samples were derived. Hence test the equality of the population variances.

40 A commercial wagon of small coal contains material from dust up to particles of the size of small apples, and it needs careful procedures of sampling to obtain a representative selection of material for the assessment of some characteristic of interest, such as the percentage ash content of the coal. In some early experiments carried out at the Fuel Laboratories of Imperial Chemical Industries, the general method of sampling small coal from a wagon was as follows.

The wagon was tipped into a hopper whence the coal flowed on to a moving conveyor which was divided into a number of compartments. The sampler removed an increment (that is a small sample of a few pounds) from every nth compartment, the first increment being taken from the centre of a compartment, the second from the right-hand side, the third from the left-hand side, and so on. The gross sample, that is the sum of the increments, was then crushed. A single small sample was then taken from the mixture for purposes of chemical analysis.

This method was used to sample 100 wagons of coal, each containing 10 tons, and the ash content of each wagon was determined. The measurements given overleaf are of the percentage ash content of the wagon loads.

12·30	14·60	17·08	13·82	13·86
9·36	15·42	16·40	14·56	11·98
6·84	17·07	16·25	12·74	13·27
11·45	15·37	19·86	13·60	18·55
16·62	15·17	12·60	17·75	15·58
15·39	17·82	17·75	12·79	14·19
17·29	17·70	9·54	12·59	16·48
10·80	14·03	15·54	10·54	15·73
16·20	20·37	17·14	19·00	18·22
11·69	12·61	11·56	13·01	13·32
8·18	13·64	14·10	10·16	15·67
14·65	16·78	11·68	16·66	15·09
12·07	13·84	16·62	11·97	19·72
13·52	14·32	16·69	10·16	16·80
14·00	20·40	16·66	12·20	11·70
14·95	15·46	16·88	18·24	12·94
14·20	15·33	14·37	16·02	13·29
13·81	13·95	12·54	12·24	9·97
16·54	16·53	12·09	17·64	16·62
11·07	12·98	13·74	13·24	17·77

Source   E.S. Grumell (1935), *Journal of the Royal Statistical Society* (Supplement), Vol. 2, p. 1.

If it may be assumed that the coal in the wagons was of comparable quality, analyse these data to determine a 95 per cent confidence interval for the true percentage ash content of this grade of coal, and a 98 per cent "symmetrical" confidence interval for the true error variability of the percentage ash content of the coal.

41   The table below shows a series of ten Rich Mixture ratings by the British 3C procedure of each of two fuels designated A and B. The ratings were obtained on ten engines, and each rating of fuel A is related to the rating alongside it for fuel B. These figures are taken from the M.A.P. monthly correlation data for October 1944.

Engine no.	Rating of fuel A	Rating of fuel B
1	100·5	98·3
2	99·8	98·5
3	101·0	100·0
4	100·5	99·5
5	101·0	101·8
6	101·4	100·5
7	102·4	102·3
8	99·5	99·8
9	102·8	100·8
10	103·6	102·0

Source   H.M. Davies (1946), *Journal of the Institute of Petroleum*, Vol. 32, p. 465.

(i) Analyse these data to show that, on the average, the difference between the ratings of the fuels is not significant at the five per cent level of significance if the variation between engines is ignored, but the observed result is significant if the engine variation is eliminated from the estimate of error variability.

(ii) Evaluate a 95 per cent confidence interval for the true difference between ratings of the two fuels.

42 In order to carry out intelligently the fortification of cereal products, it is necessary to assess the variations of the quantities of vitamins in the whole grain. An experimental investigation was therefore carried out at the Central Laboratories, General Foods Corporation, Hoboken, New Jersey, to determine the thiamin and riboflavin content of fifteen hard and sixteen soft American wheat varieties. The following table gives the measurements made in μgrams per gram.

Hard wheat variety	Thiamin	Riboflavin	Soft wheat variety	Thiamin	Riboflavin
Amber Durum	5·80	1·20	Albit	3·91	0·89
Blackhull	5·84	1·09	Currell	3·25	1·48
Ceres	4·35	1·06	Dicklow	3·63	1·24
Cheyenne	4·88	1·09	Federation	3·12	0·81
Chiefkan	4·57	1·14	Fulhio	3·39	0·85
Early Baart	6·90	1·31	Fultz	2·65	1·02
Kanred	5·20	1·41	Gladden	3·65	1·24
Marquis	4·35	1·31	Kawvale	3·39	1·06
Montana Marquis	4·98	0·96	Leap	3·69	1·30
Nebraska No. 60	4·00	1·91	Purdue No. 1	2·43	1·02
Nebred	4·98	1·27	Purplestraw	3·69	1·30
Ridit	3·65	0·94	Rex	4·77	1·10
Tenmarq	5·30	1·03	Rudy	3·51	0·82
Turkey	5·72	1·16	Thorne	3·67	1·10
Turkey Red	6·05	0·89	Triplet	4·00	1·10
			Trumbull	3·71	0·85

Source   R.T. Conner and G.T. Straub (1941), *Cereal Chemistry*, Vol. 18, p.671.

(i) Analyse these data to determine 95 per cent confidence intervals for the true differences in the thiamin and riboflavin contents of the two types of wheat varieties. Hence test the null hypotheses of no difference in the vitamin contents of hard and soft wheats. If a null hypothesis is acceptable at the five per cent level, determine a 95 per cent confidence interval for the true overall measure of the corresponding vitamin content of the two types of wheat varieties.

(ii) Also, test separately for the equality of the variances of the vitamin contents for the hard and soft wheat varieties.

43   In 1931 the American Institute of Baking, Chicago, Illinois made analyses of bread from representative bakeries located in different parts of the country. The samples of bread analysed included whole wheat, rye, raisin, salt rising and potato, but the majority of the samples were of white bread made with milk, both sliced and unsliced. The results of the analysis of 25 different samples of white bread are given in the table below. The measurements are percentages calculated on dry basis and refer to the following characteristics of the bread:

(i) Protein (N × 6·25) content;

(ii) Crude fibre content;

(iii) Carbohydrates (nitrogen-free extract);

(iv) Ether extract (fat); and

(v) Ash content.

Protein (N × 6·25)	Nitrogen-free extract	Crude fibre	Fat	Ash
15·87	76·33	0·50	4·25	3·05
15·69	76·50	0·44	4·33	3·04
15·19	76·70	0·46	4·95	2·70
15·44	76·39	0·40	4·58	3·19
15·45	76·80	0·43	4·56	2·76
15·98	75·65	0·24	4·82	3·31
15·01	77·67	0·40	4·26	2·66
14·81	77·30	0·42	4·88	2·59
15·71	75·61	0·58	5·10	3·00
15·63	76·68	0·44	4·33	2·92
14·05	78·22	0·52	4·46	2·75
14·74	77·32	0·53	4·55	2·86
15·06	76·93	0·47	4·80	2·74
14·94	76·47	0·27	5·60	2·72
14·97	76·88	0·61	4·81	2·73
14·19	77·93	0·53	4·83	2·52
14·42	77·36	0·77	4·71	2·74
13·43	77·56	0·64	5·60	2·77
14·54	77·20	0·83	4·88	2·55
14·77	77·07	0·74	4·78	2·64
14·42	76·11	1·41	5·18	2·88
14·65	77·16	0·71	4·69	2·79
16·36	75·09	0·88	4·75	2·92
14·66	78·01	0·34	4·37	2·62
14·07	78·15	0·61	4·32	2·85

Source   C.B. Morison (1931), *Cereal Chemistry*, Vol. 8, p. 415.

Earlier, in 1913, the Connecticut Agricultural Experiment Station published an extensive report of the analysis of 201 loaves representing the product of 79 Connecticut, one Springfield and three New York bakeries. The average composition of these samples of bread, calculated to the dry basis, was as follows:

(i) Protein (N × 6·25): 14·69 per cent;

(ii) Carbohydrates (nitrogen-free extract) including crude fibre: 81·33 per cent;

(iii) Ether extract (fat): 1·83 per cent; and

(iv) Ash: 2·15 per cent.

Analyse the 1931 measurements to test, at the one per cent level, the significance of the differences between them on the average and the 1913 measurements, on the assumption that

(a) the Connecticut averages are sensibly without error; and

(b) the individual observations on which the Connecticut averages were based have the same intrinsic variability as the Chicago observations.

44 Meadow foxtail grass (*Alopecurus pratensis*) is used in seed mixtures for pasture and meadow land. The seed of the grass is obtained from florets in the heads of the grass,. but the number of florets in a head of grass varies very considerably. In order to estimate this variation over the country, ten heads were chosen at random from samples of 100 heads each obtained from different areas, and the number of florets in each head was counted. The samples were obtained from meadow foxtail grass growing in the wild in Great Britain and Ireland during the summer months of 1927. The table below gives the average number of florets per head, each average being based on a count of ten heads.

283	243	308	331	202	259	282	189	226
176	228	239	261	175	283	218	163	209
287	209	201	168	176	276	179	280	265
390								

Source H.F. Barnes (1930), *Annals of Applied Biology*, Vol. 17, p. 339.

Analyse these data to estimate the mean number of florets per head of meadow foxtail grass, and determine the standard error of the estimate. Hence evaluate a 95 per cent confidence interval for the true mean number of florets per head.

45 Variation in the tensile strength of cement-mortar briquettes made from the same mixing of the material depends partly on the cement, but largely on the technique of mixing and forming and the conditions under which the briquettes are matured. From one mixing to another there may be slight differences in quality of material or changes in temperature conditions which will alter the average strength. Nevertheless, if the production is under statistical control such variation in strength will not be significant as compared with the intrinsic error variability.

The table below gives the measurements in tensile strength (measured in lb per sq. in.) of 20 samples of six strength tests, each sample being obtained from a different mixing of the same cement-mortar under similar conditions.

Sample no. 1	Sample no. 2	Sample no. 3	Sample no. 4	Sample no. 5
534	508	554	555	536
518	530	598	567	492
560	574	520	550	528
538	528	570	550	572
510	534	538	535	582
544	538	544	540	506

Sample no. 6	Sample no. 7	Sample no. 8	Sample no. 9	Sample no. 10
544	570	530	590	542
502	578	592	554	556
548	532	564	530	590
562	562	536	560	546
534	524	540	572	564
542	548	530	526	522

Sample no. 11	Sample no. 12	Sample no. 13	Sample no. 14	Sample no. 15
570	574	560	586	564
540	536	544	550	550
546	558	576	540	544
532	570	490	542	556
580	540	572	546	508
556	546	578	570	560

Sample no. 16	Sample no. 17	Sample no. 18	Sample no. 19	Sample no. 20
518	570	562	526	564
564	512	492	564	548
546	524	550	556	538
574	570	546	596	546
548	530	538	550	552
526	546	552	574	504

Source   O.L. Davies and E.S. Pearson (1934), *Journal of the Royal Statistical Society* (Supplement), Vol. 1, p. 76.

(i) Analyse these data to test the null hypothesis that the average tensile strengths of the briquettes produced by the different mixings are the same. Hence evaluate a 98 per cent confidence interval for the true tensile strength of the briquettes.

(ii) Determine a symmetrical 95 per cent confidence interval for the true error variability of the tensile strength of the briquettes.

46   O.L. Davies *et al.* in *Statistical Methods in Research and Production* (Oliver and Boyd, Edinburgh: 1958; p. 104) give the results of an investigation to find out how much of the variation from batch to batch in the quality of an intermediate product (H-acid) contributed to the variation in the yield of a dyestuff (Naphthalene Black 12B) made from it. In the experiment six samples of the intermediate, representing different batches of works manufacture, were obtained, and five preparations of the dyestuff were made in the laboratory from each sample. The following table gives the equivalent yield of each preparation as grams of standard colour determined by dye-trial.

Sample no. of H-acid	1	2	3	4	5	6
	1545	1540	1595	1445	1595	1520
Individual yield in	1440	1555	1550	1440	1630	1455
grams of standard	1440	1490	1605	1595	1515	1450
colour	1520	1560	1510	1465	1635	1480
	1580	1495	1560	1545	1625	1445

Reproduced by kind permission of the authors, the publishers, Oliver and Boyd, Ltd., Edinburgh, and Imperial Chemical Industries Ltd., England.

(i) Make an appropriate analysis of the data to test the null hypothesis than no real differences exist in the quality of the H-acid from batch to batch.

(ii) Determine the 95 per cent symmetrical confidence limits for the experimental error standard deviation, under the usual assumptions of normality.

47   The following table obtained from E.S. Keeping's *Introduction to Statistical Inference* (Van Nostrand, New York: 1962; p. 242) gives 30 measurements of the tensile strength (in kg/cm^2) made on specimens of rubber. The rubber tested was produced by five different processes and six tests were made on each kind of rubber.

Specimen no.	Process no.				
	1	2	3	4	5
1	177	116	170	181	177
2	172	179	156	190	186
3	137	182	188	210	199
4	196	143	212	173	202
5	145	156	164	172	204
6	168	174	184	187	198

Reproduced from Keeping: *Introduction to Statistical Inference*, copyright 1962 by Litton Educational Publishing Inc., New York, with permission.

Analyse these data to test the null hypothesis that, on the average, there is no difference in the tensile strength of the rubber produced by the five processes.

Suppose that the sample averages corresponding to the given processes are $\bar{y}_r$ ($r = 1, 2, \ldots , 5$), and it may be assumed that

$$E(\bar{y}_r) = \alpha + \beta(r - 3),$$

where α and β are independent parameters. Determine the least-squares estimates of α and β. Also, evaluate the 95 per cent confidence interval for β.

48   A variety of guayule (a rubber-producing plant) was planted in a field. After the plants were approximately one year old, a random sample of 54 plants was obtained. Of these 27 were normals, 15 offtypes, and 12 aberrants. The percentage rubber content was determined from each plant, and the data, quoted by W.T. Federer in *Experimental Design* (Macmillan, New York: 1955; *Problems*, p. 1), are given in the table below.

Type	Percentage rubber content						
Normal	6·97	7·11	7·26	6·80	7·01	7·00	6·35
	6·37	7·29	7·31	6·86	6·81	6·43	7·43
	6·68	7·29	7·12	6·68	7·34	5·15	6·41
	6·45	6·32	6·82	6·86	6·48	7·28	
Offtype	6·21	5·70	6·04	4·47	5·22	5·55	4·45
	4·84	5·88	5·82	6·09	5·59	6·06	5·59
	6·74						
Aberrant	4·28	7·71	6·48	7·71	7·37	7·20	7·06
	6·40	8·93	5·91	5·51	6·36		

Reproduced by kind permission of the author and The Macmillan Co., New York.

(i) Make an analysis of variance to test whether the data are consistent with the hypothesis that there is no difference in rubber content for the three types of plants.

(ii) If θ_n, θ_o, and θ_a are the true mean percentage rubber content for the normal, offtype, and aberrant plants respectively, use Student's t-statistic to test the null hypotheses $H(\theta_n - \theta_a = 0)$ and $H(2\theta_a - \theta_n - \theta_a = 0)$.

(iii) Comment on the conclusions based on the tests of (i) and (ii).

49   In an experimental study of the mineral metabolism of pullets, four White Wyandotte pullets of the same strain and hatching were used. During the period of investigation, two pullets (referred to as C_1 and C_2) were given ration C which had a high CaO content. The other two pullets (referred to as NC_3 and NC_4) were given ration NC. This ration was generally comparable with ration C except that its CaO content was low. In other respects the pullets were treated alike, and an attempt was made

to regulate the daily food consumption of the pullets. The experimental
period began on 11th July, 1934, and extended for about 70 days. During
this period pullet C_1 laid 14 eggs, pullet C_2 9 eggs, pullet NC_3 6 eggs
and pullet NC_4 8 eggs.

The following table gives the weights of CaO (in grams) found in the
total number of eggs laid by the pullets. Only six observations are available for the eggs of pullet NC_4 as the last two eggs were broken and so
rendered useless for chemical analysis.

C_1	C_2	NC_3	NC_4
2·013	2·005	2·366	2·094
2·195	1·977	2·276	2·152
2·435	2·163	2·147	1·690
2·545	2·088	1·821	1·685
2·542	2·136	1·805	1·354
2·749	2·071	1·758	0·823
2·723	1·895		
2·706	1·870		
2·686	1·824		
2·509			
2·673			
2·721			
2·655			
2·708			

Source   R.H. Common (1936), *The Journal of Agricultural Science*, Vol. 26, p.85.

(i) Analyse these data to test for the differences in the CaO content
of the eggs laid by the four pullets.

(ii) Also, obtain the 95 per cent confidence interval for the true
difference in the CaO content of the eggs due to the two rations and use
it to test the null hypothesis that this effect is not significantly different
from zero.

50   The following data are from a preliminary investigation made at the
Fleischmann Laboratories, Standard Brands Incorporated, New York, to
measure the thiamin content of seven types of American cereals — wheat,
barley, maize (corn), oats, rye, buckwheat, and millet. The samples
assayed were of different varieties and they were also obtained from different areas of the United States. It is known that there are considerable
varietal and environmental differences in the thiamin content of cereals,
and so the data cannot be regarded as completely representative of the
various grains. However, since most, though not all, of the samples were
reasonably well distributed in terms of growing regions and varieties, the
non-representativeness may be ignored, at least as a first approximation.
It is therefore possible to analyse these data to compare the thiamin
content of the seven cereals assayed. The thiamin measurements are in
micrograms per gram.

Cereal	Thiamin content										
Wheat (x_{1i})	7·1	7·3	6·9	4·5	5·0	5·2	4·5	5·7	4·5	6·7	4·7
	6·0	5·2	4·5	6·0	5·9	6·1	5·9	6·7	6·1	6·1	5·9
	5·5	6·3	6·0	5·8	5·2	4·2	4·5	4·7	5·1		
Barley (x_{2i})	6·3	7·2	7·1	8·7	6·5	8·2	7·7	7·5	7·1	7·1	6·2
	6·5	7·5	9·2	5·7	5·5	6·1	6·1	6·1	5·6	5·6	5·1
	5·6	5·1	5·3	4·7	5·3	6·6	5·8	5·5	5·1	5·1	3·8
	4·7	5·0	5·7	4·7							
Maize (x_{3i})	5·6	5·8	5·0	6·1	5·5	5·1	6·0	5·5	6·0	5·3	4·6
	4·1	4·6	4·1	5·8	4·7	5·0	8·0	6·0	6·0	4·8	4·5
	4·7										
Oats (x_{4i})	9·6	10·0	10·3	8·2	9·7	8·0	8·2	6·0	7·5	6·5	7·0
	6·0	6·0	6·0	7·0	8·8	5·2	4·8	5·0	5·7	5·7	
Rye (x_{5i})	5·7	4·0	5·7	4·7	4·5	5·0	5·0	4·5	4·1	5·0	
Buckwheat (x_{6i})	5·0	4·7	4·2	7·5	8·5						
Millet (x_{7i})	8·2	7·5	6·0								

Source   A.S. Schultz, L. Atkin, and C.N. Frey (1941), *Cereal Chemistry*, Vol. 18, p. 106.

Analyse these data to test for the differences, if any, in the thiamin content of the seven types of cereals. Also, evaluate 95 per cent confidence intervals for the true thiamin content of the cereal types and the experimental error variance of the thiamin measurements.

51   The data in the accompanying table refer to the nitrogen content of proteins found in the albumen of eggs laid by hens on diets differing only in the source of protein concentrate fed. The Barred Rock chicks used in the experiment were grown in a brooder house, and for the first few months from hatching they were reared under similar conditions except that the chicks of different groups varied, as far as possible, only in the source of the protein supplement. When the chicks were four months old ten pullets from each of the four initial groups were removed to four specially constructed pens in the laboratory. Some minor modifications were made in the diet on which the birds had previously subsisted in an endeavour to make the only variable, as nearly as possible, the quality of the protein from the different protein supplements. The four diets are referred to as buttermilk powder (Pen 28), fish meal (Pen 29), tankage (Pen 30), and meat meal (Pen 31). Another group of ten pullets was added (Pen 14) and this was fed on cod-liver meal diet.

The birds commenced laying in January 1929 and the experiment was continued until 30th August, 1929. The chemical analysis is based on eggs laid in early summer. To give sufficient material for analysis, a composite of the albumen proteins from five eggs laid by each of seven birds in each pen (excluding Pen 30) was prepared. In the case of Pen 30, the number of eggs laid was limited. However, a five-egg sample was

obtained from the eggs laid by each of six hens, but duplicate and tripli-
cate measurements of the nitrogen percentage were made on the samples
of eggs laid by two hens. For the purposes of the present analysis it may
be assumed that the nine observations made from the eggs laid by hens
in Pen 30 are independent and comparable.

The measurements given below are of nitrogen percentage of dry
weight in the albumen proteins.

Buttermilk powder Pen 28 (x_{1j})	Fish meal Pen 29 (x_{2j})	Tankage Pen 30 (x_{3j})	Meat meal Pen 31 (x_{4j})	Cod-liver meal Pen 14 (x_{5j})
15·13	14·89	15·28	15·31	14·90
15·16	15·18	15·55	15·15	15·32
14·90	15·60	15·29	15·15	15·15
15·30	14·99	14·91	15·14	15·28
15·51	15·38	15·20	15·30	15·21
15·38	15·58	15·50	15·15	15·19
15·05	15·10	14·85	14·77	15·17
		14·76		
		15·15		

Source   W.D. McFarlane, H.L.Fulmer, and T.H. Jukes (1930), *Biochemical
Journal*, Vol. 24, Part 2, p. 1601.

(i) Analyse these data to test for any differences in the nitrogen
content of the proteins in the albumen of the eggs laid by pullets given
different protein supplements in their diets.

(ii) Also, determine the 95 per cent confidence interval for the true
difference in the nitrogen content of eggs laid by pullets in Pen 28 and
the other four pens taken together.

52   The "maturation period" of the tomato plant is defined as the
number of days between the date a flower opened and the date the fruit
from the same flower was picked for market. Obviously this period cannot
be determined exactly because of the possible error at each end. In
general, tomato flowers showing the first signs of opening on any particular day are fully open at some time during the following morning and
the error at this end of the period is only a matter of a few hours. On the
other hand, the date of picking allows of much greater error, although,
whenever possible, fruits are picked when the first tinge of pink appears.
It may happen, however, that fruits too green for picking on Saturday
should have been taken on Sunday, but instead were left until Monday. It
is estimated, therefore, that the error in picking may be as much as two
days. This error is considered almost negligible in relation to the long
maturation period of about 60 days. Besides, in a large comparative study
this error will, on an average, tend to balance out.

Table for Exercise 52

Variety: Balch's Ailsa Craig

Plant no.																									
1	58	55	59	57	58	57	59	58	55	61	62	60	61	59	62	61	62	99	63	62	64	66	67	67	69
	68	71	108	64	64	67	66	68	67	70	93	95	97	100	63	61	61	64	72	78	78	94	99	75	63
	66	60	83	82	63																				
2	63	61	62	63	66	67	64	66	67	61	62	63	64	61	62	60	71	70	58	60	71	66	60	68	64
	70	66	83	67	67	107	58	65	58	65	67	63	68	74	82	80	78	58	60	69	71	75	73	71	88
	71	72	73	79	77	75	78																		
3	65	62	63	67	68	67	67	73	66	67	67	69	67	73	74	105	67	65	66	70	73	62	63	64	67
	66	79	109	60	60	63	63	58	82	97	102	80	72	71	84										
4	62	62	66	66	68	69	69	70	72	66	67	64	66	66	65	67	68	63	65	65	64	63	63	65	69
	70	69	88	57	62	63	66	65	65	67	82	97	97	63	63	67	69	103	102	92					
5	56	54	55	52	55	56	58	58	64	63	62	62	57	60	62	65	65	64	63	65	64	65	60	64	67
	64	65	56	58	58	60	61	64	66	93	95	95	63	63	65	95	98	100	100	92	98	64	60	64	
	80																								
6	61	62	62	63	67	66	68	67	67	62	66	64	64	66	62	66	65	67	67	67	70	71	69	88	94
	60	62	63	63	66	69	77	77	91	60	60	65	65	62	63	65	70	73	87	86	78	67	83		
7	62	62	62	62	63	64	67	74	65	64	64	64	66	65	65	77	65	63	63	60	68	67	69	68	99
	61	61	62	62	63	66	69	63	74	70	85	73	98	63	63	70	73	109	63	80	88	72	101	66	82	79
8	63	63	65	63	68	63	79	65	115	67	69	67	65	65	65	67	66	62	62	62	62	65	64	65	72
	96	65	66	68	68	66	70	73	99	104	106	63	63	68	62	84	105	85	64	70					

Table for Exercise 52 (concluded)

Variety: ES1

Plant no.																							
1	53 53 56 61 55 53 55 57 57 94 97 54 52 54 55 55 56 59 85 88 103
	49 50 54 55 56 52 56 65 66 61 63 66 67 64 63 63 61 66 68
	62 73 63
2	51 49 55 55 56 54 54 54 54 57 62 53 103 51 55 59 66 55 92 94
	98 91 132 92 137 58 89 106 53 54 55 74 66 62 68 66 67 64 69 68
	67 67 68 66 78 73
3	48 46 51 55 60 54 59 58 51 50 57 57 58 60 106 74 54 53 52 54 108
	54 69 72 92 89 89 53 55 82 67 68 64 62 63 100 75 87 63 61 66 73
	65 68 74 73 73
4	50 49 48 52 57 56 82 48 51 55 55 57 86 56 102 57 56 60 59 91 89
	89 53 56 53 86 133 95 106 113 54 55 63 86 93 88 64 63 71 70 70 68
	104
5	50 52 51 51 53 58 58 68 51 50 52 55 58 56 54 73 99 98 56 59 57 79
	85 93 93 105 104 52 89 57 85 80 80 57 139 70 78 84 61 61 65 67 64 63 62
	59 68 72
6	54 56 59 57 58 61 64 62 74 54 56 57 56 60 59 98 96 145 53 57 56 58 88 87
	154 142 141 139 55 88 80 83 80 84 80 79 146 131 85 76 79 82 86 130 58 60 57 63 64 68
	63 74 79 79 64 90 64 75 63 62 61 67 71

Source W.F. Bewley and W. Corbett (1930), *Annals of Applied Biology*, Vol. 17, p. 267.

The data quoted (pp. 168–9) are from an investigation carried out at the Cheshunt Experimental and Research Station, Hertfordshire. Eight plants of Balch's Ailsa Craig variety were grown in the same house under the same manurial and cultivation treatment. In addition, six plants of the variety ES1 were similarly grown in another house. The ES1 was a hybrid selected from the progeny of Ailsa Craig × Blaby and was raised at the Experimental Station, Cheshunt. The data refer to the observed maturation periods (in days) determined from all the fruit of the Ailsa Craig plants, but in the case of the ES1 plants a few trusses were not recorded. However, this complication may be ignored for the present analysis.

(i) Analyse the data for the two varieties separately to obtain estimates of the intrinsic error variability of the maturation period.

(ii) Test the significance of the null hypothesis that the intrinsic error variability of the maturation period of the two varieties is the same.

(iii) Obtain a 95 per cent confidence interval for the difference between the true maturation periods of the two varieties.

53   The quality of milk depends largely on the composition of its solid content, but there is considerable variation in quality, due, amongst other things, to the breed of cow, the age and lactation period, and the standard of feed and rearing. The total solid content of milk may be broadly classified as fat and non-fatty solids, though the latter are themselves a rather heterogeneous mixture.

A comprehensive study was carried out at the Midland Agricultural College, Sutton Bonington, Loughborough, to investigate the composition of milk, keeping in view regional and breed differences. The study was started in February 1923 and, with some variation, the weekly samples for morning and evening milk were collected from 15 herds of differing breeds located in different counties. The data collection continued till February 1926, but during this period the number of cows in the herds varied considerably. The herds were selected so as to include as great a variation as possible as regards size of herd, climatic conditions, soil types, and general management.

The milk samples of the 15 herds were analysed separately, and the the table opposite gives the frequency distribution of the percentage fat content of the milk, separately for each of the selected herds.

Since the data were collected over a considerable period of time and covered a variety of regions, breeds, and farm maintenance practices, they may be regarded as representative of a wide range of milk quality.

(i) Analyse these data to obtain a pooled estimate of the error variability of the percentage fat content of the milk. Hence evaluate a 99 per cent confidence interval for the true overall mean percentage fat content of the milk.

(ii) Also, test whether, on the average, there is any real difference in the mean percentage fat content of the milk of the different herds.

Table for Exercise 53

Mean percentage of fat content	1	2	3	4	5	6	7	8	9	10	11	12	13	14	15
													Herd no.		
2·3					1	1	1								
2·5		3			1	0	2							1	
2·7	1	2			2	3	2				1	1		0	
2·9	1	7		2	4	6	6				0	1		7	1
3·1	3	6	2	3	6	14	7	1	4	1	1	4	4	12	1
3·3	8	1	9	9	16	19	4	6	3	2	2	3	6	9	1
3·5	13	2	14	9	14	15	0	3	3	1	6	1	5	11	3
3·7	10	10	10	3	23	16	0	5	5	2	2	2	5	18	3
3·9	9	10	10	7	15	6	3	5	10	2	1	2	2	4	1
4·1	7	7	9	11	10	8	3	6	2	4	4	6	1	5	0
4·3	3	3	2	6	4	3	6	4	1	8	1	3			1
4·5	3	3	4	2	1	4	7			2	3	2			0
4·7	1	1	3	1	1	2	1			4	5	2			0
4·9	0	2	0	1		1	0			0	2	1			0
5·1	0	7	0	0			1			2	0	1			1
5·3	0	2	0	0			1				0	0			
5·5	0		1	1							0	1			
5·7	0										1				
5·9	1														
Total	60	66	64	55	98	98	44	30	28	28	29	30	23	67	12

Source H.T. Cranfield, D.G. Griffiths, and E.R. Ling (1927), *Journal of Agricultural Science*, Vol. 27, pp. 62 *and* 72.

54 "Since 1954, the number of girls aged 14, 15 and 16 having babies in Manchester has more than doubled, Dr C. Metcalfe Brown, the Medical Officer of Health, said today, when he gave the health statistics for the last year to the health committee. He described the increase as 'a disturbing fact'. In 1954 in Manchester five girls between 14 and 16 had babies; in 1955, seven; 1956, six; 1957, ten; and last year up to the end of November, thirteen." (From *The Times*, 11th February, 1959).

Assuming that no cases occurred in December 1958, test whether there is any evidence of a steady increase in the number of cases with time. If there is evidence of such an increase at the five per cent level, calculate the least number of cases that would have been necessary in 1958 for this conclusion still to hold. Also, if there is no evidence of such an increase at the one per cent level, calculate the least number of cases necessary in 1959 to provide such evidence.

Discuss briefly the validity for these particular data of the type of analysis used.

55 C.A. Bennett and N.L. Franklin in *Statistical Analysis in Chemistry and the Chemical Industry* (Wiley, New York: 1954; p. 218) give the following results obtained from measurements of the carbon content of 36 samples of ball clays from South Devon.

Sample no.	x	y	Sample no.	x	y
1	1·53	2·46	19	4·18	6·14
2	0·87	1·54	20	0·22	0·52
3	0·28	0·70	21	0·38	0·40
4	0·27	−0·40	22	0·24	0·46
5	3·07	4·82	23	1·79	2·80
6	0·25	0·30	24	0·58	2·09
7	0·25	0·64	25	6·55	9·68
8	0·29	0·78	26	2·54	4·08
9	0·12	0·12	27	1·43	2·80
10	1·50	2·36	28	2·74	3·93
11	1·31	2·14	29	6·08	8·22
12	0·31	0·08	30	0·75	0·28
13	0·14	−0·01	31	0·16	0·35
14	2·98	4·53	32	5·06	7·49
15	6·84	9·94	33	0·86	1·41
16	2·15	3·68	34	0·16	−0·50
17	1·35	1·84	35	11·43	15·89
18	0·40	0·97	36	0·19	0·18

x = Carbon determined by combustion
y = Carboniferous material by "difference" method

Reproduced by kind permission of the authors and John Wiley & Sons, Inc., New York.

The carbon content was determined by the classical combustion method (x) and by a "difference" method from the ultimate analysis of the clays (y). It is known that the combustion method is of high precision, whereas the measurement of the carboniferous material by the "difference" method is less precise as the method involves assumptions which are known to be only approximately correct. Accordingly, the x values can be taken to be without random error, and it may be assumed that

$$E(y) = \alpha + \beta x \quad \text{and} \quad \text{var}(y) = \sigma^2.$$

Determine, correct to four decimal places, the least-squares estimates of the parameters α and β; and under the usual assumptions of normality obtain the 95 per cent confidence interval for the parameter α.

56 An experimental investigation was conducted at the Bibury Trout Farm, Gloucestershire, to determine the differences, if any, between the respiratory rates of the fry of the brown trout (*Salmo fario*) raised in swift- and slow-flowing water. The fry were reared for three months in two wooden tanks which were supplied with water directly from a spring. The temperature of the water was maintained at 10°C in both tanks but the rates of flow of water in them were different. At the end of the period, the fish for the respiratory measurements were removed from the tanks in batches of ten each and the oxygen consumption measured under standard conditions. The results of the experiments are shown in the table below, the total hourly oxygen consumption (reconstructed) being given in cubic millimetres at N.T.P. and the weight of the fish in grams.

Habitat	Experiment no.	x	y
Swift-water tank	1	7·1	766·8
	2	7·0	854·0
	3	7·5	1080·0
	4	7·4	954·6
	5	7·5	802·5
	6	7·5	862·5
	7	4·4	501·6
	8	4·3	417·1
	9	6·2	595·2
	10	8·2	1033·2
Slow-water tank	1	7·2	612·0
	2	6·2	942·4
	3	4·4	365·2
	4	4·0	276·0
	5	7·7	731·5
	6	5·6	487·2
	7	5·2	369·2
	8	5·3	498·2
	9	5·6	464·8
	10	7·1	667·4

x = Wet weight of 10 fishes. y = Total oxygen consumption.

Source   R. Washbourn (1936), *Journal of Experimental Biology*, Vol. 13, p. 145.

Assuming that the respiratory rate varies linearly with wet weight, determine the regression equations for the fish in the two habitats. Also, test for the equality of the linear regression coefficients by calculating the 95 per cent confidence interval for the true difference between them.

57   E.W.H. Cruikshank obtained the following measurements for the coronary flow (y) and the auricular pressure (x) in experiments with two cats.

Cat 1		Cat 2	
Coronary flow (y)	Auricular pressure (x)	Coronary flow (y)	Auricular pressure (x)
8	7·0	3	8·5
11	9·6	9	9·8
14	12·2	10	11·2
10	10·5	14	11·2
9	8·0	14	11·2
7	6·5	14	11·2
4	5·0	10	11·2
3	3·7	7	11·2
5	6·8	9	11·2
3	6·8	13	13·5
2	6·8	22	15·5
0	6·8	10	11·8
		5	8·5
		5	5·8
		8	9·8

Source   M.H. Quenouille, *Associated Measurements*, 1952; p.76. Reproduced by kind permission of the author and Butterworths Ltd, London.

   (i) Determine the equation for the linear regression of y on x for each of these cats.

   (ii) In testing for the equality, or otherwise, of two regression coefficients, it is assumed that the variances of the dependent variables are equal. Perform the necessary calculations to show that this assumption is justified in this case, and test whether the regression coefficient of Cat 1 differs significantly from the regression coefficient of Cat 2.

58   In a biological investigation, the growth of rats was the subject of study. Experimental animals were kept in separate cages and were divided into three groups containing 10, 7, and 10 rats respectively (the second group contained fewer rats than the other two, due to an accident at the beginning of the experiment). The animals in the first group were kept as control, those in the second group had thyroxin, and those in the third group thiouracil added to their drinking water. The table opposite gives, in grams, the initial weights and the weights after one week of treatment of the rats in the three groups.

Table for Exercise 58

Control group		Thyroxin group		Thiouracil group	
Initial weight	Weight after one week	Initial weight	Weight after one week	Initial weight	Weight after one week
57	86	59	85	61	86
60	93	54	71	59	80
52	77	56	75	53	79
49	67	59	85	59	88
56	81	57	72	51	75
46	70	52	73	51	75
51	71	52	70	56	78
63	91			58	69
49	67			46	61
57	82			53	72

Source   G.E.P. Box (1950), *Biometrics*, Vol. 6, p. 362.

(i) Assuming that the weights after one week are linearly related with the initial weights, determine the three linear regressions and test for the equality of the regression coefficients.

(ii) Determine a 95 per cent confidence interval for the difference between the true regression coefficients of the thyroxin- and thiouracil-treated groups.

(iii) Analyse the gains in weight recorded to test the null hypothesis that, on the average, there is no difference in the gains in weight in the the three groups.

(iv) Test the null hypothesis that there is no difference in the variances of the final weights in the thyroxin- and thiouracil-treated groups.

59   In crop fields where two or more crops are grown in mixture, the yield rate of a crop differs according to the intensity of mixture. The following table gives the mean yield rate of wheat by intensity of mixture (intensity being measured in terms of the proportion of the net area occupied by the crop) obtained from 189 and 247 fields respectively in the Shahabad and Monghyr districts of Bihar State, India.

Shahabad			Monghyr		
Intensity	Frequency	Yield rate	Intensity	Frequency	Yield rate
3/32	9	4·0	3/32	11	5·6
7/32	17	5·7	7/32	23	4·5
11/32	30	5·1	11/32	24	4·2
15/32	32	5·7	15/32	55	5·4
19/32	35	7·1	19/32	35	6·6
23/32	19	6·9	23/32	35	6·4
27/32	10	7·1	27/32	14	9·6
31/32	37	7·5	31/32	50	8·1

For each of the districts, determine the linear regression of yield rate on intensity of mixture, and also test for the equality of the two district regression coefficients. If the test of significance gives a significant result at the one per cent level, obtain a 95 per cent confidence interval for the true difference between the district regression coefficients, otherwise obtain a 95 per cent confidence interval for the common regression coefficient.

60   R.M. Jacobs (*Quality Control in Central New York Industry*: 1952) cites a problem faced by a particular company manufacturing air-conditioning units due to its failure to meet the finished weight requirements of a connecting rod used in the compressors. The compressor was a major part of the air-conditioning units, and rod weight was critical in a good compressor. The wastage caused by the rejection of finished connecting rods which did not meet the weight requirement seemed considerable, and, in addition, there was the expense of testing the weight of a finished connecting rod.

To reduce this cost, it was felt that an estimate of the rough casting could be used to give enough information on the weight of the finished rod. The table below gives the rough weight in lb (x) of the rough casts and the finished weight in lb (y) of 25 connecting rods.

Rod no.	Rough weight (x)	Finished weight (y)	Rod no.	Rough weight (x)	Finished weight (y)
1	2·745	2·080	14	2·635	1·990
2	2·700	2·045	15	2·630	1·990
3	2·690	2·050	16	2·625	1·995
4	2·680	2·005	17	2·625	1·985
·5	2·675	2·035	18	2·620	1·970
6	2·670	2·035	19	2·615	1·985
7	2·665	2·020	20	2·615	1·990
8	2·660	2·005	21	2·615	1·995
9	2·655	2·010	22	2·610	1·990
10	2·655	2·000	23	2·590	1·975
11	2·650	2·000	24	2·590	1·995
12	2·650	2·005	25	2·565	1·955
13	2·645	2·015			

Assuming that y is linearly related with x so that, for any given x,

$$E(y) = \alpha + \beta(x - \bar{x}), \quad \text{var}(y) = \sigma^2,$$

where α, β, and σ^2 are parameters, and $\bar{x}$ is the average weight of the rough casts of the sampled rods, determine the least-squares estimates of these parameters. Hence, evaluate, correct to four decimal places, the estimated value of y corresponding to $x = 2 \cdot 60$, and calculate, correct to six decimal places, the standard error of the estimate.

61 O.L. Davies *et al.* (*Statistical Methods in Research and Production*, Oliver and Boyd, Edinburgh: 1958; p. 155) give the results of an investigation carried out on the properties of a synthetic rubber. Varying amounts and types of compounding materials were used in the preparation of a number of specimens of this rubber and a wide range in the physical properties was obtained. The table below gives the abrasion loss and hardness of 30 different specimens.

Abrasion loss (g/hp hour)	Hardness (degrees Shore)	Abrasion loss (g/hp hour)	Hardness (degrees Shore)
372	45	196	68
206	55	128	75
175	61	97	83
154	66	64	88
136	71	249	59
112	71	219	71
55	81	186	80
45	86	155	82
221	53	114	89
166	60	341	51
164	64	340	59
113	68	283	65
82	79	267	74
32	81	215	81
228	56	148	86

Reproduced by kind permission of the authors, the publishers, Oliver and Boyd Ltd., Edinburgh, and Imperial Chemical Industries Ltd., England.

The determination of hardness is simple but that of abrasion loss requires elaborate experimentation and is much more difficult to carry out. There would therefore be considerable advantage if abrasion loss could be predicted with sufficient accuracy from the measurements of hardness. If it may be assumed that abrasion loss is linearly dependent upon hardness, estimate the regression equation.

Also, under the usual assumptions of normality, evaluate a 95 per cent confidence interval for the expected abrasion value corresponding to a hardness value of 90. Comment briefly on the practical value of the result obtained.

62 The moisture content of wheat and flour is a characteristic of great importance. Not only are wheat and flour bought and sold on a weight basis, but the water content of the material has a very intimate connection with the keeping quality. However, the water content of wheat and flour is extremely variable; it is known to vary from 5 to 25 per cent, sometimes even more. This variation is due to the hygroscopicity of the material, that is, the power of the material to absorb moisture from, or give up moisture to, the surrounding atmosphere according to the temperature and humidity of the latter.

An experimental investigation was carried out by the Research Association of British Flour Millers, St Albans, to measure the rate at which flour loses moisture under controlled conditions. The method consisted in exposing flour over concentrated sulphuric acid in a closed vessel at a constant temperature of 25°C. The apparatus was so arranged that the flour could be weighed at intervals without removing it from the drying vessel or in any way disturbing the drying process or the drying conditions. In the study, two separate samples, each of a patent blended flour obtained from City Imperial wheat and No. 1 Northern Manitoba wheat, were used. The initial moisture contents of the four samples were 13·07, 14·66, 14·53, and 13·49 per cent respectively. The table opposite gives these initial values and the percentage water remaining in the samples at various stages (measured in minutes) of the drying process up to approximately 200 minutes.

Over the time-interval considered, the water content declines approximately exponentially, and it may be assumed that the logarithm of the percentage water content decreases linearly with time. Analyse these data on this basis to determine the four linear regression coefficients, separately for each sample. Hence test the equality of the regression coefficients and, if this is tenable at the one per cent level of significance, determine the pooled regression equation and a 95 per cent confidence interval for the common regression parameter.

Note   Calculate the common logarithms of the percentage water content correct to two decimal places.

63   The process of generation of electricity consists essentially in setting in motion a set of coils mounted in a magnetic field. When this motion is produced by either steam or water-driven turbines, the essentials of the process of generation are roughly as follows. Combustible materials are fed into furnaces and the heat produced is used to evaporate water and produce superheated steam under pressure. The steam in turn passes into a turbine, which may be imagined as a series of glorified windmills mounted on a single shaft and rotated by the pressure of steam passing along them. The energy of rotation is in turn passed on to the generator, where it is converted into electrical energy.

In practice, the process is visualised as occurring in two stages. In the first stage, chemical energy contained in the fuel is transformed into heat energy in the form of superheated steam, and the plant accomplishing this transformation is classed as boiler plant. In the second stage, heat energy contained in the steam is converted into electrical energy, the plant accomplishing this transformation being known as generating plant. The electrical engineer is particularly interested in the relationship between the input of chemical energy and the output of electrical energy.

(i) The table at top of p. 180 gives the monthly heat consumption, measured in millions of British thermal units (Btu), and the output, measured in

Table for Exercise 62

City Imperial				No. 1 Northern Manitoba			
Sample no. 1		Sample no. 2		Sample no. 1		Sample no. 2	
Time in minutes	Percentage water content	Time in minutes	Percentage water content	Time in minutes	Percentage water content	Time in minutes	Percentage water content
0	13·07	0	14·66	0	14·53	0	13·49
15·50	11·93	8·00	13·99	15·00	13·05	11·75	12·39
40·00	10·48	15·00	13·28	28·00	11·96	26·25	11·28
51·50	9·84	22·50	12·60	41·50	11·00	39·50	10·46
76·50	8·79	30·00	11·99	68·75	9·45	65·75	9·06
86·50	8·45	36·50	11·53	108·00	7·85	105·00	7·54
136·50	6·95	49·00	10·66	125·50	7·30	122·75	7·06
153·25	6·61	68·50	9·53	138·00	7·00	134·75	6·80
166·75	6·40	90·00	8·52	151·00	6·70	149·50	6·51
187·75	6·09	101·00	8·05	165·75	6·50	164·75	6·30
200·50	5·96	113·50	7·59	179·00	6·33	177·75	6·19
		137·50	6·92	201·00	6·02	198·25	5·84
		152·50	6·55				
		169·00	6·18				
		183·50	5·90				
		195·00	5·72				

Source E.A. Fisher (1927), *Cereal Chemistry*, Vol. 4, p. 184.

Units supplied	Heat consumed	Units supplied	Heat consumed
3·173	59 636	11·632	173 965
3·389	55 266	13·474	196 530
8·771	142 814	11·584	172 064
7·703	124 807	10·949	162 246
5·661	94 118	14·520	212 716
11·097	167 094	4·056	59 639
11·515	175 448	14·344	211 407
13·598	201 571	13·316	194 961
12·587	182 266	15·852	233 603
11·878	179 476	13·260	194 932
13·087	190 724	14·690	213 024

millions of kilowatt hours (kWh) of electricity supplied, of a generating station having one generating set and three boilers, these boilers having identical characteristics.

(ii) As a more general situation, the monthly heat consumption and output of a generating station having six generating sets and thirteen boilers of widely differing capacities was studied. The table below records the measurements obtained.

Units supplied	Heat consumed	Units supplied	Heat consumed
12·200	227 549	13·417	253 478
10·204	193 564	10·889	210 477
9·763	187 027	12·431	238 524
4·629	89 127	9·781	201 061
5·476	109 189	8·716	184 149
4·468	85 772	6·696	138 066
6·543	129 639	5·788	121 290
5·251	108 650	5·529	113 687
7·495	152 119	8·741	173 590
9·552	186 376	12·617	244 861
10·926	207 024	13·365	258 060
13·656	257 456	13·309	260 603

Source   J.L. Ineson (1939), *Journal of the Royal Statistical Society* (Supplement), Vol. 6, p. 149.

Assuming that a linear relationship holds between heat input and electricity output, analyse these data separately to determine the regressions from (i) and (ii). Hence compare the slope parameters of the two regressions.

64   During 1931–3, a comprehensive investigation of lamp-making processes was carried out at the Research Laboratories of the General Electric Co., Ltd., Wembley. One aspect of the study was the quality of the fine tungsten wires of which coils used in the lamps were made. The aim was to obtain wire which would give the longest possible life when

operating in a lamp at a given efficiency of light production, or inversely, the highest possible efficiency for a given life. At that time the life-test, which was destructive, involved a period of 1000 hours and an expenditure of approximately £1 per 100-watt lamp tested. It was therefore considered desirable to devise wire quality tests which required much less time and expense. As a result of some experimental work a short-period test was suggested. The measurements made on the wires by this test were seen to be negatively correlated with the performance of the wire in the lamps.

The following table gives the measurements made by the short-period test on various wires produced in 1933. These observations were made on eight different dates during the period from April to June, 1933.

304	300	281	305	324	310	326
299	302	281	304	318	304	327
299	303	282	302	324	310	327
301	288	281	309	324	310	324
301	302	285	305	318	310	322
294	309	300	275	300	318	299
292	314	301	278	294	318	292
289	313	301	275	300	315	292
291	309	282	274	302	316	291
291	310	301	274	303	324	289
315	302	292	292	309	308	344
315	301	296	295	316	303	345
315	301	294	290	313	299	344
315	302	299	292	310	302	341
314	301	298	294	314	302	341
281	298	295	307	265	316	297
279	298	291	302	267	317	300
279	299	292	307	266	317	297
277	300	296	302	268	317	296
280	295	292	302	266	314	293
282	273	296	306	296	295	322
282	272	298	307	293	299	319
284	272	296	304	301	299	319
279	272	299	307	294	295	319
278	274	300	306	293	299	318
304	272	288	324	294	323	322
302	271	289	331	294	324	328
304	271	292	326	290	326	323
304	271	295	323	296	333	324
304	271	291	326	291	327	323
280	278	292	319	313	266	323
280	280	291	324	314	268	318
283	281	290	320	312	270	320
285	282	290	322	313	267	327
284	280	285	324	313	265	320

(i) Tabulate these observations as a frequency distribution having class-intervals with class-limits 265–; 270–; 275–; ...; 345–. Evaluate the mean and variance of the distribution. Hence determine a 99 per cent confidence interval for the true mean of the measurements made on the wires by the short-period test.

A few wires of similar quality were tested by the short-period test and the lifetimes (in hours) of lamps using coils made from them were also measured. These paired observations are given in the table below.

Short-period test	Lifetime (in hours)	Short-period test	Lifetime (in hours)
355	855	293	1120
332	1015	288	1320
350	1070	315	1225
360	890	305	1055
357	1095	315	1390
363	900	306	1385
355	905	286	1700
345	950	289	2070
356	1170	296	1395
276	1605	335	1105

(ii) Use these observations to determine the linear regression of lifetime on the measurement made by the short-period test. Hence evaluate the estimate of the average lifetime of the lamps made using the wires tested in the first sample. Also, calculate the standard error of this estimated average lifetime.

Source   W.J. Jennett and B.P. Dudding (1936), *Journal of the Royal Statistical Society* (Supplement), Vol. 1, p. 1.

65   Yield and protein content are the most important characters of wheat, the former because of its direct effect on the farmer's revenue and the latter because of its relation to baking quality. Unfortunately, when wheat is grown under climatic conditions favouring high yield the protein content is usually low. During the years 1930 and 1931, certain experiments were carried out in Alberta, Canada, to measure the effect of seasonal variation on the yield and protein content of wheat. In both years the field selected was reasonably level and did not show any obvious variations in soil. The crop was a pure line of Red Bobs 222 in 1930 and of Marquis in 1931, so that variation in inherited characters was reduced to a minimum. Just prior to the harvesting of the main crop, fifty 18-foot rows were cut by hand at locations scattered over the field. The field used in 1930 was approximately 14 by 34 rods, and that used in 1931 about 8 by 62 rods in size. Each 18-foot row was threshed separately, the yield recorded and the grain preserved for protein determinations. The following table gives the yields in grams and the protein content, expressed as percentages calculated at 13·5 per cent moisture content.

Red Bobs 222: 1930				Marquis: 1931			
Yield	Protein	Yield	Protein	Yield	Protein	Yield	Protein
243	14·1	294	15·3	223	13·3	164	15·4
424	14·4	232	15·5	373	14·3	239	14·6
305	14·8	246	14·9	303	14·1	90	15·0
322	14·4	253	14·3	342	14·8	353	14·4
413	14·3	318	14·7	298	14·7	342	14·0
323	14·7	445	13·4	373	13·6	304	14·0
235	15·5	339	14·9	426	14·5	306	13·5
182	15·3	314	14·6	419	14·0	178	14·4
137	16·2	230	15·3	376	13·6	294	14·1
255	12·9	235	15·1	326	12·9	236	14·1
344	14·9	218	15·1	501	13·7	212	14·5
349	14·4	262	15·2	175	13·9	273	14·8
228	15·7	238	14·8	349	12·9	226	14·8
377	14·9	259	15·1	303	14·2	418	14·4
300	15·0	249	14·5	297	15·1	390	14·1
252	15·4	463	14·1	232	14·7	514	14·5
286	14·9	390	14·5	45	14·8	229	14·7
273	14·2	433	14·5	143	14·7	151	14·7
228	15·6	483	12·5	357	14·4	193	15·4
145	16·5	330	14·2	363	14·1	181	14·6
400	13·5	294	14·6	393	14·3	178	14·2
444	13·0	310	15·3	319	14·2	234	14·8
300	15·1	208	16·7	314	14·4	262	14·6
296	15·3	270	15·8	185	14·5	230	15·2
360	15·0	321	15·7	94	15·0	316	15·3

Source   J.G. Malloch and R. Newton (1934), *Canadian Journal of Research*, Vol. 10, p.774.

(i) Analyse these data to compare separately the yields and protein contents for the two years, assuming that the two wheat varieties were of comparable quality.

(ii) Determine, separately for the two years, the linear regression of yield on protein content and hence evaluate a 95 per cent confidence interval for the difference between the true regression coefficients. Is there evidence for the belief that the linear association between yield and protein content is the same for the two years considered?

66   The data given below pertain to the total protein content and the percentage glutenin determined in different flour samples. The glutenin was determined by the method of Blish and Sandstedt, but by different investigators in three independent studies. In the first two studies the wheat varieties were generally representative, but in the third study durum wheats only were used.

Study no. 1		Study no. 2		Study no. 3	
Protein	Glutenin	Protein	Glutenin	Protein	Glutenin
11·00	4·67	9·73	3·48	14·89	5·72
10·66	4·46	13·99	5·31	14·95	4·88
11·69	5·06	13·07	5·02	14·17	5·23
12·31	5·47	12·53	4·85	14·31	5·50
11·46	5·11	14·29	5·62	11·87	3·72
9·69	4·73	12·34	4·51	12·58	4·57
11·23	4·30	10·61	3·96	14·94	5·32
10·92	4·20	10·63	4·04	12·76	4·66
8·66	3·39	12·99	5·13	14·59	5·87
11·51	4·44	14·10	5·50	10·94	4·10
14·65	5·73	11·56	4·50	16·59	5·62
15·44	5·75	12·24	4·78	10·51	3·95
14·08	5·70	10·09	3·64	12·40	4·70
14·14	5·70	12·76	4·58	12·41	4·79
14·31	5·75	12·49	4·35	12·22	4·85
15·61	6·30	14·45	5·52	12·69	4·71
16·89	7·21	13·64	5·53	12·97	4·99
				13·16	3·82

Sources M.J. Blish, R.C. Abbott, and H. Platenius (1927), *Cereal Chemistry*, Vol. 4, p. 129; E. Grewe and C.H. Bailey (1927), *Cereal Chemistry*, Vol. 4, p. 230; H. Vogel and C.H. Bailey (1927), *Cereal Chemistry*, Vol. 4, p. 137.

Assuming that a linear relation exists between the glutenin content and total protein content of wheat, analyse the data, separately for each study, to estimate the regression equations.

Also, show that in the light of the sample information, it is reasonable to assume that the slopes of the three regression lines are, in fact, equal. Hence obtain a 95 per cent confidence interval for the true value of the common slope parameter, and the equation of the estimated common regression line.

67 Some experiments were carried out at the Rothamsted Experimental Station, Harpenden, to study the effect of aeration on denitrification — the reduction of nitrates to gaseous nitrogen, sometimes mixed with oxides of nitrogen. Two non-identifiable species of denitrifying bacteria, classified under the genus *Pseudomonas*, were used in these experiments. A simple synthetic liquid medium was used, and its composition was as follows: glucose 0·35 per cent, KNO_3 0·1 per cent, C/N ratio 10, initial pH 7·0.

A heavy suspension of bacterial growth from a young agar slope[*] was made in 10–25 ml of sterile basis solution, and the number of cells counted. The volume of this suspension required to give an initial count of about 10 million bacteria/ml was then pipetted into flasks containing

[*]An "agar slope" is a special chemical surface used for rearing bacterial cultures.

the medium. All flasks contained the same amount of the medium and were incubated at room temperature and in the light, but they were maintained at three levels of air supply, referred to as aerated, control, and anaerobic conditions respectively. Control flasks were plugged with cotton-wool and were undisturbed except for sampling. In the aerated flasks the cotton-wool was replaced by a sterilized rubber stopper carrying inlet and outlet air tubes, and a small steady stream of air, filtered through cotton-wool, was bubbled through the medium. Anaerobic flasks, smaller in size, were stoppered with cotton-wool, and were incubated in a closed vacuum desiccator containing alkaline pyrogallal, under reduced pressure. Samples were taken from all flasks at the beginning and end of incubation, and usually during its course.

I   A part of the data collected from these experiments and quoted below refers to the growth of bacterial numbers at different ages for the three experimental aeration levels. The age of the cultures is in days and the bacterial numbers in millions/ml.

Anaerobic cultures		Control cultures		Aerated cultures	
Age	Bacterial number	Age	Bacterial number	Age	Bacterial number
2	248	1	336	1	578
3	576	1	242	1	332
7	1260	2	1058	2	2288
14	2240	2	1014	2	1522
14	1490	2	648	2	1240
14	1520	2	1048	2	1748
16	1410	3	1348	3	3700
		7	2072	7	3600
		7	2925	7	4100
		7	2240	7	4450
		14	2825	14	4125
		14	2560	16	4400
		14	4900		
		16	3550		

These observations were obtained from different experiments and so there is an inter-experimental source of variation, but for present purposes this may be included in the component of error variability. It is seen that under all the three levels of aeration there is a steady increase in bacterial numbers with age. If it may be assumed that for each level of aeration a linear relation exists between number and age, estimate these regression lines and test for the equality of their slopes.

II   Another part of the data refers to the bacterial numbers (measured in 100 millions/ml) and the estimated "protein" nitrogen (measured in mg/l) under the three levels of aeration. This nitrogen is that component of the total which is presumably locked up in the cells of the bacteria. The accuracy of these values is limited because they are based on the differences of two determinations – that of total nitrogen and that of "non-protein nitrogen" precipitated with basic lead acetate after the end of the experiments.

Anaerobic cultures		Control cultures		Aerated cultures	
Bacterial number	"Protein" nitrogen	Bacterial number	"Protein" nitrogen	Bacterial number	"Protein" nitrogen
14·10	14	28·25	20	36·00	57
22·40	18	20·72	30	41·00	39
21·10	31	29·25	27	44·00	53
15·55	34	35·50	29	41·25	22
24·95	39	28·25	25	30·40	66
39·70	53	25·60	42		
55·40	83	25·55	53		
		52·00	45		

Assuming that for each level of aeration there is a linear regression between "protein" nitrogen and bacterial number, estimate these regressions and test for the equality of the slopes.

Also, in both I and II evaluate 95 per cent confidence intervals for the individual regression coefficients.

Source J. Meiklejohn (1940), *Annals of Applied Biology*, Vol. 27, p. 558.

68   Plant seed in the resting condition may have all the materials necessary for germination except water and oxygen. The process of germination, essentially a speeding up of the physiological processes of the young plant, is preceded and accompanied by the absorption of water. Some empirical evidence based on laboratory practice showed that durum wheats (*Triticum durum*, Desf.) required less water than common wheats (*Triticum vulgare*, Vill.) in germination. It was thought that this difference might be related to the variations found in germination. An investigation was therefore planned to measure quantitatively the absorption of some common and durum wheat varieties under various conditions and to determine whether significant differences in absorption were to be found between varieties of the two species.

The samples of wheat used in the investigation were obtained from the Dominion Experimental Stations at Swift Current, Indian Head and Rosthern in Saskatchewan, Brandon in Manitoba, and the Division of Agronomy and Plant Genetics, University of Minnesota. The common wheat varieties were Marquis and Reward, and the durum varieties were Mindum and Kubanka. Not all the varieties were grown at each centre from which the seeds were obtained and, in order to make comparisons eliminating regional differences, the following wheat varieties could be used in pairs:

(i) Marquis and Mindum based on the average of seven samples;

(ii) Marquis and Reward based on the average of five samples; and

(iii) Mindum and Kubanka based on the average of two samples.

Unfortunately, these comparisons are not wholly independent since there was some overlap of the samples used.

In each sample, duplicate lots of 100 seeds were counted out, cracked and broken kernels being discarded. The seed lots were weighed and then put to soak. The absorption of water as a percentage of the weight was determined at different time-intervals from 4 hours to 168 hours. The average absorption of the two lots gave the percentage absorption for each sample. The temperature in these experiments varied form 19° to 30°C.

The following table gives the total percentage absorption in the paired groups (i), (ii), and (iii). No observations are recorded in group (i) for 28 hours because four of the samples in this group were inadvertently subjected to increased temperature in the 24—28 hour period.

Immersion in hours	Group (i)		Group (ii)		Group (iii)	
	Marquis	Mindum	Marquis	Reward	Mindum	Kubanka
4	23·13	20·24	23·39	22·45	18·41	20·35
8	34·52	30·43	35·55	34·17	30·44	34·08
12	41·98	37·13	42·04	40·09	37·17	40·92
16	45·86	41·05	46·59	44·48	41·63	45·50
20	48·71	44·36	49·54	47·73	44·35	49·18
24	·51·86	47·64	52·08	51·10	48·19	52·92
28	−	−	55·40	53·70	52·41	55·49
32	56·07	51·89	56·80	55·18	53·42	57·78
36	58·04	54·33	58·87	57·83	55·31	59·03
40	59·89	56·08	60·59	59·56	57·49	61·38
44	61·32	57·65	61·93	61·04	58·89	62·37
48	63·97	59·63	63·99	63·34	61·27	65·42
60	66·94	63·03	67·28	67·08	66·59	70·18
72	69·08	65·00	69·41	69·42	68·42	71·69
96	74·12	71·20	72·75	74·20	75·24	78·02
120	76·26	73·60	74·33	76·57	78·65	81·99
144	76·36	76·27	74·47	78·05	84·34	87·02
168	75·45	76·86	73·18	78·42	85·16	87·28

Source   T.W.L. Burke (1930), *Scientific Agriculture*, Vol. 10, p. 367.

If p denotes the percentage absorption after t hours of immersion, transform the observations and the time-scale by the transformations

$$y = \sin^{-1} \sqrt{p} \quad \text{and} \quad x = \log_{10} t.$$

It is then seen that y is approximately linearly related to x. Analyse the transformed data on this basis to compare the linear rates of water absorption in the groups (i), (ii) and (iii). Also, obtain 95 per cent confidence intervals for the differences in the linear rates in the groups.

Note   Calculate the x values correct to two decimal places and the y values correct to one decimal place.

69   Bread-making is a complex phenomenon, and it is not easy to predict
the volume of bread from the properties of the flour-dough. Nevertheless,
it was believed that a simple and general relation existed between the
capacity of a flour-dough to be stretched into a thin membrane and the
specific volume of bread obtainable with this flour. A statistical relation
of this kind would make it possible to predict rather simply the baking
value of a flour.

I   In a series of experiments, a special apparatus known as an "exten-
simeter" was used to measure the extensibility of doughs, the observations
made being expressed in terms of a "coefficient of extension" of the
dough (E). The flours were also used to form the loaves of dough of 175
grams in weight baked in an oven for 18 minutes at a temperature of 270°C.
The specific volume of the bread was determined by a special procedure,
and it was expressed in cc per 100 grams of bread. The table below gives
the values of the coefficient of extension and the corresponding values of
the specific volume of the bread obtained in 30 independent experiments.

E	Specific volume	E	Specific volume
234	359	254	338
112	274	169	270
199	341	270	363
199	331	229	341
90	259	241	341
264	380	270	358
77	228	126	238
186	326	284	348
206	313	284	358
239	339	244	343
250	371	50	217
239	338	146	302
220	314	212	328
116	214	277	372
116	230	176	317

If it is assumed that the specific volume of bread (y) is linearly related to the square root of the coefficient of extension (x), analyse the
data to estimate this regression, and evaluate a 95 per cent confidence
interval for the regression coefficient.

Note   Calculate the x values correct to one decimal place.

II   It was felt that the practical usefulness of the above method for
assessing the baking value of flour would depend considerably on the
reliability of the measurements of extensibility. Although the methods of
mixing the dough and of making the extensibility test were standardised,
it was recognised that in mixing a dough it would be difficult to duplicate
all the conditions affecting the properties of a substance that is so

Table for Exercise 69

Dough no.	Test no. 1	Test no. 2	Test no. 3	Test no. 4	Test no. 5	Test no. 6	Test no. 7	Test no. 8	Test no. 9	Test no. 10
1	20·1	18·3	17·7	16·0	17·4	18·0	18·6	16·4	15·7	20·3
2	17·9	19·2	19·6	18·7	19·9	17·9	19·8	19·8	18·7	18·4
3	19·3	17·8	17·8	18·3	16·1	17·8	19·8	18·0	20·2	19·3
4	16·8	16·8	17·8	17·3	16·6	17·8	17·8	19·2	16·8	17·7
5	17·0	19·2	18·1	19·4	20·8	19·4	18·3	15·4	19·5	19·6
6	18·3	18·8	18·5	17·3	18·1	16·2	17·2	17·1	16·5	15·5
7	19·2	18·7	18·7	18·5	18·2	20·1	18·2	18·3	18·4	18·2
8	18·2	18·3	16·1	16·7	15·7	16·4	15·1	17·8	15·3	19·7
9	17·2	18·3	20·9	17·4	18·5	17·3	19·0	18·9	19·9	16·8
10	17·3	18·5	20·1	18·2	18·7	17·4	17·5	20·1	18·4	20·5
11	14·7	16·4	15·0	14·2	14·1	14·7	14·5	14·6	14·7	13·4
12	16·9	17·2	15·7	16·2	15·2	14·0	17·2	15·7	18·4	17·3
13	18·1	16·3	17·7	19·2	17·1	16·7	17·8	15·4	15·4	16·3
14	16·3	16·8	16·0	18·1	17·5	16·7	16·8	15·8	16·7	15·9
15	15·7	16·3	18·8	15·2	15·4	18·5	16·3	18·5	15·4	14·4
16	16·7	16·1	15·6	16·4	16·3	16·1	17·9	16·8	16·6	16·5
17	14·3	15·5	15·6	15·2	15·6	14·7	14·9	15·7	14·3	13·7
18	19·3	20·2	18·7	18·3	17·8	18·4	18·6	20·3	19·7	19·8
19	19·4	17·3	18·3	17·2	17·2	17·2	20·6	18·6	17·8	17·5
20	16·8	19·8	18·3	18·3	19·6	18·2	19·8	19·0	19·7	18·5
21	20·7	17·8	18·3	18·0	17·2	18·0	18·8	17·0	17·9	18·3
22	17·8	16·6	17·7	17·7	17·2	17·2	16·9	18·2	17·7	16·2
23	16·7	16·8	17·2	16·8	17·2	16·4	16·2	17·0	16·9	17·7
24	19·2	17·9	19·0	18·7	18·8	18·0	17·5	18·8	18·7	19·7
25	19·7	18·5	20·5	21·5	18·0	17·5	18·2	17·8	19·8	18·8

sensitive to treatment. In order to assess the variability in the measurement of extensibility of doughs, 25 doughs from the same flour were
tested, ten tests being made on each dough. The table on page 189 records
the square roots of the coefficients of extension.

Analyse these data to test for differences in extensibility of the
doughs. Obtain a "symmetrical" 95 per cent confidence interval for the
experimental error variance and a 99 per cent confidence interval for the
true value of the square root of the coefficient of extension of the flour
used.

Source   C.H. Bailey and A.M. Le Vesconte (1924), *Cereal Chemistry*, Vol. 1, p. 38

70   The following data pertain to the gains in weight of baby chicks
reared on four tropical feedstuffs for a certain period of time. Although
the details of the layout of the experiment are not given, it may be
assumed that the chicks in the four experimental groups were selected
randomly, so that the observed differences in the gains in weight between
the groups represent only those due to feeding treatments.

	Treatment		
1	2	3	4
55	61	42	169
49	112	97	137
42	30	81	169
21	89	95	85
52	63	92	154

Source   Query 70, *Biometrics* (1949), Vol. 5, p. 250.

(i) Analyse these data to test the null hypothesis that, on the
average, there is no difference in the gains in weight attributable to the
four treatments. Also, determine, correct to four decimal places, the best
estimate of the experimental error variance σ^2.

(ii) If the above null hypothesis is rejected at the one per cent level
it may be assumed that $\bar{y}_i$, the observed mean of the ith group, is such
that

$$E(\bar{y}_i) = \alpha + \beta x_i, \quad \text{for } i = 1, 2, 3, 4,$$

where α and β are independent parameters, and the x_i are non-random
variables which take the values $-1 \cdot 5$, $-0 \cdot 5$, $0 \cdot 0$, and $2 \cdot 0$ corresponding
to the four treatments. Evaluate the least-squares estimates of α and β
and a 95 per cent confidence interval for β. Reconstruct the estimates of
the treatment means on this basis. Hence test for the adequacy of the
above linear regression hypothesis for the means.

71   The following table, derived from data given by C.H. Goulden in
Methods of Statistical Analysis (Wiley, New York: 1952; p. 120), pertains
to the carotene content determined by two methods for 20 wheat varieties.

Variety no.	Carotene in flour	Carotene in wheat	Variety no.	Carotene in flour	Carotene in wheat
1	2·39	1·18	11	2·60	1·58
2	3·11	2·13	12	2·11	1·45
3	2·15	1·41	13	2·30	1·74
4	1·96	1·42	14	1·80	1·42
5	2·02	1·50	15	2·00	1·45
6	1·76	1·25	16	2·05	1·87
7	2·10	1·65	17	2·09	2·00
8	2·12	1·24	18	2·33	1·65
9	2·28	1·48	19	2·29	1·64
10	1·86	1·35	20	2·30	1·62

Reproduced by kind permission of the author and John Wiley & Sons, Inc., New York.

By one method carotene was determined on the whole wheat, and by the other method on flour. The measurements for carotene in the wheat are lower than for carotene in the flour, which is the reverse of the actual condition. This was due to a different method of extraction for the whole wheat which gave lower but comparable results.

(i) Make an appropriate analysis of the data to test the null hypothesis that there is no difference in the experimental error variance of the two methods for the determination of carotene content.

(ii) Estimate the true difference in the carotene measurements obtained by the two methods used, and determine the standard error of the estimate.

(iii) Determine, under suitable assumptions, the linear rates at which the carotene content of flour varies with that of wheat and vice versa. Hence, or otherwise, obtain an estimate of the correlation between the measurements of carotene content made by the two methods.

72 The gasoline colour test is a reliable and objective method for determining the colour of flour. This test uses the fact that a considerable portion of the colour imparted to flour originates in the bran and germ. After extracting these materials with a suitable solvent, the colour is measured quantitatively by comparing the gasoline extract with a potassium chromate solution of known strength. The method, however, required considerable time; for example, the extraction of a 20-gram sample needed 16 or more hours with 100 cc of clear colourless gasoline. It was believed that comparable results in colour determination could be achieved by using smaller samples with constant stirring for a short period of time.

A study was, therefore, made to determine the extraction of 5 grams of flour with 100 cc of gasoline, and 20 grams of flour with 100 cc of gasoline stirred with a mechanical stirrer for 15 and 30 minutes, respectively. These two methods are referred to as Methods I and II. The third method was that of the Association of Official Agricultural Chemists (A.O.A.C.); this required 20 grams of flour and 100 cc of gasoline, and took 24 or more hours to complete. Forty-eight samples of flour having a considerable range of variation in gasoline colour values were selected. The results obtained by the three methods are given in the table below.

Method I	Method II	A.O.A.C. Method	Method I	Method II	A.O.A.C. Method
1·99	1·92	2·02	1·20	1·11	1·15
1·99	1·89	2·02	0·82	0·72	0·67
2·13	2·16	2·10	1·04	1·05	0·97
2·25	2·17	2·14	1·00	0·88	0·82
1·90	2·04	1·95	1·91	1·86	1·75
0·76	0·81	0·80	1·27	1·22	1·17
1·23	1·29	1·33	0·94	0·85	0·69
1·16	1·14	1·07	1·06	0·97	0·88
1·53	1·50	1·47	2·32	2·33	2·23
1·26	1·21	1·26	1·10	0·89	0·97
1·32	1·24	1·22	2·13	2·13	2·12
1·42	1·39	1·35	2·42	2·34	2·37
1·74	1·71	1·80	1·33	1·18	1·17
0·76	0·77	0·80	1·64	1·46	1·40
1·47	1·36	1·32	1·16	1·03	1·00
1·35	1·26	1·09	1·42	1·39	1·42
1·62	1·51	1·43	1·23	1·16	1·15
1·36	1·31	1·20	1·37	1·25	1·25
1·23	1·17	1·06	1·29	1·19	1·16
1·15	1·06	0·95	1·19	1·15	1·00
1·57	1·53	1·38	1·19	0·92	0·90
1·18	1·17	1·14	1·50	1·41	1·40
1·18	1·14	0·98	1·28	1·17	1·04
1·05	1·03	1·05	1·51	1·45	1·26

Source   D.A. Coleman and A. Christie (1926), *Cereal Chemistry*, Vol. 3, p. 84.

(i) Analyse the data to compare the relative accuracy of the measurements of the gasoline colour value of flour made by Methods I and II.

(ii) Also, is it reasonable to infer that the measurements made by Methods I and II are of equal accuracy?

(iii) By using separately the measurements obtained by Methods I and II as concomitant observations, determine the linear regressions of the measurements made by the A.O.A.C. method on the two sets of concomitant observations. Hence evaluate 95 per cent confidence intervals for the two linear regression coefficients.

73   The standard technique for determining the total moisture content of wheat flour is the oven-drying method, which involves the exposure of the flour to a temperature of 130°C for one hour. The measurements obtained by this method are known to be extremely accurate with a negligibly small experimental error. A quick and relatively simple alternative method is based on the volumetric assessment of the acetylene produced by the interaction of calcium carbide and the flour under investigation. A measurement by this method can be made in about five minutes, but the accuracy of the measurements so made appears to be less than those obtained by the oven-drying method.

An experimental investigation was carried out at the Department of Agricultural Chemistry, University of Nebraska, Lincoln, U.S.A., to determine the differences, if any, between the assessments of the percentage moisture contents of wheat flours of various types and origin made by the two methods. Forty-two flours of varying grades, types, and moisture content were obtained from more than twenty different mills. The samples varied form 0·37 to 1·09 per cent in ash content, and included both hard and soft wheat flours. Samples of one gram each were used to determine the percentage moisture content by the oven-drying and the carbide methods. The measurements are given in the table below.

Percentage moisture by oven method	Percentage moisture by carbide method	Manometric pressure in millimetres	Percentage moisture by oven method	Percentage moisture by carbide method	Manometric pressure in millimetres
12·57	12·65	91·5	10·09	9·99	58·0
13·98	13·85	103·0	10·20	9·93	57·0
9·80	9·75	55·0	10·38	10·06	58·5
9·57	9·57	53·0	8·65	8·50	42·5
14·02	13·85	103·0	9·66	9·55	55·0
13·65	13·75	103·5	10·16	10·19	62·5
9·47	9·35	50·5	9·75	9·57	53·0
9·86	9·77	55·5	11·75	11·80	82·0
11·92	11·95	81·0	10·30	10·61	67·5
12·53	12·47	87·0	10·54	10·35	64·5
8·25	8·35	39·0	10·24	9·95	59·5
8·76	8·65	42·5	14·72	14·95	118·0
8·38	8·47	40·5	14·27	14·45	112·0
7·76	7·95	36·5	13·23	12·91	94·0
14·29	14·35	111·0	12·83	12·95	95·0
14·10	14·25	109·5	11·88	12·19	86·0
9·87	9·59	55·5	13·36	13·55	101·5
11·30	11·25	75·0	7·17	7·45	30·5
11·52	11·35	76·0	7·05	7·29	28·5
14·16	13·95	106·0	7·91	8·13	38·0
10·22	10·10	59·0	7·91	8·13	38·0

Source   M.J. Blish and B.D. Hites (1930), *Cereal Chemistry*, Vol. 7, p. 99.

(i) Analyse these data to test whether, on the average, there is any difference between the two sets of comparative measurements of percentage moisture content. Also, test whether these observations are of equal intrinsic accuracy.

(ii) Since the experimental flours were reasonably representative of the bulk of commercial quality, use the above data to determine the best linear relation for evaluating the percentage moisture content of wheat flour from the observations made by the carbide method.

(iii) The third column of the above table gives the manometric pressure readings in millimetres. This pressure is proportional to the amount of acetylene gas produced, and is therefore an index to the quantity of water that has reacted with the carbide. However, not all of the flour moisture reacts with the carbide, though the pressure readings are generally closely correlated with the percentage moisture in the flour. Use these pressure readings as a concomitant variable to determine the linear regression of moisture content (by oven method) on manometric pressure. Hence estimate the percentage moisture content of wheat flour which gave a manometric pressure of 98·6. Also, evaluate a 99 per cent confidence interval for the true percentage moisture content corresponding to a manometric pressure of 98·6.

7.4   In a statistical study of smallpox, Karl Pearson suggested that the severity of attack may be measured by the two tests of days (a) from onset to bath and (b) from eruption to bath. The term "bath" refers here to the stage when a patient has the first bath after recovery from illness. The table opposite gives the frequency distributions according to the two criteria, which were obtained from certain cases provided by Dr John Brownlee.

Of these, 779 were vaccinated, 55 unvaccinated and 21 doubtful. The deaths recorded before recovery were excluded from the frequency distributions.

(i) Evaluate, correct to four decimal places, the mean and variance of each of the frequency distributions.

(ii) Assuming that the two sample distributions are from populations having means θ_1, θ_2 and variances σ_1^2 and σ_2^2, determine the standard errors of the best estimates of θ_1 and θ_2. Hence calculate the 95 per cent large-sample confidence intervals for the parameters θ_1 and θ_2.

(iii) Explain very briefly whether or not the sample information given is adequate to evaluate a large-sample confidence interval for the difference $\theta_1 - \theta_2$.

Table for Exercise 74

Onset to bath		Eruption to bath	
Days	Frequency	Days	Frequency
4–5	2	4–5	10
6–7	13	6–7	55
8–9	40	8–9	164
10–11	131	10–11	174
12–13	192	12–13	156·
14–15	152	14–15	96
16–17	99	16–17	71
18–19	73	18–19	39
20–21	40	20–21	26
22–23	24	22–23	13
24–25	13	24–25	21
26–27	17	26–27	6
28–29	10	28–29	8
30–31	6	30–31	2
32–33	4	32–33	7
34–35	6	34–35	3
36–37	2	36–37	3
38–39	1	38–39	1
40–41	0		
42–43	1		
Total	826	Total	855

Source K. Pearson (1906), *Biometrika*, Vol. 5, p. 505

75 The following table gives the index (y) of the total taxable incomes in the U.K. during the years 1948–65, with the index for the base year 1948 taken to be 100.

Year	Index (y)	Year	Index (y)
1948	100	1957	180
1949	106	1958	185
1950	112	1959	196
1951	124	1960	214
1952	130	1961	231
1953	136	1962	243
1954	146	1963	251
1955	159	1964	272
1956	172	1965	301

Source Private communication from H.M.A. Fazel, University of Leicester (1968).

If x denotes time and it is assumed that

$$E(y_\nu) = \alpha + \beta(x_\nu - \bar{x}) + \gamma(x_\nu - \bar{x})^2,$$

where $\bar{x}$ is the average of the observed x values, use the method of least squares to obtain, correct to six decimal places, the estimates of the parameters α, β, and γ. Hence evaluate, correct to two decimal places, the best fitting values of the index for the years 1948–65. Comment on the results obtained.

76   In the performance of a blast furnace, the output of pig iron per ton of coke, which is a measure of operating efficiency, is affected for quite different reasons by the percentage of available carbon in the coke (because it is largely the carbon which does the work of smelting), and by the hearth temperature of the furnace. The latter cannot be measured directly but it is believed to be reflected, other things being equal, in the lowness of the ultimate percentage of silicon in the pig iron coming out of the furnace. The following table, obtained from a coking plant of the United Steel Companies Ltd., Workington, gives the records for 74 weeks of the coded weekly output of pig iron per ton of coke (z), the available percentage of carbon in the coke (x), and the percentage of silicon in the pig iron (y).

Serial no.	x	y	z	Serial no.	x	y	z	Serial no.	x	y	z
1	50	12	56	26	8	42	62	51	11	34	61
2	52	15	59	27	26	40	50	52	14	35	70
3	31	26	66	28	29	35	70	53	33	23	45
4	43	50	65	29	44	26	65	54	26	22	45
5	41	35	61	30	38	35	76	55	48	28	71
6	41	24	59	31	43	41	53	56	33	12	67
7	43	36	59	32	39	47	46	57	28	27	63
8	38	53	59	33	49	42	59	58	19	33	59
9	51	33	66	34	52	28	53	59	33	35	61
10	54	34	97	35	35	47	48	60	28	43	51
11	39	32	90	36	28	50	43	61	46	37	44
12	45	34	35	37	14	35	57	62	48	25	56
13	35	41	62	38	24	47	48	63	36	20	57
14	44	18	63	39	22	44	43	64	16	48	36
15	56	19	67	40	18	66	10	65	34	50	40
16	54	34	49	41	19	70	34	66	35	37	44
17	49	55	41	42	13	54	35	67	37	45	63
18	51	42	39	43	23	24	45	68	53	62	63
19	38	40	55	44	14	48	40	69	41	70	73
20	33	41	74	45	19	74	14	70	34	30	75
21	28	53	75	46	18	41	6	71	27	47	46
22	36	31	59	47	15	67	14	72	31	21	67
23	38	41	76	48	12	63	16	73	36	26	70
24	8	45	64	49	28	45	32	74	63	35	54
25	8	43	16	50	47	47	21				

Source   W.M. Gibson and G.H. Jowett (1957), *Applied Statistics*, Vol. 6, p. 189.

(i) Assuming a relation of the form

$$E(z_\nu) = \alpha + \beta(x_\nu - \bar{x}) + \gamma(y_\nu - \bar{y}), \quad \text{for } \nu = 1, 2, \ldots, 74,$$

where $\bar{x}$, $\bar{y}$, and $\bar{z}$ are the means respectively of the x, y, and z values given, use the method of least squares to obtain estimates of the parameters α, β, and γ correct to four decimal places.

(ii) If it is assumed that $\gamma = 0$, test the null hypothesis $H(\beta = 0)$ under the usual assumptions of normality.

77 The table overleaf gives the lengths (x) and breadths (y), measured in millimetres of 243 eggs of the cuckoo (*Cuculus canorus*) from the collections in the Charterhouse Museum and the British Museum of Natural History, South Kensington. Of the eggs measured, 223 were known to have been deposited in the nests of 42 different species of birds, while the foster-parents of the remaining 20 were not ascertainable.

Assuming that the bivariate normal model is suitable for the length and breadth measurements, analyse the data —

(i) to obtain a 95 per cent confidence interval for the true difference in the mean length and breadth of the eggs;

(ii) to test whether the length and breadth measurements are of equal intrinsic variability; and

(iii) to test whether the length and breadth measurements are uncorrelate

Also, determine the linear regression of length on breadth and that of breadth on length.

78 By means of a technique based on the use of a projection microscope and a sensitive electric photometer, it was possible to measure both the areas and transparencies of red blood-cells stained with acid fuchsine. The table on p. 199 gives the bivariate frequency distribution obtained from 400 observations made on a sample of normal blood. The horizontal array represents areas, arranged in class-intervals of 3 mm deflection as shown by a galvanometer. The vertical array indicates the relative transparency, the class-interval being 0·03. The class-marks used in the table for both the horizontal and vertical arrays are midpoints of the class-intervals.

(i) Calculate, correct to six decimal places, the product-moment correlation coefficient between relative transparency and area of the red blood-cells.

(ii) It is known that, on a slide of normal blood, as ordinarily seen through a microscope, the larger cells appear paler than the smaller ones. If it is assumed that the regression of transparency on cell area is linear, analyse the above data to estimate this regression. Also, evaluate a 95 per cent confidence interval for the regression coefficient.

(iii) Represent graphically the means of the transparency arrays and the estimated regression line of transparency and cell area.

Table for Exercise 77

x	y	x	y	x	y	x	y	x	y	x	y	x	y
22·5	17·0	22·3	16·3	21·8	16·7	22·0	16·3	23·2	16·9	21·2	15·7	21·9	16·9
20·1	14·9	20·6	16·2	21·1	16·5	22·6	17·0	22·0	17·1	20·9	16·0	24·0	17·2
23·3	16·0	22·1	16·8	23·4	16·2	22·0	16·0	22·2	17·0	22·7	14·5	22·3	16·8
22·9	17·4	21·9	17·0	23·8	16·3	22·1	16·4	21·2	16·1	22·8	16·7	22·6	17·0
23·1	17·4	23·0	16·9	23·3	16·7	21·1	16·4	21·6	16·5	22·1	16·9	22·0	17·0
22·0	16·5	22·0	17·0	24·0	17·5	23·0	17·0	21·6	16·5	23·4	17·0	22·7	16·9
22·3	17·2	22·0	17·0	23·5	17·3	21·3	16·1	21·9	16·1	21·2	16·2	22·3	17·3
23·6	17·2	22·1	17·3	23·2	16·4	19·9	16·0	22·0	16·5	22·5	17·0	22·5	16·9
24·7	18·0	22·0	16·8	24·0	17·3	22·9	16·0	22·9	17·9	23·9	17·7	21·2	15·9
23·7	17·8	19·6	15·8	22·4	16·0	23·3	16·1	22·8	16·5	24·0	16·0	22·4	17·2
24·0	18·0	22·8	17·1	23·9	16·4	22·1	17·8	22·7	16·7	22·8	17·2	22·2	16·7
20·4	15·0	22·0	16·9	22·0	17·0	20·9	15·3	21·0	16·0	23·2	16·0	22·2	16·4
21·3	16·0	23·4	16·4	23·9	16·9	21·9	17·0	22·5	16·2	22·1	15·8	23·0	16·3
22·0	16·5	23·8	16·4	20·9	15·8	22·9	16·4	21·9	16·8	22·4	16·5	19·8	15·0
24·2	17·3	23·3	16·8	23·8	17·3	22·4	16·6	24·0	17·7	23·0	16·8	22·1	16·0
21·7	16·9	22·5	17·1	25·0	17·5	19·1	14·0	23·2	16·2	22·0	17·0	21·5	16·2
21·0	16·1	22·3	17·0	24·0	17·5	20·9	15·3	22·3	16·7	22·1	17·1	20·9	15·7
20·1	15·8	21·9	17·1	21·7	16·2	23·0	16·0	23·0	17·0	20·8	15·7	22·0	16·2
21·9	15·9	22·0	17·2	23·8	16·5	21·0	16·1	23·1	17·1	22·2	18·1	21·0	15·5
21·9	16·2	21·7	16·2	22·8	16·2	22·0	16·3	23·2	16·9	21·2	15·6	22·3	16·0
21·7	16·1	23·3	16·7	23·1	17·1	21·3	16·5	20·9	15·9	22·5	16·4	21·0	15·9
22·6	17·0	22·2	16·8	23·1	16·1	22·0	16·2	22·5	16·0	21·1	15·9	20·3	15·5
20·9	16·2	22·3	16·2	23·5	16·9	22·0	15·2	21·9	16·0	23·0	16·3	20·9	15·9
21·6	16·2	22·8	16·4	23·0	16·7	21·3	15·8	22·5	16·1	23·4	16·7	22·0	16·0
22·2	16·9	22·9	17·2	23·0	17·0	21·1	15·8	23·3	17·2	24·0	17·0	20·0	15·7
22·5	16·9	23·7	17·0	21·8	16·0	23·1	16·6	23·2	16·5	23·3	16·3	20·8	15·9
22·2	17·3	22·0	17·2	23·0	15·9	21·0	16·0	23·0	17·0	23·1	16·7	21·2	16·0
24·3	16·8	21·9	17·0	23·3	17·1	23·0	16·2	21·1	17·0	22·4	16·5	21·0	16·0
22·3	16·8	22·2	16·2	22·4	16·6	23·1	16·8	22·9	17·0	21·8	16·0	24·2	16·9
22·6	17·0	24·4	16·2	22·4	16·9	22·9	16·0	23·3	16·8	21·8	16·0	22·8	16·6
20·1	16·5	22·7	16·3	23·0	16·1	23·2	17·0	20·1	15·8	24·9	16·8	24·7	16·3
22·0	16·9	23·3	16·6	23·0	17·2	24·4	17·9	22·9	17·0	24·0	15·8	24·0	18·8
22·8	16·5	24·0	17·0	23·0	16·2	21·5	16·0	23·3	16·1	22·1	16·2	22·9	17·1
22·0	17·0	23·6	16·9	23·9	16·9	23·3	16·4	22·3	16·2	21·0	17·1		
22·4	17·0	22·1	16·3	22·3	15·2	22·3	16·3	22·9	17·3	22·6	16·0		

Source   O.H. Latter (1901), *Biometrika*, Vol. 1, p. 164.

Table for Exercise 78

Transparency	Area											
	28	31	34	37	40	43	46	49	52	55	58	61
0·3155			1	1								
0·3455												
0·3755		1	3	1	1			1				
0·4055	2		3	3	3	3	1					
0·4355			5	12	7	2						
0·4655		3	5	8	8	8	4					
0·4955	1	2	6	10	17	10	4	4				
0·5255		2	4	13	29	21	12	3	2			
0·5555			4	8	14	21	15	6	3			
0·5855			4	5	13	16	11	12	3	1		
0·6155			1	3	3	6	5	4	2	2	3	
0·6455				1	1	3	1	1				1
0·6755												
0·7055										1		

Source A. Savage, C.H. Goulden and J.M. Isa (1935), *Canadian Journal of Research*, Vol. 12, p. 803.

Table for Exercise 79

Regular weekly wage of father	Weekly rent											
	2·0	2·5	3·0	3·5	4·0	4·5	5·0	5·5	6·0	6·5	7·0	7·5
7 –		1		1								
9 –												
11 –			2	1								
13 –			1	2	1	1						
15 –	1	2	5	7	8	2						
17 –	3	4	22	23	17	12	2	2				
19 –	4	12	32	75	88	55	18	6	1			
21 –	1	8	30	62	98	53	11	10				1
23 –		6	39	81	131	85	25	14	5			1
25 –	1	7	33	74	130	135	25	19	3	1		1
27 –	1	4	6	24	47	51	16	16	2	3		
29 –	1	1	8	26	36	66	32	29	5	9	1	1
31 –		1		1	1	3	2	2	1			
33 –						3	2	2	2	1		
35 –		1		6	8	14	11	7	3	1	1	1
37 –					1	1	1	1				
39 –				1		2	2	2	4	2	2	
41 –												
43 –												
45 –					2							
47 –												
49 –								1				

Source E.M. Elderton (1925–6), *Annals of Eugenics*, Vol. 1, p. 277.

79   The lower table on page 199 gives the bivariate frequency distribution of the regular wage of father (in shillings*) and the weekly rent (in shillings*) paid by regular wage-earners in Bradford during 1911–12. The data were originally supplied by Dr W.A. Evans, Medical Officer of Health, Bradford.

(i) Calculate, correct to six decimal places, the product-moment correlation coefficient between weekly wage of father and weekly rent. Also, test the hypothesis that the two variables are uncorrelated.

(ii) Determine the two linear regression equations between weekly wage of father and weekly rent. Represent graphically the array means and the regression lines.

80   Table 1 gives the frequency distribution of age (in years) and body temperature (in degrees Fahrenheit) of convicts, based on some data collected by Dr Charles Goring in Parkhurst prison.

Table 2 gives the frequency distribution of head length (in inches) and reaction time to sight (in 1/100 of a second) of 4690 males, based on data collected by Sir Francis Galton, and corrected for age.

Table 3 gives the frequency distribution of maximum and minimum temperatures (in degrees Fahrenheit) at Rothamsted for the month of August during the years 1876–1926. The table is condensed from that of the original paper by doubling the length of the class-intervals of the arrays.

Table 4 gives the frequency distribution of the number of pistils and stamens in 268 early flowers of the lesser celandine (*Ranunculus Ficaria*), which appeared between 27th February and 17th March, 1899.

Table 5 gives the frequency distribution of the number of pistils and stamens in 373 late flowers of the lesser celandine (*Ranunculus Ficaria*), which appeared between 17th and 23rd April, 1899.

The two sets of observations were made by Professor Julius Macleod and first published in *Botanisch Jaarboek*, Jaargang XI, 1899.

Table 6 gives the frequency distribution of the length (in cm) and longitudinal girth (in cm) of 955 eggs of the common tern (*Sterna fluviatilis*) laid at Blakeney Point during the laying season of 1920.

For each of the six bivariate frequency distributions given, calculate the product-moment correlation coefficient and the two estimated linear regressions. Represent graphically the array means and the regression lines.

---

* i.e. 12 old pence, or 5 new pence.

Table 1 for Exercise 80

Age (Midpoints)	Temperature (Midpoints)																	
	96·5	96·7	96·9	97·1	97·3	97·5	97·7	97·9	98·1	98·3	98·5	98·7	98·9	99·1	99·3	99·5	99·7	99·9
22					1		1		2		3	4	3	2				
25				1	1	3	6	5	6	9	11	8	5	4	3	1	1	
28			1		5	4	4	9	10	14	15	14	3	5			1	
31					2	5	4	8	15	15	23	17	7	7				
34					2	5	5	7	9	12	21	14	7	4	4	2	1	
37		1			3	4	6	9	10	13	12	14	7	5	1			
40							5	9	7	12	14	9	12	6	2	1		
43					2	2	6	7	8	12	8	9	5	2	1	1		
46							1	5	8	7	4	8	3		1			1
49						1	5	4	5	7	3	4	2	2				
52			1			1	5	4	2	5	1	4	3					
55					4	5	1	6	6	7	1	2	2	1	3	1		
58							5		5	5	4	4	1		1			
61		1		1	1		4	5	5	6	7	3	5	2				
64				1		3	2	7	8	12	6	9	3	1				
67	1					1	5	3	6	7	1	3						
70						3	2	2	5	4	4	4	4					
73					1		2	3	4	5		1	1		1	1		
76						2	1	2	2		2							
79							1	2	2									
82							1					1						
85							1	1		1								

Source M.H. Whiting (1915), *Biometrika*, Vol. 11, p. 1.

Table 2 for Exercise 80

Head length (Midpoints)

Reaction time (Midpoints)	6·805	6·905	7·005	7·105	7·205	7·305	7·405	7·505	7·605	7·705	7·805	7·905	8·005	8·105	8·205	8·305	8·405	8·505	8·605
5·995								1											
7·995						1		1											
9·995			1	2	1	5	5	10	1										2
11·995				1	1	1	5	9	10	12	12	7	6	4	1	2	1		
13·995			2	4	6	10	23	29	10	15	15	5	8	6	4	1		2	
15·995			4	2	14	37	67	115	30	49	39	27	19	9	7	6	1	2	
17·995		5	4	11	22	43	76	143	131	138	145	121	71	39	17	11	1		
19·995	2	1	5	13	37	58	122	148	191	201	177	146	83	52	22	9	3	1	
21·995		1	1	7	11	25	41	53	176	196	215	138	81	63	31	5		1	
23·995			1		7	10	24	22	76	56	78	49	28	25	8	1			
25·995				3	3	5	13	13	58	45	42	24	19	11	5				
27·995			1		1		1	5	13	12	14	5	5	6	2				
29·995					1		1		5	2	4	2	2	2					
31·995									1	4	2	1	2	1					
33·995							1	1			1								
35·995											1								
37·995																			
39·995						1													
41·995																			
43·995										1									

Source   G.E. Harmon (1926), *Biometrika*, Vol. 18, p. 207.

Table 3 for Exercise 80

Maximum temperature	Minimum temperature													
	36–	38–	40–	42–	44–	46–	48–	50–	52–	54–	56–	58–	60–	62–
50–					1									
52–														
54–			1	3		1	3	2	3					
56–		1		3	4	2	4	3	4	1	1			
58–	1	1	4	2	7	10	11	11	13	2	4	3		
60–		2	3	10	11	13	24	22	35	8	8	3		1
62–		2	3	6	9	24	29	37	54	16	12	7		
64–		1	3	10	16	25	43	39	40	28	18	12	2	
66–			2	9	9	16	28	38	35	37	33	13	2	
68–	2		1	6	16	19	20	33	14	35	20	14	4	2
70–	1	2	1	3	9	9	10	15	19	32	25	10	5	1
72–			3	3	6	10	9	14	13	19	7	7	3	1
74–					4	7	12	6	7	11	8	2	3	
76–				2	5	3	8	4	9	8	3	5	3	
78–					2	3	5	7	5	9	3	2	1	
80–					1	2		3	4	8	3	2		
82–				1			2	1	1	1	2			
84–						1			1		1		1	
86–											1		1	
88–								1		2	1	1		
90–											1		1	
92–														

Source R.A. Fisher and T.N. Hoblyn (1928), *Geografiska Annaler*, Vol. 3, p. 267.

Table 4 for Exercise 80

Number of pistils	Number of stamens																				
	18	19	20	21	22	23	24	25	26	27	28	29	30	31	32	33	34	35	36	37	38
2																				1	
3																					
4																					
5																					
6																					
7												1									
8																					
9												1									
10		1						1													
11				2				1													
12			3	1	2	1	1	3	1	1											
13		1	1	1	4	1	1	1			1	1									
14		4	3		1	2	4	1	2	3	1	1									
15			1	2	4	3	7	4	4	5	2	1		1	1						
16	1			2			1	5	3	5	4	5	3	2							
17			'		2	2	1	4	2	3	1	2	5	2	1						
18				1		2	4	3	1	7	1	3	2	1		2					
19						1	2		5	4	4	1	1	1	1	1					
20				2			1	1	2	2		1	3	4	2	1					
21				1					2	2			2	4	1	1					
22								1	2	3		1	2	3	1	2					
23									1	1		2	1	1	2			2			
24									1	1		2									
25														1	2						1
26													1		2						
27								1		1								2			
28																	1				
29																					
30																					
31												1									

Table 5 for Exercise 80

Number of stamens

Number of pistils	8	9	10	11	12	13	14	15	16	17	18	19	20	21	22	23	24	25	26	27	28	29
5				1																		
6			1	1	4	1		3														
7				2		4	2	4														
8	1					3	10	14														
9			1		1	2	3	7														
10			2			1	3	4	2								1					
11				1			2	5	10	1	3	1	1									
12				1			2	4	6	2	7	1	1									
13							1		8	5	13	4	5	3								
14								1	4	6	6	7	6	2	2	1				1		
15								1	4	3	5	7	6	4	3		1					
16										7	13	9	12	1	5	1	1		1			
17									1	2	8	7	2	3	1	1		2				
18										2	5	4	5	4	2	2		2	3			1
19										2	3		2	2	2	1	1	1	1	1	1	
20										1	3	3	1		1							
21													1			1						
22												1			1							
23																			1			
24																		1				

Source W.F.R. Weldon (1901), *Biometrika*, Vol. 1, p. 125.

Table 6 for Exercise 80

Longitudinal girth	Length									
	3·55–	3·60–	3·65–	3·70–	3·75–	3·80–	3·85–	3·90–	3·95–	4·00–
9·80–	1									
9·90–										
10·00–										
10·10–										
10·20–										
10·30–				1				1		
10·40–					1		1			
10·50–					2	2	1			
10·60–					1	6	2	4	2	
10·70–						2	4		3	1
10·80–					1	3	3	8	9	8
10·90–					1	1	1	11	7	11
11·00–				1			1	2	20	20
11·10–							1	3	6	20
11·20–							1	1	7	6
11·30–										
11·40–										2
11·50–										
11·60–									1	
11·70–										
11·80–							1			
11·90–										1
12·00–										
12·10–										
12·20–										
12·30–										

Source   "A Third Co-operative Study" (1923), *Biometrika*, Vol. 15, p. 294.

Table 6 for Exercise 80 (concluded)

Length												
4·05—	4·10—	4·15—	4·20—	4·25—	4·30—	4·35—	4·40—	4·45—	4·50—	4·55—	4·60—	4·65—
1												
2	1											
5	1											
23	10	1	1									
20	25	8	3	1	2	1						
21	37	29	11	5			1					
7	38	25	23	13	5	1	1	1				
3	10	25	29	21	9	2						
1	5	11	20	27	22	9	1	2				
	1	2	8	15	30	12	5	1	1			
2			2	9	25	16	10	2				
					9	10	15	11	1	1		
			1		4	2	5	4	6	6		
						1	2	4	5	3	1	
									1	1	1	3
									1	1		1
												1

81   Karl Pearson (*Philosophical Magazine*, 1900) quotes the following results of Weldon's classic experiment in which he tossed twelve dice 26 306 times and each time observed the number of dice which showed a 5 or a 6 on the uppermost face.

Number of dice with 5 or 6 uppermost	Frequency
0	185
1	1 149
2	3 265
3	5 475
4	6 114
5	5 194
6	3 067
7	1 331
8	403
9	105
10	14
11	4
12	0
Total	26 306

Assuming that the dice were unbiased, use the appropriate binomial expansion to calculate the expected frequency in each of the thirteen classes of Weldon's data. Test the agreement between the observed and expected frequencies.

82   The accompanying table of 707 digits from 0 to 9 is claimed to be a random sample from an infinite population in which the ten digits are all equally likely.

Make a frequency distribution of the digits in the sample and use it to test the hypothesis that the sample is truly random.

Further, assuming the randomness of the sample, use approximate large-sample procedures for testing the hypotheses that

(i) the proportion of 5's and 7's are the same in the population; and

(ii) the pooled proportion of 5's and 7's in the sample does not differ significantly from that obtained on the hypothesis that the digits have equal relative frequencies in the population.

Explain the contradictions, if any, in the conclusions obtained from the above three tests of significance.

83   H. Cramér (*Mathematical Methods of Statistics*, Princeton: 1946; p. 436) quotes some data obtained by N.G. Holmberg on the number of red blood-corpuscles in the 169 cells of a haemocytometer (see page 210).

Table for Exercise 82

(i) Test in two different ways whether these observations may reasonably be supposed to follow a Poisson distribution, stating the conclusion drawn from each of these tests. If the two tests lead to contradictory conclusions, state which one is to be preferred, and why.

(ii) Assuming that the observed distribution can be represented by a Poisson model with mean μ, obtain a large-sample 95 per cent confidence interval for μ.

(iii) Calculate a large-sample 95 per cent confidence interval for μ which does not depend upon the Poisson assumption.

Table for Exercise 83

Number of corpuscles	Frequency	Number of corpuscles	Frequency
0	0	11	15
1	0	12	21
2	0	13	18
3	0	14	17
4	1	15	16
5	3	16	9
6	5	17	6
7	8	18	3
8	13	19	2
9	14	20	2
10	15	21	1
Total			169

Reproduced by kind permission of the author, the publishers, Princeton University Press, New Jersey, and Almqvist and Wiksell—Gebers, Stockholm.

84   The following data relate to the sentence-length (in words) of a sample from *Regimen in Acute Diseases*, one of the books of the *Hippocratic Corpus* of the Greek physician Hippocrates (c. 460–377 B.C.). The logarithmic distribution is grouped with class-intervals of 0·10, and sentences ten words long falling on the dividing line between the class-intervals 0·90–1·00 and 1·00–1·10 (midpoints 0·95 and 1·05) are apportioned equally between the two class-intervals.

Sentence-length	Mean log length	Frequency	Sentence-length	Mean log length	Frequency
2	0·35	1	16–19	1·25	48
3	0·45	4	20–25	1·35	56
4	0·55	6	26–31	1·45	24
5	0·65	12	32–39	1·55	12
6	0·75	15	40–50	1·65	11
7	0·85	17	51–63	1·75	2
8–10	0·95	41	64–79	1·85	2
10–12	1·05	50	80–100	1·95	2
13–15	1·15	52			
Total					355

Source   W.C. Wake (1957), *Journal of the Royal Statistical Society, Series A*, Vol. 120, p. 331.

Draw a histogram of the logarithms of sentence-length and fit a normal distribution to this histogram. Test for the goodness of fit. Comment upon these data and your results.

85 The religion and sex distribution of the children (aged 0–15 years) who were in- and out-patients at the Bethlem Royal and Maudsley Hospitals during 1952–4 was as follows:

Religion	Boys	Girls	Total
Church of England (CE)	553	295	848
Roman Catholic (RC)	86	39	125
Nonconformist (NC)	26	21	47
Jewish (J)	16	7	23
Others (O)	19	9	28
Total	700	371	1071

Source *The Bethlem Royal Hospital and the Maudsley Hospital Triennial Statistical Report 1952–4.*

(i) Test these data for any differences in religion between the boys and girls.

(ii) The Catholic year book gives the percentage of Roman Catholics in England for the same period as 7 per cent. Test whether the proportion of Catholics among these patients differs from that in the general population, and comment.

86 The following table gives the numbers of patients treated for various diseases of the digestive system in the Falkirk Royal Infirmary during the years 1893–1914, classified by sex and occupation.

Sex	Occupation	Disease				
		Appendicitis	Hernia	Peptic ulcer	Other diseases	Total
Female	Women "gainfully employed"	51	13	28	12	104
	Other women	49	56	17	34	156
Male	Iron-moulders	23	45	2	12	82
	Other metal-workers	19	26	5	6	56
	Transport and communication workers	13	29	2	6	50
	Coal miners	29	8	2	7	46
Total		184	177	56	77	494

Analyse these data to test

(i) whether there is any difference in the disease distribution between "gainfully employed" women and other women;

(ii) whether there is any difference in the disease distribution between the different male occupation groups; and

(iii) whether there is, on the average, any sex difference in the disease distribution.

(iv) Obtain an approximate 95 per cent confidence interval for the difference in proportions of appendicitis patients between males and females, and hence, or otherwise, conclude whether these proportions differ significantly.

87 For the Lancashire cotton industry, the nineteenth century was a period of steady growth, which was made possible by technological advances and the increasing imports of raw cotton from both India and Egypt. The Balfour Committee on the Survey of Textile Industries gave the following table showing the number of persons employed in the cotton factories in the United Kingdom during the years 1839–98.

Year	Males	Females	Total
1839	112 941	146 395	259 336
1847	134 091	182 236	316 327
1850	141 501	189 423	330 924
1856	157 186	222 027	379 213
1861	182 556	269 013	451 569
1870	178 397	271 690	450 087
1874	187 626	291 895	479 521
1878	185 472	297 431	482 903
1885	196 378	307 691	504 069
1890	208 187	320 608	528 795
1895	205 230	333 653	538 883
1898	197 701	328 406	526 107

Source   W. Clare-Lees (1933–34), *Transactions of the Manchester Statistical Society*.

During this period the total number of persons employed in the cotton factories more than doubled, and this growth was reflected in the increase of both male and female workers. Analyse these data to test whether the sex proportion of the workers remained the same during the years considered. If so, evaluate a large-sample 95 per cent confidence interval for the overall true proportion of male workers employed in the cotton factories Alternatively, assuming that the proportion of male workers declined linearly with time, determine the best-fitting trend line to the observed proportions (calculated correct to four decimal places) of male workers.

Comment on the results obtained.

88 S.D. Wicksell (*Kungl. Fysiogr. Sällsk. Handl.*, 1926, Vol. 37) gives the distribution according to age of mothers (in years) of the sex of 928 570 children born in Norway during the years 1871 to 1900 (see table on opposite page).

(i) Analyse these data to test separately for each age-group that there is no difference in the sex ratio of the births.

(ii) Irrespective of the age of the mothers, test in two different ways that there is no overall difference in the proportions of boys and girls born.

Age of mother	Boys	Girls	Total
Below 20	5 496	5 365	10 861
20 –	67 987	63 743	131 730
25 –	122 119	116 194	238 313
30 –	120 077	112 979	233 056
35 –	96 353	91 392	187 745
40 –	53 855	50 354	104 209
45 –	11 646	11 010	22 656
Total	477 533	451 037	928 570

(iii) Also, test the heterogeneity of the sex ratio in the age-groups as a whole.

(iv) Determine an approximate large-sample 95 per cent confidence interval for the overall true proportion of male births.

(v) Obtain an approximate large-sample 95 per cent confidence interval for the true difference in the proportion of boys born to mothers who were more than and less than 20 years of age respectively. Is it reasonable to conclude that this difference is zero?

89 H. Cramér in *Mathematical Methods of Statistics* (Princeton University Press: 1946; p. 447) gives the following sex distribution of children born in Sweden in 1935.

Month	Boys	Girls	Total	Month	Boys	Girls	Total
1	3743	3537	7280	7	3 964	3 621	7 585
2	3550	3407	6957	8	3 797	3 596	7 393
3	4017	3866	7883	9	3 712	3 491	7 203
4	4173	3711	7884	10	3 512	3 391	6 903
5	4117	3775	7892	11	3 392	3 160	6 552
6	3944	3665	7609	12	3 761	3 371	7 132
Total					45 682	42 591	88 273

Analyse these data to test that

(i) there is no difference in the sex proportion of births in each of the twelve months of the year;

(ii) there is no difference in the overall sex proportion of births in the year; and

(iii) the monthly figures consistently show the same sex proportion.

Also, determine the large-sample 95 per cent confidence interval for the true proportion of male births in the population, and comment on the results of your analysis.

90 The table overleaf gives the distribution of the number of entries and the number of passes in different subjects in the A-level examinations held during the summer of 1965. The examinations were conducted by the following eight Examining Boards:

Associated Examining Board for the General Certificate of Education.
Northern Universities' Joint Matriculation Board.
Oxford Delegacy of Local Examinations.
Oxford and Cambridge Schools Examination Board.
Southern Universities' Joint Board for School Examinations.
University of Cambridge Local Examinations Syndicate.
University of London School Examinations Council.
Welsh Joint Education Committee.

The figures do not include overseas candidates who sat for the G.C.E. examinations specifically and separately organised by the Cambridge and London Boards for overseas candidates. But candidates who took G.C.E. examinations at overseas centres for the home certificate awarded by these and other Boards are included in the table, as are candidates from Scotland, Northern Ireland and elsewhere in the United Kingdom.

Subject	Number of entries	Number of passes
Mathematics	58 811	41 894
Physics	43 396	29 948
Chemistry	32 017	21 810
Biological subjects	26 314	17 802
Other science and technical	5 570	3 836
Total	166 108	115 290

Source   Statistics of Education, 1965 (Department of Education and Science, 1967).

(i) Analyse these data to test that there is no difference in the success rates in the different subjects.

(ii) Also, test the null hypothesis that the success rate in mathematics is no different from that of the other science subjects taken as a whole.

(iii) Determine 95 per cent large-sample confidence intervals for the true success rate in mathematics, and for the difference between the success rates in mathematics and all the other subjects considered jointly.

(iv) Discuss briefly any reasons which may, in your opinion, invalidate the above statistical analysis.

91   In the autumn of 1932, some breeding experiments were carried out at the U.S. Animal Husbandry Experiment Farm, Beltsville, Md., for a study of colour mutation in the Rhode Island Red fowl. The following data pertain to the number of chicks of different colour which resulted from three different types of parents, down colour being a reliable index of adult plumage colour.

Type of parents	Number of chicks		
	Coloured down	White down	Total
A	236	58	294
B	159	59	218
C	35	9	44
Total	430	126	556

Source J.P. Quinn (1934), *Journal of Genetics*, Vol. 29, p. 75.

(i) Test whether there is any significant difference between the relative frequencies of chicks with white down produced by the different types of parents.

(ii) If there is no significant difference between chicks of the different types of parents, obtain an approximate 95 per cent confidence interval for the total proportion of chicks having white down.

(iii) A genetic hypothesis predicts that this proportion should be $\frac{1}{4}$. Use the confidence interval to test this hypothesis.

(iv) Use a suitable χ^2 test to test the hypothesis in (iii).

92 The following extract is from a paper describing the pedigree over five generations of a piebald family. The family was domiciled in the Birmingham area, and there was a tradition that the first piebald in the family was a Frenchman who settled in England about 100 years ago. The family was in the subgroup of piebalds with a white frontal blaze and dorsal surface pigmented. The characteristic is a genetic abnormality and, according to the laws of Mendelian inheritance, it is expected that half the family would be affected by the characteristic and half unaffected.

"The total number of members in the five generations is 38, of which 21 were affected and 17 were not. It must, however, be taken into account that six of the latter could not, according to Mendelian laws, have exhibited the abnormality, and so if one deducts also the first two piebalds about whose sibs nothing is known, this gives a proportion of 19 piebalds to 11 who did not exhibit this character, which is close enough to the expected ratio of 1 : 1. The departure of 19 from the expected value 15 has a standard error of 2·74, and therefore cannot be considered significant.

"Among 30 individuals we find:
  (a) 11 affected and 2 unaffected males, and
  (b) 8 affected and 9 unaffected females.

"There is thus a considerable preponderance of affected persons among the males. Since 11 differs from the expected value 6·5 by 4·5, and the appropriate standard error is 1·80, the difference may therefore be significant. There is thus a considerable preponderance of piebalds among the males."

Source   A.M. Nussey (1938), *Biometrika*, Vol. 30, p. 65.

(i) Explain carefully what has been done by the author, repeating any calculations he has made in his approximate significance tests and commenting on his methods and any conclusions he has drawn. The comments should be supported by the necessary calculations and tests.

(ii) Use exact formulae to show that the males in the family differ significantly from expectation as regards this characteristic at the 2·5 per cent level of significance.

93 The following table gives the age distribution of the male populations of Birmingham and Liverpool in 1911. Although age is a metrical character, the ten age-groups in the table are distinct, and a year is the unit of measurement used.

Age-group	Birmingham		Liverpool	
	Population	Deaths	Population	Deaths
0 –	32 552	2003	45 889	3117
5 –	58 653	161	78 518	326
15 –	48 431	162	62 751	309
25 –	47 212	252	58 216	476
35 –	37 897	382	47 711	632
45 –	25 431	454	32 664	775
55 –	15 384	575	20 198	944
65 –	7 535	511	10 215	904
75 –	1 944	321	2 194	335
85 –	173	58	191	62
Total	275 212	4879	358 547	7880

Source   K. Pearson and J.F. Tocher (1915–17), *Biometrika*, Vol. 11, p. 159.

(i) Analyse these data, separately for each age-group, to test whether there is any difference in the mortality rates of the two cities. Hence evaluate a heterogeneity χ^2 to test whether the age-groups show the same pattern of differential mortality in the two cities.

(ii) Estimate the true difference in the overall mortality in the two cities and evaluate, correct to six significant figures, the standard error of the estimate. Hence calculate a large-sample 95 per cent confidence interval for the true difference in overall mortality in Liverpool and Birmingham per 1000 males.

94 In a statistical study of the effects of vaccination on the incidence of smallpox, the following data on the mortality due to the disease were obtained from several areas in the United Kingdom during different periods in the latter half of the nineteenth century.
   (i) Gloucester (1895–6);
   (ii) Leicester (1892–3);
   (iii) Sheffield (1887–8);
   (iv) City of Glasgow Smallpox Hospital, Belvedere (1892–5);

(v) Homerton Hospital (1873–84) and Fulham Hospital (1880–5), London; and

(vi) Cases reported in *The Times* for London in 1901.

(i) Gloucester

	Recoveries	Deaths	Total
Vaccinated	1091	120	1211
Unvaccinated	454	314	768
Total	1545	434	1979

(ii) Leicester

	Recoveries	Deaths	Total
Vaccinated	197	2	199
Unvaccinated	139	19	158
Total	336	21	357

(iii) Sheffield

	Recoveries	Deaths	Total
Vaccinated	3951	200	4151
Unvaccinated	278	274	552
Total	4229	474	4703

(iv) Glasgow

	Recoveries	Deaths	Total
Vaccinated	622	21	643
Unvaccinated	31	26	57
Total	653	47	700

(v) London hospitals

	Recoveries	Deaths	Total
Vaccinated	8207	692	8 899
Unvaccinated	1424	1103	2 527
Total	9631	1795	11 426

(vi) London (1901)

	Recoveries	Deaths	Total
Vaccinated	847	153	1000
Unvaccinated	126	158	284
Total	973	311	1284

Source   W.R. Macdonell (1901–2), *Biometrika*, Vol. 1, p. 375.

Analyse these data, separately for each table, to test the null hypothesis that there is no association between vaccination and mortality due to smallpox. Hence evaluate a heterogeneity χ^2 to test that the results obtained from the six tables are in agreement. Also, determine a large-sample 95 per cent confidence interval for the overall true difference between the percentage mortality amongst vaccinated and unvaccinated persons attacked by smallpox.

95   Two independent binomial populations give rise to observations in the two distinct classes A and a but with probabilities $1 - \theta, \theta$ and $1 - \theta^2, \theta^2$ respectively, where θ is an unknown parameter such that $0 < \theta < 1$. Random samples of sizes n_1 and n_2 respectively are taken from the populations, and the observed frequencies in the A and a classes are:

	A	a	
1st sample	x_{11}	x_{12}	$x_{11} + x_{12} = n_1$
2nd sample	x_{21}	x_{22}	$x_{21} + x_{22} = n_2$

Derive the equation for $\hat{\theta}$, the joint maximum-likelihood estimate of θ obtained from the two samples. Hence indicate how the heterogeneity χ^2, that is the goodness-of-fit χ^2, which tests jointly the agreement between the observed and estimated frequencies in the two samples, can be evaluated.

A genetical hypothesis of some interest postulates that $\theta = 0\cdot5$. Show that the χ^2 which tests jointly the deviations of the observations from their expectation based on the null hypothesis $H(\theta = 0\cdot5)$ is

$$\chi_m^2 = \frac{[(x_{12} - x_{11}) + 2(3x_{22} - x_{21})/3]^2}{n_1 + 4n_2/3} \quad \text{with 1 d.f.}$$

In two independent breeding experiments with some varieties of nasturtium (*Tropaeolum majus*), the observed frequencies were:

$$x_{11} = 81, \ x_{12} = 104; \quad x_{21} = 75, \ x_{22} = 28.$$

Source   A.A. Moffett (1936), *Journal of Genetics*, Vol. 33, p. 151.

Evaluate, correct to six decimal places, the maximum-likelihood estimate of θ and test the significance of the heterogeneity χ^2. Also, test the null hypothesis $H(\theta = 0\cdot5)$.

96   Three independent experiments were conducted on some wheat crosses with respect to the character called "fired" which causes the leaf-blades to die prematurely, changing their colour soon after flowering through a greenish grey to brownish red and finally to the ordinary straw colour. But if the change of colour happens to begin somewhat late, or if the ripening sets in early, so that the normal straw colour-change is taking place at the same time, it is difficult to make a clear distinction between fired and normal plants.

The table below gives the observed frequencies obtained in the three crosses investigated. There is no error in the classification of the fired plants, but there is a possibility of misclassification with normal plants.

Experiment no.	Cross	Observed frequency		Total
		Fired	Normal	
I	Shepherd × Cadia	57	88	145
II	Shepherd × Cleveland	60	108	168
III	Shepherd × Federation	44	80	124
Total	—	161	276	437
IV	Test families	5	55	60

Source   H.F. Smith (1937), *Annals of Eugenics*, Vol. 8, p. 94.

Accordingly, 60 normal-looking plants were tested in a fourth breeding experiment and it was observed that 55 of these were true normals but the remaining 5 were fired.

If there is no misclassification, then in the first three experiments the fired and normal plants should be in the expected ratio of $27 : 37$. But because of misclassification, the expected proportions of fired and normal plants may be assumed to be $27\lambda/64$ and $1 - (27\lambda/64)$ respectively, where λ is an unknown parameter in the range $0 < \lambda < 1$. Also, the expected proportions of fired and normal plants in the fourth experiment are

$$1 - \frac{37}{64 - 27\lambda} \quad \text{and} \quad \frac{37}{64 - 27\lambda} \quad \text{respectively.}$$

(i) Determine the maximum-likelihood equation for $\hat{\lambda}$, the estimate of λ, and hence evaluate $\hat{\lambda}$ numerically correct to six decimal places.

(ii) Calculate, correct to four decimal places, the expected frequencies in the first three experiments, and then test whether the data of these experiments are in agreement with the null hypothesis that the true proportions of fired and normal plants are $27\lambda/64$ and $1 - (27\lambda/64)$.

(iii) Finally, by pooling the observations of Experiments I to III, test whether these data and those obtained from the test families are in agreement with the null hypothesis that the true proportions of fired and normal plants are $27\lambda/64$ and $1 - (27\lambda/64)$.

97   If a_1, a_2, a_3, a_4 ($\Sigma\, a_i \equiv n$) are the observed frequencies in the four mutually exclusive classes of a multinomial distribution, and the corresponding expected proportions are $\frac{1}{4}(2 + \theta)$, $\frac{1}{4}(1 - \theta)$, $\frac{1}{4}(1 - \theta)$, $\frac{1}{4}\theta$, where $0 < \theta < 1$ is a parameter, determine the variance of the linear function

$$X = \theta(1 - \theta)a_1 - \theta(2 + \theta)(a_2 + a_3) + (2 + \theta)(1 - \theta)a_4.$$

Hence verify that

$$\frac{X^2}{\text{var}(X)} = \frac{2[\theta(1 - \theta)a_1 - \theta(2 + \theta)(a_2 + a_3) + (2 + \theta)(1 - \theta)a_4]^2}{n\theta(1 - \theta)(2 + \theta)(1 + 2\theta)}.$$

W. Bateson, in *Mendel's Principles of Heredity* (1909), gives the following observed frequencies obtained in two separate experiments with the sweet pea, the two characters considered being long *v.* round pollen grains, and purple *v.* red petals.

Experiment no.	Purple, long	Purple, round	Red, long	Red, round	Total
I	296	19	27	85	427
II	583	26	24	170	803

Source   K. Mather (1935), *Annals of Eugenics*, Vol. 6, p. 399.

According to genetical considerations, the true proportions in the four classes of each experiment should be the same as the expected proportions of the multinomial distribution given above.

(i) Irrespective of the value of θ, test separately whether the observed frequencies in each experiment of the two characters long v. round and purple v. red are in the ratio 3 : 1.

(ii) Determine, correct to six decimal places and separately for each experiment, the maximum-likelihood estimates of the parameter θ, and evaluate the estimates of the large-sample variances of these estimates. Hence use a large-sample approximation to test the null hypothesis that the true values of θ are the same for the populations sampled in the two experiments.

(iii) Also, after pooling the two classes having the same expectation in each experiment, evaluate the χ^2's to test the goodness of fit.

(iv) Evaluate, correct to six decimal places, the joint maximum-likelihood estimate of θ on the assumption that the true value of the parameter is the same for the two experimental populations sampled. Use this estimate to evaluate χ^2 for testing the equality of the parameter values for the two populations.

(v) Use the maximum-likelihood estimate obtained in (iv) to evaluate a goodness-of-fit χ^2 by pooling in each experiment the classes having the same expectation, and then test the significance of the heterogeneity χ^2.

98   The following data are adapted from experimental results obtained in three separate studies carried out by E.J. Collins, A.P. Lunden and I. Jørstad, and R.N. Salaman and J.W. Lesley, respectively, on the problem of immunity to wart disease [*Synchytrium endobioticum*, (Schilb.) Perc.] in the potato. Different immune and susceptible varieties of potato were used in the experimental crosses, and those considered here are classified in the five classes:

Class I: Immune, selfed;   Class II: Immune × Immune;
Class III: Susceptible, selfed;   Class IV: Immune × Susceptible;
Class V: Susceptible × Immune.

It is believed that the susceptible varieties of potato fall into two groups, namely, those that produce only susceptible seedlings on selfing, and those that produce susceptible and resistant seedlings on selfing. The figures quoted here in Class III pertain to crosses with the latter type of susceptible varieties.

The frequencies given in the table opposite are of seedlings obtained from the crosses considered in the five classes.

Class	Resistant (R)	Susceptible (S)
I	331	94
II	291	88
III	36	41
IV	146	148
V	130	133

Source   E.J. Collins (1935), *Annals of Botany*, Vol. 49, p. 480.

Analyse these data to test —

(i) whether the frequencies in Classes I and II are in agreement with the hypothesis that resistant and susceptible seedlings are in the ratio 3 : 1;

(ii) whether the frequencies in Classes III, IV, and V are in agreement with the hypothesis that resistant and susceptible seedlings are in the ratio 1 : 1.

Also, test whether the data are jointly in agreement with the expected proportions.

99   A multinomial distribution has k distinct classes, and in a random sample of N observations from the distribution the observed frequencies in the k classes are $a_1, a_2, \ldots, a_k$ respectively. The corresponding expected frequencies are $m_1, m_2, \ldots, m_k$ ($\Sigma\, a_i = \Sigma\, m_i = N$). Assuming that the m_i's are all functions of an unknown parameter p, prove that the equation for $\hat{p}$, the maximum-likelihood estimate of p, is

$$\sum_{i=1}^{k} \left[\frac{a_i}{m_i} \times \frac{dm_i}{dp} \right]_{p=\hat{p}} = 0.$$

Also, show that the linear function of the a_i's

$$X = \sum_{i=1}^{k} \frac{a_i}{m_i} \times \frac{dm_i}{dp}$$

has zero expectation and that

$$\text{var}(X) = \sum_{i=1}^{k} \frac{1}{m_i} \left(\frac{dm_i}{dp} \right)^2.$$

The table overleaf gives the observed frequencies and the expected proportions of the four distinct types of progenies obtained in six independent experiments with different varieties of *Papaver rhoeas*, where p is an unknown parameter such that $0 < p < 1$.

Show that the equation for $\hat{p}$, the pooled maximum-likelihood estimate of p, is

$$3012 - 19\,434\hat{p} + 4930\hat{p}^2 + 3613\hat{p}^3 - 12\,974\hat{p}^4 + 12\,160\hat{p}^5 - 7496\hat{p}^6 +$$
$$+ 2365\hat{p}^7 = 0.$$

Hence use a suitable iterative method to determine $\hat{p}$ correct to five decimal places.

Experiment no.		Class A_1	Class A_2	Class A_3	Class A_4	Total
I	Observed frequency	191	36	37	203	467
	Expected proportion	$\frac{1}{2}(1-p)$	$\frac{1}{2}p$	$\frac{1}{2}p$	$\frac{1}{2}(1-p)$	1
II	Observed frequency	37	205	203	44	489
	Expected proportion	$\frac{1}{2}p$	$\frac{1}{2}(1-p)$	$\frac{1}{2}(1-p)$	$\frac{1}{2}p$	1
III	Observed frequency	298	54	34	136	522
	Expected proportion	$\frac{1}{4}(3-2p+p^2)$	$\frac{1}{4}p(2-p)$	$\frac{1}{4}p(2-p)$	$\frac{1}{4}(1-p)^2$	1
IV	Observed frequency	51	29	19	1	100
	Expected proportion	$\frac{1}{4}(2+p^2)$	$\frac{1}{4}(1-p^2)$	$\frac{1}{4}(1-p^2)$	$\frac{1}{4}p^2$	1
V	Observed frequency	9	8	2	10	29
	Expected proportion	$\frac{1}{4}(2-p)$	$\frac{1}{4}(1+p)$	$\frac{1}{4}p$	$\frac{1}{4}(1-p)$	1
VI	Observed frequency	48	20	63	5	136
	Expected proportion	$\frac{1}{4}(1+p)$	$\frac{1}{4}(1-p)$	$\frac{1}{4}(2-p)$	$\frac{1}{4}p$	1

Source   J. Philp (1937), *Journal of Genetics,* Vol. 28, p. 175.

By using $\hat{p}$ for p, evaluate the ratio $X^2/\mathrm{var}(X)$ separately for each of the above six experiments. Hence calculate the pooled goodness-of-fit χ^2 to test the agreement between the observed and expected frequencies in the six experiments.

Note   It may be assumed that for large samples $X^2/\mathrm{var}(X)$ is distributed as χ^2 with 1 d.f.

100   In a breeding experiment with certain varieties of *Antirrhinum majus*, three sister plants were crossed with the same ivory individual and the following frequencies of yellow and ivory plants were obtained in the three families:

Experiment I	Yellow	Ivory	Total
Family 1	33	12	45
Family 2	38	22	60
Family 3	27	15	42
Total	98	49	147

In two further independent experiments four yellow plants each from families 1 and 2 were self-pollinated and gave the following progeny:

Experiment II	Yellow	Ivory	Total
Family 1a	27	2	29
Family 1b	15	12	27
Family 1c	24	6	30
Family 1d	25	5	30
Total	91	25	116

Experiment III	Yellow	Ivory	Total
Family 2a	44	16	60
Family 2b	21	7	28
Family 2c	23	7	30
Family 2d	22	6	28
Total	110	36	146

Source   K. Mather (1937), *Annals of Eugenics*, Vol. 8, p. 96.

According to the simplest Mendelian hypothesis, the yellow and ivory plants in the families of Experiment I should be in the ratio 1 : 1 and those of Experiments II and III in the ratio 3 : 1.

Analyse these data to answer the following questions:

(i) Are the yellow and ivory totals in Experiment I in the ratio 1 : 1?

(ii) Are the yellow and ivory totals obtained by pooling the observations of Experiments II and III in the ratio 3 : 1?

(iii) Are the totals considered in (i) and (ii) jointly in agreement with their respective Mendelian expectations?

(iv) If the hypothesis in (iii) is rejected, it may be assumed that the true ratios are $1 - \theta : \theta$ and $1 - \theta^2 : \theta^2$ instead of 1 : 1 and 3 : 1 respectively. Estimate θ by the method of maximum likelihood, and by using this estimated value test the significance of the heterogeneity χ^2 obtained form the totals considered in (i) and (ii).

Analyse the data of the three experiments separately to answer in each case the following questions:

(v) Are the yellow and ivory totals in agreement with the Mendelian expectation?

(vi) Are the individual families consistently showing the same proportion of yellow and ivory plants?

ANSWERS AND HINTS ON SOLUTIONS
Chapter 3

1 (i) Suppose n is the required sample size. Then if w is the random variable denoting the number of persons in the sample in favour of the airport, we have

$$P(w \leqslant 0.62n) = \sum_{w=0}^{0.62n} \binom{n}{w}(0.64)^w(0.36)^{n-w} = 0.002.$$

Hence, using the normal approximation, we have

$$P(w \leqslant 0.62n) \sim \Phi\left[\frac{0.62n - 0.64n + 0.5}{\sqrt{(n \times 0.64 \times 0.36)}}\right] = 0.002,$$

so that $\quad (-0.02n + 0.5)/\sqrt{(0.2304n)} = -2.878,$

or $\quad (n - 25)^2 = 4770.94n.$

Hence, approximately, n satisfies the quadratic equation

$$n^2 - 4821n + 625 = 0.$$

The required root gives $n = 4821$.

(ii) In this case, we have

$$P(w \geqslant 0.60n) = 1 - P(w < 0.60n)$$

$$\sim 1 - \Phi\left[\frac{0.60n - 0.59n - 0.5}{\sqrt{(n \times 0.59 \times 0.41)}}\right] = 0.001,$$

so that $\quad (0.01n - 0.5)/\sqrt{(0.2419n)} = 3.09$

or $\quad (n - 50)^2 = 23\,096.85n.$

Therefore, approximately,

$$n^2 - 23\,197n + 2500 = 0,$$

and the required root gives $n = 23\,197$.

2 Suppose x is a random variable denoting the lifetime of bulbs produced by the new process. Then x is normally distributed with

$$E(x) = 2375 \quad \text{and} \quad \text{s.d.}(x) = 250.$$

If the sample size is n and $\bar{x}$ denotes the average lifetime of the bulbs sampled, then $\bar{x}$ is also a normal random variable such that

$$E(\bar{x}) = 2375 \quad \text{and} \quad \text{s.d.}(\bar{x}) = 250/\sqrt{n}.$$

The engineer will fail to adopt the new process if $\bar{x} \leqslant 2300$. Hence n is to be so determined that

$$P(\bar{x} \leqslant 2300) = \Phi\left[\frac{2300 - 2375}{250/\sqrt{n}}\right] = 0\cdot01.$$

Therefore $\qquad\qquad -(75\sqrt{n})/250 = -2\cdot326,$

whence $n = 60\cdot11$ or 61 correct to the nearest integer containing n.
    If s.d. $(x) = 320$, then the equation for n is

$$-(75\sqrt{n})/320 = -2\cdot326,$$

whence $n = 98\cdot49$ or 99 correct to the nearest integer containing n.

3   Let p be the true proportion and W the total number of adults in the city who are in favour of giving financial support to the theatre.
    The sampling fraction is $8500/750\,000 = 17/1500$.
    The maximum-likelihood estimate of p is

$$\hat{p} = 6893/8500 = 0\cdot8109.$$

The unbiased estimate of $\text{var}(\hat{p})$ is

$$\frac{0\cdot8109 \times 0\cdot1891(1 - 17/1500)}{8499} = 10^{-4} \times 0\cdot178\,378,$$

whence $\qquad\qquad \text{s.e.}(\hat{p}) = 10^{-2} \times 0\cdot4223.$

    The maximum-likelihood estimate of W is

$$\hat{W} = (6893 \times 750\,000)/8500 = 608\,206$$

and $\qquad\qquad \text{s.e.}(\hat{W}) = 750\,000 \times \text{s.e.}(\hat{p}) = 3167\cdot25.$

    If the population is "effectively" infinite, then the unbiased estimate of $\text{var}(\hat{p})$ is

$$(0\cdot8109 \times 0\cdot1891)/8499 = 10^{-4} \times 0\cdot181\,490,$$

whence $\qquad\qquad \text{s.e.}(\hat{p}) = 10^{-2} \times 0\cdot4260.$

4   Suppose p_1 and p_2 are the true proportions of motorists and non-motorists respectively who are in favour of the tests. Then the maximum-likelihood estimate of p_1 is

$$\hat{p}_1 = 6092/11\,850 = 0\cdot5141$$

and that of p_2 is $\quad \hat{p}_2 = 14\,968/17\,532 = 0\cdot8538.$
The unbiased estimate of $\text{var}(\hat{p}_1)$ is
$$(0\cdot5141 \times 0\cdot4859)/11\,849 = 10^{-4} \times 0\cdot210\,820,$$

whence   s.e. $(\hat{p}_1) = 10^{-2} \times 0.4592$.

Similarly, the unbiased estimate of var $(\hat{p}_2)$ is

$$(0.8538 \times 0.1462)/17\ 531 = 10^{-5} \times 0.712\ 028,$$

whence   s.e. $(\hat{p}_2) = 10^{-2} \times 0.2668$.

The best estimate of u is

$$\hat{u} = (11\ 850\hat{p}_1 + 17\ 532\hat{p}_2)/29\ 382 = 0.7168,$$

and that of v is $\hat{v} = \frac{1}{2}(\hat{p}_1 + \hat{p}_2) = 0.6840$.

Also, $\text{var}(\hat{u}) = \dfrac{1}{(29\ 382)^2} [(11\ 850)^2 \text{var}(\hat{p}_1) + (17\ 532)^2 \text{var}(\hat{p}_2)],$

whence the unbiased estimate of var $(\hat{u})$ is

$$\frac{1}{(29\ 382)^2}[(11\ 850)^2 \times 10^{-4} \times 0.210\ 820 + (17\ 532)^2 \times 10^{-5} \times 0.712\ 028]$$

$$= 10^{-5} \times 0.596\ 426.$$

Therefore   s.e. $(\hat{u}) = 10^{-2} \times 0.2442$.

Again, $\text{var}(\hat{v}) = \frac{1}{4}[\text{var}(\hat{p}_1) + \text{var}(\hat{p}_2)],$

and the unbiased estimate of var $(\hat{v})$ is

$$\tfrac{1}{4}[10^{-4} \times 0.210\ 820 + 10^{-5} \times 0.712\ 028] = 10^{-5} \times 0.705\ 057\ 5.$$

Therefore   s.e. $(\hat{v}) = 10^{-2} \times 0.2655$.

5   The probability of obtaining the observed sample is by the multinomial distribution

$$\frac{4164!}{2972!\ 171!\ 190!\ 831!} [\tfrac{1}{4}(2 + \theta)]^{2972} [\tfrac{1}{4}(1 - \theta)]^{361} [\tfrac{1}{4}\theta]^{831}.$$

Therefore the log likelihood is

$$\log L = \text{constant} + 2972 \log(2 + \theta) + 361 \log(1 - \theta) + 831 \log \theta,$$

whence the equation for the maximum-likelihood estimate $\hat{\theta}$ is

$$\frac{2972}{2 + \hat{\theta}} - \frac{361}{1 - \hat{\theta}} + \frac{831}{\hat{\theta}} = 0$$

or $1662 + 1419\hat{\theta} - 4164\hat{\theta}^2 = 0$.

The appropriate root of this equation gives $\hat{\theta} = 0.824\ 734$. Again,

$$\frac{d^2 \log L}{d\theta^2} = -\frac{2972}{(2 + \theta)^2} - \frac{361}{(1 - \theta)^2} - \frac{831}{\theta^2},$$

so that $\qquad -E\left[\dfrac{d^2 \log L}{d\theta^2}\right] = \dfrac{n}{4}\left[\dfrac{1}{2 + \theta} + \dfrac{2}{1 - \theta} + \dfrac{1}{\theta}\right]$, where $n = 4164$,

$$= \frac{n(1 + 2\theta)}{2\theta(1 - \theta)(2 + \theta)}.$$

Therefore, for large samples,

$$\operatorname{var}(\hat{\theta}) = \frac{2\theta(1 - \theta)(2 + \theta)}{n(1 + 2\theta)},$$

and a consistent estimate of this variance is

$$\frac{2\hat{\theta}(1 - \hat{\theta})(2 + \hat{\theta})}{n(1 + 2\hat{\theta})} = 10^{-4} \times 0 \cdot 740\ 201.$$

Therefore $\qquad$ s.e. $(\hat{\theta}) = 10^{-2} \times 0 \cdot 8603$.

The expected frequencies are

$$\frac{n}{4}(2 + \hat{\theta}) = 2940 \cdot 55; \qquad \frac{n}{4}(1 - \hat{\theta}) = 182 \cdot 45;$$

$$\frac{n}{4}(1 - \hat{\theta}) = 182 \cdot 45; \qquad \frac{n}{4}\hat{\theta} = 858 \cdot 55.$$

6 We have

$$p(x) = \frac{1}{1 + e^{-(a + x)}} \qquad \text{and} \qquad 1 - p(x) = \frac{e^{-(a + x)}}{1 + e^{-(a + x)}}.$$

Let r_1, r_2 and r_3 be the number of insects killed at strengths corresponding to $x = -1, 0, 1$ respectively, so that $r_1 + r_2 + r_3 = r$. Then, since the three experiments are independent, the joint probability of obtaining the observed results is

$$L = \binom{n}{r_1}\left[\frac{1}{1 + e^{-(a - 1)}}\right]^{r_1}\left[\frac{e^{-(a - 1)}}{1 + e^{-(a - 1)}}\right]^{n - r_1} \times \binom{n}{r_2}\left[\frac{1}{1 + e^{-a}}\right]^{r_2}\left[\frac{e^{-a}}{1 + e^{-a}}\right]^{n - r_2}$$

$$\times \binom{n}{r_3}\left[\frac{1}{1 + e^{-(a + 1)}}\right]^{r_3}\left[\frac{e^{-(a + 1)}}{1 + e^{-(a + 1)}}\right]^{n - r_3}.$$

Therefore the log likelihood is

$$\log L = \text{constant} - a(3n - r) - n\log[1 + e^{-(a - 1)}] - n\log[1 + e^{-a}] - n\log[1 + e^{-(a + 1)}],$$

so that $\qquad \dfrac{d \log L}{da} = -(3n - r) + \dfrac{n\ e^{-(a - 1)}}{1 + e^{-(a - 1)}} + \dfrac{n\ e^{-a}}{1 + e^{-a}} + \dfrac{n\ e^{-(a + 1)}}{1 + e^{-(a + 1)}},$

whence the equation for $\hat{a}$ is

$$-\frac{3n - r}{n} + \frac{1}{1 + e^{\hat{a} - 1}} + \frac{1}{1 + e^{\hat{a}}} + \frac{1}{1 + e^{\hat{a} + 1}} = 0.$$

Hence the equation for $y = e^{-\hat{\alpha}}$ is found to be

$$y^3 + (1 - \phi)(e + 1 + e^{-1})y^2 + (1 - 2\phi)(e + 1 + e^{-1})y + 1 - 3\phi = 0.$$

For the numerical solution, set $\lambda \equiv e + 1 + e^{-1}$, so that $\lambda = 4 \cdot 086\ 161$ correct to six decimal places. Then for $\phi = 11/6$, the equation for y may be written as

$$f(y) \equiv 6y^3 - \lambda y(5y + 16) - 27 = 0.$$

Since $f(y) < 0$ for $0 \leqslant y \leqslant 5$ and $f(y) > 0$ for $y = 6$, the required positive root lies between 5 and 6. An appropriate difference table of the values of $f(y)$ is as follows:

y	$f(y)$	Δ	Δ^2	Δ^3
5·45	− 18·887 475			
		12·523 456		
5·50	− 6·364 019		392 845	
		12·916 301		4502
5·55	6·552 282		397 347	
		13·313 648		4498
5·60	19·865 930		401 845	
		13·715 493		
5·65	33·581 423			

If $y_0 = a + \theta h$ is the required root, then $a = 5 \cdot 50$ and $h = 0 \cdot 05$. Inverse interpolation then gives $\theta = 0 \cdot 4965$ correct to four decimal places. Therefore

$$y_0 = 5 \cdot 5248 \quad \text{and} \quad \hat{\alpha} = -\log_e y_0 = -1 \cdot 7092.$$

7   Leicester, 1968.

The joint likelihood of the three independent experiments is

$$L = \frac{1395!}{397!\ 297!\ 289!\ 412!} [\tfrac{1}{2}(1 - p)]^{397} [\tfrac{1}{2}p]^{297} [\tfrac{1}{2}p]^{289} [\tfrac{1}{2}(1 - p)]^{412}$$

$$\times \frac{414!}{78!\ 136!\ 120!\ 80!} [\tfrac{1}{2}p]^{78} [\tfrac{1}{2}(1 - p)]^{136} [\tfrac{1}{2}(1 - p)]^{120} [\tfrac{1}{2}p]^{80}$$

$$\times \frac{1267!}{461!\ 161!\ 515!\ 130!} [\tfrac{1}{4}(1 + p)]^{461} [\tfrac{1}{4}(1 - p)]^{161} [\tfrac{1}{4}(2 - p)]^{515} [\tfrac{1}{4}p]^{130}.$$

Therefore

$$\log L = \text{constant} + 1226 \log (1 - p) + 874 \log p + 461 \log (1 + p) + {} + 515 \log (2 - p).$$

Hence the equation for $\hat{p}$ is

$$-\frac{1266}{1 - \hat{p}} + \frac{874}{\hat{p}} + \frac{461}{1 + \hat{p}} - \frac{515}{2 - \hat{p}} = 0,$$

which reduces to

$$1748 - 2919\hat{p} - 4357\hat{p}^2 + 3076\hat{p}^3 = 0,$$

so that $\qquad \hat{p} = (1748 - 4357\hat{p}^2 + 3076\hat{p}^3)/2919$

Iteration now gives $\hat{p} = 0\cdot4162$ correct to four decimal places.

8   The joint probability of the three independent samples is

$$L = \binom{147}{98}\left(\frac{2-\theta}{3}\right)^{98}\left(\frac{1+\theta}{3}\right) \times \binom{29}{27}\left(\frac{3-2\theta+\theta^2}{3}\right)\left(\frac{2\theta-\theta^2}{3}\right)^2$$

$$\times \binom{27}{15}\left(\frac{2+\theta^2}{3}\right)^{15}\left(\frac{1-\theta^2}{3}\right)^{12}.$$

Therefore the joint log likelihood is

$$\log L = \text{constant} + 98\log(2-\theta) + 49\log(1+\theta) + 27\log(3-2\theta+\theta^2) +$$
$$+ 2\log(2\theta-\theta^2) + 15\log(2+\theta^2) + 12\log(1-\theta^2),$$

so that

$$\frac{d\log L}{d\theta} = -\frac{98}{2-\theta} + \frac{49}{1+\theta} + \frac{27(-2+2\theta)}{3-2\theta+\theta^2} + \frac{2(2-2\theta)}{\theta(2-\theta)} + \frac{30\theta}{2+\theta^2} - \frac{24\theta}{1-\theta^2},$$

whence the equation for the maximum-likelihood estimate $\hat{\theta}$ is

$$24 - 256\hat{\theta} - 654\hat{\theta} + 1608\hat{\theta}^3 - 1905\hat{\theta}^4 + 1611\hat{\theta}^5 - 831\hat{\theta}^6 + 259\hat{\theta}^7 = 0.$$

For iteration, this equation may be rewritten as

$$\hat{\theta} = \frac{24 - \hat{\theta}^2(654 - 1608\hat{\theta} + 1905\hat{\theta}^2 - 1611\hat{\theta}^3 + 831\hat{\theta}^4 - 259\hat{\theta}^5)}{256}$$

$$= 0\cdot093\,750 - 2\cdot554\,687\,5\hat{\theta}^2(1 - 2\cdot458\,716\hat{\theta} + 2\cdot912\,844\hat{\theta}^2 -$$
$$- 2\cdot463\,303\hat{\theta}^3 + 1\cdot270\,642\hat{\theta}^4 -$$
$$- 0\cdot396\,024\hat{\theta}^5).$$

Hence iteration gives $\hat{\theta} = 0\cdot080\,254$.

In general, suppose n_1, n_2, n_3 are the sample sizes and p_1, p_2, p_3 the true probabilities of yellow plants in the three experiments. Then the goodness-of-fit χ^2 is

$$\sum_{i=1}^{3} (x_i - n_i\hat{p}_i)^2/n_i\hat{p}_i(1-\hat{p}_i) \quad \text{with 2 d.f.,}$$

where x_1, x_2, x_3 are the observed frequencies of yellow plants in the three experiments and $\hat{p}_1$, $\hat{p}_2$, $\hat{p}_3$ are estimates of the true probabilities which are functions of θ jointly estimated from the data. Hence

$$\hat{p}_1 = \frac{1}{3}(2-\hat{\theta}) = 0\cdot639\,915; \qquad \frac{(x_1 - n_1\hat{p}_1)^2}{n_1\hat{p}_1(1-\hat{p}_1)} = 0\cdot4566;$$

$$\hat{p}_2 = \frac{1}{3}(3 - 2\hat{\theta} + \hat{\theta}^2) = 0\text{·}948\ 644; \qquad \frac{(x_2 - n_2\hat{p}_2)^2}{n_2\hat{p}_2(1 - \hat{p}_2)} = 0\text{·}1846;$$

$$\hat{p}_3 = \frac{1}{3}(2 + \hat{\theta}^2) = 0\text{·}668\ 814; \qquad \frac{(x_3 - n_3\hat{p}_3)^2}{n_3\hat{p}_3(1 - \hat{p}_3)} = 1\text{·}5636.$$

Hence the goodness-of-fit $\chi^2 = 2\text{·}2048$ with 2 d.f. This observed value is obviously not significant and so the fit is good.

9 For the frequency distribution of the depth of alburnum,

$$\text{mean} = 2\text{·}914\ 088; \qquad \text{variance} = 0\text{·}645\ 112.$$

If the normal population has mean μ and variance σ^2, then the least-squares estimates are

$$\mu^* = 2\text{·}914\ 088 \quad \text{and} \quad \sigma^{2*} = 0\text{·}645\ 112.$$

Estimate var $(\mu^*) = \sigma^{2*}/1370 = 10^{-3} \times 0\text{·}470\ 885.$
Therefore $\qquad\qquad$ s.e. $(\mu^*) = 10^{-1} \times 0\text{·}2170.$

$$P(\mu^* - \mu > 0\text{·}045) = P\left[\frac{\mu^* - \mu}{\text{s.e.}(\mu^*)} > \frac{0\text{·}045}{0\text{·}021\ 70}\right]$$

$$= 1 - \Phi(2\text{·}074) = 0\text{·}0190.$$

10 For the weight distribution of the eggs which hatched

$$\text{mean} = 73\text{·}3428; \qquad \text{variance} = 43\text{·}217\ 973;$$

and for the weight distribution of the eggs which did not hatch

$$\text{mean} = 73\text{·}8881; \qquad \text{variance} = 59\text{·}639\ 571.$$

The above are the least-squares estimates of $\theta_1, \sigma_1^2, \theta_2, \sigma_2^2$ respectively.

(i) The best estimate of $\theta_1 - \theta_2$ is

$$\theta_1^* - \theta_2^* = 73\text{·}3428 - 73\text{·}8881 = -0\text{·}5453;$$

and $\qquad$ s.e. $(\theta_1^* - \theta_2^*) = \left[\dfrac{43\text{·}217\ 973}{1202} + \dfrac{59\text{·}639\ 571}{688}\right]^{\frac{1}{2}} = 0\text{·}3502.$

If $\sigma_1 = \sigma_2 = \sigma$, say, then the best estimate of σ^2 is

$$\sigma^{2*} = \frac{1201 \times 43\text{·}217\ 973 + 687 \times 59\text{·}639\ 571}{1888} = 49\text{·}193\ 417.$$

Therefore $\qquad\qquad \sigma^* = 7\text{·}013\ 802.$

Hence $\qquad$ s.e. $(\theta_1^* - \theta_2^*) = 7\text{·}013\ 802\left[\dfrac{1}{1202} + \dfrac{1}{688}\right]^{\frac{1}{2}} = 0\text{·}3353.$

(ii) If p is the true proportion of fertile eggs which hatch, then the estimate $p^* = 1202/1890 = 0 \cdot 635\ 979 = 0 \cdot 6360$ to 4 s.f.

The unbiased estimate of var (p^*)

$$= \frac{0 \cdot 635\ 979 \times 0 \cdot 364\ 021}{1889} = 10^{-3} \times 0 \cdot 122\ 557.$$

Therefore   s.e. $(p^*) = 10^{-1} \times 0 \cdot 110\ 705 = 10^{-1} \times 0 \cdot 1107$ to 4 s.f.

11   Leicester, 1961.
    The total number of sample points corresponding to the division of the 8 cups into two groups is

$$\binom{8}{4} = 70.$$

Therefore the probability of success is $1/70$, if the lady has no power of discrimination and she allocates the cups into the two groups randomly. Then in 10 independent experiments the probability of the lady achieving r successes is

$$P(r) = \binom{10}{r} \left(\frac{1}{70}\right)^r \left(\frac{69}{70}\right)^{10-r}, \quad 0 \leqslant r \leqslant 10.$$

Hence $P(r \geqslant 2) = 1 - P(r \leqslant 1) = 0 \cdot 008\ 509.$

This probability is less than one per cent. This means that if the lady has no ability to discriminate then a significant result has occurred. Therefore the lady establishes her claim at the assigned level of significance if she achieves two or more successes in the 10 experiments.
    On the other hand, in the second experiment the total number of sample points is

$$\binom{12}{6} = 924.$$

There is only one way of correctly dividing the 12 cups. In addition, the lady can achieve a success by choosing correctly 5 cups of each kind. This can be done in

$$\binom{6}{5}\binom{6}{5} = 36 \text{ ways.}$$

Therefore the probability of success is $37/924$ which is $> 1/70$. But now, in order to establish her claim, the lady has to achieve successes in 3 or more of the 10 independent experiments. The probability for this is

$$1 - \sum_{r=0}^{2} \binom{10}{r} \left(\frac{37}{924}\right)^r \left(\frac{887}{924}\right)^{10-r} = 0 \cdot 006\ 230.$$

This probability is also less than one per cent and so, if the lady achieves 3 or more successes in the 10 repetitions of the second experiment, then she again establishes her claim. But the second procedure is slightly more stringent than the first, although the probability of a single success in the second experiment is greater than in the first. This increase in the probability of an individual success is more than compensated for by the requirement that, in order to establish her claim, the lady has to achieve three or more successes in the second experiment instead of two or more successes as in the first experiment.

12   Suppose p is the probability that the new model irons have an improved safety performance. Then the null hypothesis is $H(p = 0·5)$ with the alternatives $p \neq 0·5$. Therefore, for a two-sided test, the required probability is

$$P(|w - 6| \geq 2) = P(w \geq 8) + P(w \leq 4),$$

where w is the random variable denoting the number of new model irons in the sample which gave an improved safety performance. Hence, by symmetry, we have

$$P(|w - 6| \geq 2) = 2 \sum_{r=0}^{4} \binom{12}{r} (\tfrac{1}{2})^{12} = \frac{794}{2048} = 0·3877.$$

This probability is large and so the null hypothesis is acceptable. Therefore the experiments do not provide sufficient evidence for the improved safety performance of the new model irons.

In the second case, the null hypothesis is again $H(p = 0·5)$ with the alternatives $p \neq 0·5$. Therefore the required probability is

$$P(|w - 60| \geq 11) = 2P(w \geq 71) = 2[1 - P(w < 70)]$$

$$\sim 2\left[1 - \Phi\left(\frac{70 - 60 + 0·5}{\sqrt{30}}\right)\right] = 0·055\ 24.$$

The observed result is now approaching significance and so there is some suspicion about the credibility of the null hypothesis $H(p = 0·5)$.

If it is believed that there is an improvement in the safety performance of the new model irons, then the null hypothesis is again $H(p = 0·5)$ but with the alternatives $p > 0·5$. Hence the required probabilities based on the two observed results are

$$P(w \geq 8) = 0·1938 \quad \text{and} \quad P(w \geq 71) \sim 0·027\ 62.$$

Therefore the first experiment gives a non-significant result and so the null hypothesis is still acceptable, whereas the second experiment gives reasonably definite evidence for the rejection of the null hypothesis.

13   If p is the true proportion of students in favour of the proposal, then the maximum-likelihood estimate of p is

$$\hat{p} = 384/472 = 0·8136,$$

and the unbiased estimate of var $(\hat{p})$ is
$$(0 \cdot 8136 \times 0 \cdot 1864)/471 = 10^{-3} \times 0 \cdot 321\ 985.$$

Therefore $\qquad$ s.e. $(\hat{p}) = 10^{-1} \times 0 \cdot 1794.$

(i) If it is believed that less than 75 per cent of the students are in favour of the proposal, then the null hypothesis is $H(p = 0 \cdot 75)$ with the alternatives $p < 0 \cdot 75$. Hence, if the null hypothesis is true, then

$E(\hat{p}) = 0 \cdot 75$ and var $(\hat{p}) = (0 \cdot 75 \times 0 \cdot 25)/472 = 10^{-3} \times 0 \cdot 397\ 246.$

Therefore $\qquad$ s.d. $(\hat{p}) = 10^{-1} \times 0 \cdot 1993.$

If the null hypothesis is true, then
$$z = \frac{\hat{p} - 0 \cdot 75}{\text{s.d.}\,(\hat{p})} = \frac{0 \cdot 0636}{0 \cdot 019\ 93} = 3 \cdot 191 \text{ is } \sim N(0,\ 1).$$

Hence to test the significance of the null hypothesis $H(p = 0 \cdot 75)$ with the alternatives $p < 0 \cdot 75$, the required probability is
$$P(z \leqslant 3 \cdot 191) \sim \Phi(3 \cdot 191) = 0 \cdot 999\ 29.$$

This is a very large probability and therefore the null hypothesis $H(p = 0 \cdot 75)$ is acceptable against the alternatives $p < 0 \cdot 75$. Therefore the data do not support the belief that less than 75 per cent of the students are in favour of the proposal.

(ii) In the second case, the null hypothesis is, again, $H(p = 0 \cdot 75)$ but the alternatives are $p \neq 0 \cdot 75$. We thus use a two-sided test of significance, and the required probability is
$$P(|z| \geqslant 3 \cdot 191) \sim 2[1 - \Phi(3 \cdot 191)] = 0 \cdot 001\ 42.$$

This probability is less than one per cent and so we can now confidently reject the null hypothesis $H(p = 0 \cdot 75)$ as against the alternatives $p \neq 0 \cdot 75$.

(iii) If the university has 4872 students and the null hypothesis $H(p = 0 \cdot 75)$ is true, then
$$\text{var}\,(\hat{p}) = 10^{-3} \times 0 \cdot 397\ 246 \times \frac{(4872 - 472)}{4871} = 10^{-3} \times 0 \cdot 358\ 834.$$

Therefore $\qquad$ s.d. $(\hat{p}) = 10^{-1} \times 0 \cdot 1894.$

Hence to test the null hypothesis $H(p = 0 \cdot 75)$ with the alternatives $p < 0 \cdot 75$, the observed value of
$$z = 0 \cdot 0636/0 \cdot 018\ 94 = 3 \cdot 358.$$

Therefore $\qquad P(z \leqslant 3 \cdot 358) \sim \Phi(3 \cdot 358) = 0 \cdot 999\ 60.$

Thus the null hypothesis $H(p = 0 \cdot 75)$ may be confidently accepted against the alternatives $p < 0 \cdot 75$.

Similarly, to test the null hypothesis $H(p = 0.75)$ with the alternatives $p \neq 0.75$, the relevant probability is

$$P(|z| \geq 3.358) \sim 2[1 - \Phi(3.358)] = 0.000\ 80.$$

The decision to reject the null hypothesis is reinforced by the finiteness of the population of students sampled.

14 Let p_1 and p_2 respectively be the proportions of persons in the city in favour of the radio station before and after its establishment. Then the maximum-likelihood estimates of p_1 and p_2 are

$$\hat{p}_1 = \frac{7642}{15\ 892} = 0.480\ 871 \quad \text{and} \quad \hat{p}_2 = \frac{8935}{18\ 682} = 0.478\ 268.$$

Also, the unbiased estimate of $\text{var}(\hat{p}_1)$ is

$$\frac{0.480\ 871 \times 0.519\ 129}{15\ 891} = 10^{-4} \times 0.157\ 091,$$

and that of $\text{var}(\hat{p}_2)$ is

$$\frac{0.478\ 268 \times 0.521\ 732}{18\ 681} = 10^{-4} \times 0.133\ 573.$$

If the null hypothesis $H(p_1 = 0.5)$ is true, then

$$E(\hat{p}_1) = 0.5 \quad \text{and} \quad \text{var}(\hat{p}_1) = 0.25/15\ 892 = 10^{-4} \times 0.157\ 312.$$

Therefore $\qquad\qquad$ s.d. $(\hat{p}_1) = 10^{-2} \times 0.3966.$
Hence, if the null hypothesis $H(p_1 = 0.5)$ is true, we have

$$z = \frac{0.4809 - 0.5}{10^{-2} \times 0.3966} = -4.186 \text{ is } \sim N(0, 1).$$

But since the alternatives to the null hypothesis under test are $p_1 > 0.5$, therefore the relevant probability is

$$P(z \geq -4.816) \sim 1.$$

Therefore the null hypothesis $H(p_1 = 0.5)$ is almost certainly acceptable in preference to the alternatives $p_1 > 0.5$.

To test the null hypothesis $H(p_1 - p_2 = 0)$, we may first assume that if this hypothesis is true then $p_1 = p_2 = p$, say. Then the maximum-likelihood estimate of p is

$$\hat{p} = 16\ 577/34\ 574 = 0.479\ 464;$$

and the unbiased estimate of $\text{var}(\hat{p}_1 - \hat{p}_2)$ is

$$\frac{0.479\ 464 \times 0.520\ 536}{34\ 573} \times \frac{(34\ 574)^2}{15\ 892 \times 18\ 682} = 10^{-4} \times 0.290\ 648.$$

Therefore $\qquad\qquad$ s.e. $(\hat{p}_1 - \hat{p}_2) = 10^{-2} \times 0.539\ 117.$
Hence, if the null hypothesis $H(p_1 - p_2 = 0)$ is true, we have

$$z = 0.002\ 603/(10^{-2} \times 0.539\ 117) = 0.483 \text{ is } \sim N(0, 1).$$

Since the alternatives to the null hypothesis are $p_1 - p_2 \neq 0$, therefore the required probability for testing significance is

$$P(|z| \geqslant 0 \cdot 483) \sim 2[1 - \Phi(0 \cdot 483)] = 0 \cdot 629\ 10.$$

This is a large probability and so the null hypothesis is acceptable. In other words, the data are in agreement with the hypothesis that there has been no change in public opinion about the radio station.

In general, for $p_1 - p_2 \neq 0$, var $(\hat{p}_1 - \hat{p}_2) = $ var $(\hat{p}_1) + $ var $(\hat{p}_2)$, and an unbiased estimate of var $(\hat{p}_1 - \hat{p}_2)$ is

$$10^{-4} \times 0 \cdot 157\ 091 + 10^{-4} \times 0 \cdot 133\ 573 = 10^{-4} \times 0 \cdot 290\ 664.$$

Therefore   s.e. $(\hat{p}_1 - \hat{p}_2) = 10^{-2} \times 0 \cdot 539\ 133.$

(i) If the null hypothesis $H(p_1 - p_2 = 0 \cdot 05)$ is true, then

$$z = \frac{0 \cdot 002\ 603 - 0 \cdot 05}{10^{-2} \times 0 \cdot 539\ 133} = -8 \cdot 791 \text{ is } \sim N(0, 1).$$

Since the alternatives to this null hypothesis are $p_1 - p_2 < 0 \cdot 05$, therefore the required probability for testing significance is

$$P(z \leqslant -8 \cdot 791) \sim \Phi(-8 \cdot 791) \sim 0.$$

Hence the null hypothesis may almost certainly be rejected.

(ii) On the other hand, if the null hypothesis $H(p_1 - p_2 = 0 \cdot 07)$ is true, then

$$z = \frac{0 \cdot 002\ 603 - 0 \cdot 07}{10^{-2} \times 0 \cdot 539\ 133} = -12 \cdot 50 \text{ is } \sim N(0, 1).$$

The alternatives to the null hypothesis are $p_1 - p_2 > 0 \cdot 07$, and so the required probability for testing significance is

$$P(z \geqslant -12 \cdot 50) \sim 1.$$

Hence the null hypothesis is acceptable in preference to $p_1 - p_2 > 0 \cdot 07$.

15   Let p_1 and p_2 be the true proportions of girls of medium hair-colour in Edinburgh and Glasgow respectively. Then the maximum-likelihood estimates of p_1 and p_2 are

$$\hat{p}_1 = 4008/9743 = 0 \cdot 411\ 372 \quad \text{and} \quad \hat{p}_2 = 17\ 529/39\ 764 = 0 \cdot 440\ 826.$$

Hence the best estimate of $p_1 - p_2$ is $\hat{p}_1 - \hat{p}_2 = -0 \cdot 029\ 454.$
   Also, the unbiased estimate of var $(\hat{p}_1)$ is

$$\frac{0 \cdot 411\ 372 \times 0 \cdot 588\ 626}{9742} = 10^{-4} \times 0 \cdot 248\ 558,$$

and that of var $(\hat{p}_2)$ is

$$\frac{0 \cdot 440\ 826 \times 0 \cdot 559\ 174}{39\ 763} = 10^{-5} \times 0 \cdot 619\ 919.$$

Therefore the best estimate of var $(\hat{p}_1 - \hat{p}_2)$ is $10^{-4} \times 0 \cdot 310\ 550$ and

$$\text{s.e.} (\hat{p}_1 - \hat{p}_2) = 10^{-2} \times 0 \cdot 557\ 270.$$

If the null hypothesis $H(p_1 - p_2 = 0)$ is true, then we may assume that $p_1 = p_2 = p$, say. The best estimate of p is

$$\hat{p} = 21\ 537/49\ 507 = 0.435\ 029,$$

and the unbiased estimate of $\text{var}(\hat{p})$ is

$$\frac{0.435\ 029 \times 0.564\ 971}{49\ 506} = 10^{-5} \times 0.496\ 463.$$

Hence, if the null hypothesis is true, the unbiased estimate of $\text{var}(\hat{p}_1 - \hat{p}_2)$ is

$$10^{-5} \times 0.496\ 463 \times \frac{(49\ 507)^2}{9743 \times 39\ 764} = 10^{-4} \times 0.314\ 078.$$

Therefore   s.e. $(\hat{p}_1 - \hat{p}_2) = 10^{-2} \times 0.560\ 427.$

If the null hypothesis $H(p_1 - p_2 = 0)$ is true, then

$$z = - \frac{0.029\ 454}{10^{-2} \times 0.560\ 427} = -5.256 \text{ is } \sim N(0, 1).$$

Since the alternatives to the null hypothesis are $p_1 - p_2 \neq 0$, therefore the required probability for testing significance is

$$P(|z| \geqslant 5.256) \sim 2[1 - \Phi(5.256)] \sim 0.$$

Therefore the null hypothesis is rejected decisively.

If the true proportion of girls of medium hair-colour in the combined population of the two cities is p, then, as already calculated, the best estimate of p is $\hat{p} = 0.435\ 029$. Further, if the null hypothesis $H(p = 0.45)$ is true, then

$$\text{var}(\hat{p}) = (0.45 \times 0.55)/49\ 507 = 10^{-5} \times 0.499\ 929.$$

Therefore   s.d. $(\hat{p}) = 10^{-2} \times 0.223\ 591.$

Hence, if the above null hypothesis is true, we have

$$z = \frac{0.435\ 029 - 0.45}{10^{-2} \times 0.223\ 591} = -6.696 \text{ is } \sim N(0, 1).$$

(i) If the alternatives to the null hypothesis $H(p = 0.45)$ are $p \neq 0.45$, then the required probability for testing significance is

$$P(|z| \geqslant 6.696) \sim 2[1 - \Phi(6.696)] \sim 0.$$

Therefore the null hypothesis is rejected decisively.

(ii) If the alternatives to the null hypothesis $H(p = 0.45)$ are $p < 0.45$, then the required probability for testing significance is

$$P(z \leqslant -6.696) \sim 0.$$

Therefore the null hypothesis is again rejected decisively.

(iii)   If the alternatives to the null hypothesis $H(p = 0.45)$ are $p > 0.45$, then the required probability for testing significance is

$$P(z \geqslant -6.696) \sim 1.$$

Therefore the null hypothesis is now almost certainly acceptable.

16   Suppose p_1 and p_2 are the true death-rates of diphtheria patients receiving the antitoxin and ordinary treatments. Then the maximum-likelihood estimates of p_1 and p_2 are

$$\hat{p}_1 = 37/228 = 0 \cdot 162\ 281 \quad \text{and} \quad \hat{p}_2 = 28/337 = 10^{-1} \times 0 \cdot 830\ 861.$$

Therefore the best estimate of $p_1 - p_2$ is $\hat{p}_1 - \hat{p}_2 = 0 \cdot 079\ 195.$
The unbiased estimate of $\text{var}(\hat{p}_1)$ is

$$\frac{0 \cdot 162\ 281 \times 0 \cdot 837\ 719}{227} = 10^{-3} \times 0 \cdot 598\ 881$$

and that of $\text{var}(\hat{p}_2)$ is

$$\frac{10^{-1} \times 0 \cdot 830\ 861 \times 0 \cdot 916\ 913\ 9}{336} = 10^{-3} \times 0 \cdot 226\ 735.$$

Hence the unbiased estimate of $\text{var}(\hat{p}_1 - \hat{p}_2)$ is $10^{-3} \times 0 \cdot 825\ 616.$
Therefore $\text{s.e.}(\hat{p}_1 - \hat{p}_2) = 10^{-1} \times 0 \cdot 287\ 335.$
    If the null hypothesis $H(p_1 - p_2 = 0)$ is true, then we may assume that $p_1 = p_2 = p$, say. The best estimate of p is

$$\hat{p} = 65/565 = 0 \cdot 115\ 044,$$

and the unbiased estimate of $\text{var}(\hat{p})$ is

$$\frac{0 \cdot 115\ 044 \times 0 \cdot 884\ 956}{564} = 10^{-3} \times 0 \cdot 180\ 512.$$

Hence, if the above null hypothesis is true, the unbiased estimate of $\text{var}(\hat{p}_1 - \hat{p}_2)$ is

$$10^{-3} \times 0 \cdot 180\ 512 \times \frac{(565)^2}{228 \times 337} = 10^{-3} \times 0 \cdot 749\ 960.$$

Therefore $\text{s.e.}(\hat{p}_1 - \hat{p}_2) = 10^{-1} \times 0 \cdot 273\ 854.$
If the null hypothesis $H(p_1 - p_2 = 0)$ is true, then

$$z = 0 \cdot 079\ 195/(10^{-1} \times 0 \cdot 273\ 854) = 2 \cdot 892 \text{ is } \sim N(0, 1).$$

Since the alternatives to the null hypothesis $H(p_1 - p_2 = 0)$ are $p_1 - p_2 \neq 0$, therefore the required probability for testing significance is

$$P(|z| \geqslant 2 \cdot 892) \sim 2[1 - \Phi(2 \cdot 892)] = 0 \cdot 003\ 84.$$

The observed result is highly significant ($<$ one per cent) and so the null hypothesis may be confidently rejected.

    (i) The null hypothesis is $H(p_1 - p_2 = 0 \cdot 10)$ with the alternatives $p_1 - p_2 > 0 \cdot 10$. If this null hypothesis is true, then

$$z = \frac{0 \cdot 079\ 195 - 0 \cdot 10}{10^{-1} \times 0 \cdot 287\ 335} = -0 \cdot 724 \text{ is } \sim N(0, 1).$$

Hence the required probability for testing significance is

$$P(z \geqslant -0 \cdot 724) \sim 1 - \Phi(-0 \cdot 724) = 0 \cdot 765\ 46.$$

This probability is large and so the null hypothesis is acceptable. We conclude that the data do not give any evidence for believing that the difference between p_1 and p_2 is $> 0 \cdot 10$.

(ii) The null hypothesis is now $H(p_1 - p_2 = 0.07)$ with the alternatives $p_1 - p_2 < 0.07$. Hence if this null hypothesis is true, then

$$z = 0.009\ 195/(10^{-1} \times 0.287\ 335) = 0.320 \text{ is } N(0, 1).$$

Therefore the required probability for testing significance is

$$P(z \leqslant 0.320) \sim \Phi(0.320) = 0.625\ 52.$$

This probability is large and so the null hypothesis is acceptable. The data do not suggest that the difference between p_1 and p_2 is < 0.07.

17   Suppose p_1 and p_2 are the true proportions of heavy smokers in the populations of lung cancer and other patients respectively. Then the maximum-likelihood estimates of p_1 and p_2 are

$$\hat{p}_1 = 283/532 = 0.531\ 955 \quad \text{and} \quad \hat{p}_2 = 728/2791 = 0.260\ 838.$$

Therefore the best estimate of $p_1 - p_2$ is $\hat{p}_1 - \hat{p}_2 = 0.271\ 117$. The unbiased estimate of $\text{var}(\hat{p}_1)$ is

$$\frac{0.531\ 955 \times 0.468\ 045}{531} = 10^{-3} \times 0.468\ 887$$

and that of $\text{var}(\hat{p}_2)$ is

$$\frac{0.260\ 838 \times 0.739\ 162}{2790} = 10^{-4} \times 0.691\ 045.$$

Therefore the unbiased estimate of $\text{var}(\hat{p}_1 - \hat{p}_2)$ is $10^{-3} \times 0.537\ 992$ and $\text{s.e.}(\hat{p}_1 - \hat{p}_2) = 10^{-1} \times 0.231\ 947.$
    If the null hypothesis $H(p_1 - p_2 = 0)$ is true, then we may write $p_1 = p_2 = p$, say. The best estimate of p is

$$\hat{p} = 1011/3323 = 0.304\ 243$$

and the unbiased estimate of $\text{var}(\hat{p})$ is

$$\frac{0.304\ 243 \times 0.695\ 757}{3322} = 10^{-4} \times 0.637\ 204$$

and that of $\text{var}(\hat{p}_1 - \hat{p}_2)$ is

$$10^{-4} \times 0.637\ 204 \times (3323)^2/532 \times 2791 = 10^{-3} \times 0.473\ 879.$$

Therefore $\text{s.e.}(\hat{p}_1 - \hat{p}_2) = 10^{-1} \times 0.217\ 688.$

(i) To test the null hypothesis $H(p_1 - p_2 = 0)$ with the alternatives $p_1 - p_2 \neq 0$, we have

$$z = 0.271\ 117/(10^{-1} \times 0.217\ 688) = 12.454 \text{ is } \sim N(0, 1).$$

The required probability for testing significance is

$$P(|z| \geqslant 12.454) \sim 0.$$

Hence the null hypothesis can be rejected decisively. The proportions of heavy smokers in the populations of lung cancer and other patients are almost certainly not the same.

(ii) If the null hypothesis is $H(p_1 - p_2 = 0.25)$ with the alternatives $p_1 - p_2 > 0.25$, then

$$z = 0.021\ 117/(10^{-1} \times 0.231\ 947) = 0.910 \text{ is } \sim N(0, 1).$$

The required probability for testing significance is

$$P(z \geqslant 0.910) \sim 1 - \Phi(0.910) = 0.181\ 41.$$

This is a large probability and so the null hypothesis can be accepted. Therefore the data are consistent with the hypothesis that the proportion of heavy smokers in the population of lung cancer patients may be 25 per cent more than the proportion of heavy smokers in the population of other patients.

However, the results of the two tests of significance do not prove a causal relationship between heavy smoking and lung cancer. Thus, for example, it may well be that persons who are prone to lung cancer may also have a propensity for heavy smoking.

18 Since var $(p^*) = \dfrac{p(1 - p)}{n} \leqslant 1/4n$, therefore s.d. $(p^*) \leqslant 1/(2\sqrt{n})$.

Suppose p_1 and p_2 are the true proportions defective in the components produced by the two factories. If $\hat{p}_1$ and $\hat{p}_2$ are the maximum-likelihood estimates of the parameters based on the observed samples, then the best estimate of $p_1 - p_2$ is $\hat{p}_1 - \hat{p}_2 = 0.086$.

Also, $\text{var}(\hat{p}_1 - \hat{p}_2) = \dfrac{p_1(1 - p_1)}{450} + \dfrac{p_2(1 - p_2)}{680}$

$$\leqslant \tfrac{1}{4} \times 1130/(680 \times 450) = 10^{-3} \times 0.923\ 203.$$

Therefore   s.d. $(\hat{p}_1 - \hat{p}_2) \leqslant 10^{-1} \times 0.303\ 843.$

To test the null hypothesis $H(p_1 - p_2 = 0)$ with the alternatives $p_1 - p_2 \neq 0$, we have

$$z = \frac{\hat{p}_1 - \hat{p}_2}{\text{s.d.}\,(\hat{p}_1 - \hat{p}_2)} \geqslant \frac{\hat{p}_1 - \hat{p}_2}{\text{max s.d.}\,(\hat{p}_1 - \hat{p}_2)} = 2.830,$$

where z is $N(0, 1)$ if the null hypothesis under test is true. Hence the required probability for testing significance is

$$P\left[|z| \geqslant \frac{\hat{p}_1 - \hat{p}_2}{\text{s.d.}\,(\hat{p}_1 - \hat{p}_2)}\right] \leqslant P(|z| \geqslant 2.830) \sim 2[1 - \Phi(2.830)] = 0.004\ 66.$$

Thus, irrespective of the individual values of $\hat{p}_1$ and $\hat{p}_2$, we have

$$P\left[|z| \geqslant \frac{\hat{p}_1 - \hat{p}_2}{\text{s.d.}\,(\hat{p}_1 - \hat{p}_2)}\right] \leqslant 0.004\ 66.$$

This probability is small and the null hypothesis can be rejected at the one per cent level. The data do not support the hypothesis that the true proportions defective in the products of the two factories are the same.

19 By suffixing a zero in the fourth decimal place, the given proportions (p_i) may be regarded as four-digit decimal fractions. Form the cumulative distribution of these proportions. Use these cumulative proportions as class-limits and then, by regarding the given four-digit random numbers as four-digit decimal fractions, the sample frequency distribution is formed. This is given in the following table:

Age-group	Midpoint (x_i)	Proportion (p_i)	Cumulative proportion	Observed frequency
0 –	2·5	0·0650	0·0650	9
5 –	10	0·1310	0·1960	10
15 –	20	0·1480	0·3440	13
25 –	30	0·1610	0·5050	15
35 –	40	0·1520	0·6570	19
45 –	50	0·1310	0·7880	12
55 –	60	0·1090	0·8970	7
65 –	70	0·0710	0·9680	8
75 –	80	0·0280	0·9960	7
85 –	90	0·0040	1·0000	0
Total		1·0000		100

Sample average $\bar{x} = 37\cdot325$; sample standard deviation $s_x = 22\cdot4825$.

Suppose x is a random variable denoting age. Then, in the population,

$$E(x) = \sum p_i x_i = 36\cdot0025; \quad E(x^2) = \sum p_i x_i^2 = 1740\cdot206\ 25.$$

Therefore $E(x) = 36\cdot00$; s.d. $(x) = 21\cdot0719$; s.d. $(\bar{x}) = 2\cdot107\ 19$.

(i) To test that the sample average $(\bar{x})$ may have arisen from a sample from a population with mean $36\cdot00$, we have

$$z = (37\cdot235 - 36\cdot00)/2\cdot107\ 19 = 0\cdot629 \text{ is } N(0, 1).$$

Therefore, for a two-sided test, the required probability is

$$P(|z| \geqslant 0\cdot629) = 2[1 - \Phi(0\cdot629)] = 0\cdot529\ 36.$$

This probability is large and so the difference between the sample average and the population mean can be attributed to sampling error.

(ii) When var(x) is assumed unknown, then s.e. $(\bar{x}) = 2\cdot248\ 25$. Hence we now have

$$t = (37\cdot325 - 36\cdot00)/2\cdot248\ 25 = 0\cdot589 \text{ is } \sim N(0, 1).$$

Therefore, for a two-sided test, the required probability is

$$P(|t| \geqslant 0\cdot589) \sim 2[1 - \Phi(0\cdot589)] = 0\cdot555\ 88.$$

This probability is also large and so the difference between the sample average and the population mean may again be due to chance.

20   The frequency distribution of the 345 y observations is as follows:

Class-mark	Frequency	Class-mark	Frequency
0·00−	2	1·01−	44
0·11−	4	1·11−	27
0·21−	11	1·21−	30
0·31−	8	1·31−	21
0·41−	15	1·41−	18
0·51−	18	1·51−	12
0·61−	22	1·61−	9
0·71−	33	1·71−	6
0·81−	27	1·81−	3
0·91−	32	1·91−	3
Total			345

The midpoint of the first class-interval is 0·055.
The sample mean $\bar{y} = 0·9901$ and sample standard deviation $s_y = 0·3899$.
   From the theoretical distribution of y we have

$$E(y^r) = \int_0^1 y^{r+1} dy + \int_1^2 y^r (2 - y)\, dy \qquad (r \geqslant 0)$$

$$= \frac{1}{r+2} + \frac{2^{r+2} - (r+3)}{(r+1)(r+2)}.$$

Hence $E(y) = 1$; $E(y^2) = 7/6$; $\mathrm{var}(y) = 1/6$; s.d. $(y) = 1/\sqrt{6} = 0·4082$.

21   The frequency distribution of the 500 y observations is as follows:

Class-mark	Frequency	Class-mark	Frequency
0·00−	125	2·51−	4
0·26−	80	2·76−	13
0·51−	74	3·01−	4
0·76−	47	3·26−	0
1·01−	40	3·51−	4
1·26−	31	3·76−	4
1·51−	24	4·01−	0
1·76−	17	4·26−	0
2·01−	17	4·51−	7
2·26−	9	4·76−	0
Total			500

The midpoint of the first class-interval is 0·13.
The sample mean $\bar{y} = 0·9485$ and sample standard deviation $s_y = 0·9276$.
   The theoretical distribution of y has the probability density function
e^{-y}, for $y \geqslant 0$. Therefore, for any $r > -1$, we have

$$E(y^r) = \int_0^\infty e^{-y} y^r \, dy = \Gamma(r + 1).$$

Hence $E(y) = 1$; $E(y^2) = 2$; $\mathrm{var}(y) = $ s.d. $(y) = 1$.

22   The frequency distribution of the 420 y observations is as follows:

Class-mark	Frequency	Class-mark	Frequency
0·00 –	25	0·54 –	26
0·06 –	22	0·60 –	25
0·12 –	27	0·66 –	28
0·18 –	20	0·72 –	25
0·24 –	27	0·78 –	32
0·30 –	20	0·84 –	27
0·36 –	25	0·90 –	18
0·42 –	25	0·96 –	23
0·48 –	25		
Total			420

The midpoint of the first class-interval is 0·03.

The sample mean $\bar{y}$ = 0·5147 and sample standard deviation s_y = 0·2900.

The theoretical distribution of y is uniform in the (0, 1) interval.

Therefore $E(y)$ = 0·5; $\text{var}(y)$ = 1/12; s.d.(y) = 0·288 675.

Hence   s.d.$(\bar{y})$ = $0·288\ 675/\sqrt{420}$ = 0·014 09,

and z = $(0·5147 - 0·5)/0·014\ 09$ = 1·043 is $\sim N(0, 1)$.

Therefore, for a two-sided test, the required probability is

$$P(|z| \geqslant 1·043) \sim 2[1 - \Phi(1·043)] = 0·296\ 96.$$

This probability is large and so it is reasonable to infer that the difference between $\bar{y}$ and $E(y)$ may be due to sampling error.

23   The frequency distribution of the 504 observations of the sample ranges is as follows:

Class-mark	Frequency	Class-mark	Frequency
0·44 –	1	0·72 –	46
0·48 –	2	0·76 –	55
0·52 –	8	0·80 –	60
0·56 –	15	0·84 –	71
0·60 –	12	0·88 –	83
0·64 –	23	0·92 –	63
0·68 –	24	0·96 –	41
Total			504

The midpoint of the first class-interval is 0·455.

The sample mean $\bar{x}$ = 0·8188 and sample standard deviation s_x = 0·1112.

From the general distribution of the range obtained from samples of size n, we have

$$E(x^r) = n(n - 1) \int_0^1 x^{n+r-2}(1 - x)\,dx$$

$$= n(n - 1)B(n + r - 1, 2), \quad \text{for } n + r - 1 > 0.$$

Therefore $E(x) = n(n - 1)B(n, 2) = (n - 1)/(n + 1)$

and $E(x^2) = n(n - 1)B(n + 1, 2) = \dfrac{n(n - 1)}{(n + 1)(n + 2)}$,

whence $\text{var}(x) = \dfrac{2(n - 1)}{(n + 1)^2(n + 2)}$.

In particular, for $n = 10$,

$\quad E(x) = 9/11 = 0 \cdot 8182; \quad \text{var}(x) = 3/242 = 10^{-1} \times 0 \cdot 123\ 967.$

Therefore $\text{s.d.}(x) = 0 \cdot 111\ 340$

and $\text{s.d.}(\bar{x}) = 0 \cdot 111\ 340/\sqrt{504} = 10^{-2} \times 0 \cdot 4959.$

24   Suppose the ith plant has n_i seeds of which r_i are round and the rest angular. Then, assuming that the ten plants provide independent samples, the joint likelihood of the observations is

$$L = \prod_{i=1}^{10} \binom{n_i}{r_i} p^{r_i}(1 - p)^{n_i - r_i}.$$

Hence the equation for the maximum-likelihood estimate $\hat{p}$ is

$$\sum_{i=1}^{10} [r_i(1 - \hat{p}) - (n_i - r_i)\hat{p}] = 0,$$

so that $\hat{p} = \displaystyle\sum_{i=1}^{10} r_i \bigg/ \sum_{i=1}^{10} n_i.$

Also, since the r_i are independent binomial variables, we have

$$\text{var}(\hat{p}) = p(1 - p) \bigg/ \sum_{i=1}^{10} n_i.$$

For the given numerical values of n_i and r_i, we have

$$\hat{p} = 336/437 = 0 \cdot 768\ 879.$$

The unbiased estimate of $\text{var}(\hat{p})$ is

$$(0 \cdot 768\ 879 \times 0 \cdot 231\ 121)/436 = 10^{-3} \times 0 \cdot 407\ 518,$$

so that $\text{s.e.}(\hat{p}) = 10^{-1} \times 0 \cdot 201\ 886.$

The large-sample 95 per cent confidence interval for p is

$$\hat{p} \pm 1 \cdot 96 \times \text{s.e.}(\hat{p}),$$

which gives the limits $0 \cdot 8084$, $0 \cdot 7293$.

Since the point $0 \cdot 75$ lies within the above confidence interval, it follows that the null hypothesis $H(p = 0 \cdot 75)$ is acceptable at the 5 per cent level. This test is two-sided and the alternatives to the null hypothesis are $p \neq 0 \cdot 75$.

Alternatively, if the null hypothesis $H(p = 0 \cdot 75)$ is true, then

$$\text{var}(\hat{p}) = (0 \cdot 75 \times 0 \cdot 25)/437 = 10^{-3} \times 0 \cdot 429\ 062,$$

so that $\qquad$ s.d. $(\hat{p}) = 10^{-1} \times 0\cdot207\ 138$.

Hence, if the null hypothesis under test is true, we have

$$z = (\hat{p} - 0\cdot75)/\text{s.d.}\,(\hat{p}) = 0\cdot911 \text{ is } \sim N(0, 1).$$

Since the alternatives to the null hypothesis are $p \neq 0\cdot75$, the required probability for testing significance is

$$P(|z| \geqslant 0\cdot911) \sim 2[1 - \Phi(0\cdot911)] = 0\cdot362\ 30.$$

This probability is large and so the null hypothesis can be accepted.

25 The probability distribution based on the outcomes of the experiment is as follows:

$$P(H^3) = p^3; \quad P(H^2T) = 3p^2(1 - p); \quad P(HT^2) = 3p(1 - p)^2; \quad P(T^3) = (1 - p)^3.$$

In general, if the observed frequencies in the four classes are n_1, n_2, n_3, n_4 ($\Sigma n_i = N$) respectively, then the sample likelihood is

$$L = \frac{N!}{\prod\limits_{i=1}^{4} n!} [p^3]^{n_1}[3p^2(1 - p)]^{n_2}[3p(1 - p)^2]^{n_3}[(1 - p)^3]^{n_4},$$

so that

$$\log L = \text{constant} + (3n_1 + 2n_2 + n_3)\log p + (n_2 + 2n_3 + 3n_4)\log(1 - p).$$

Hence the equation of the maximum-likelihood estimate $\hat{p}$ is

$$(3n_1 + 2n_2 + n_3)(1 - \hat{p}) - (n_2 + 2n_3 + 3n_4)\hat{p} = 0,$$

whence $\qquad$ $\hat{p} = (3n_1 + 2n_2 + n_3)/3N$.

Again, $\qquad$ $\dfrac{d^2\log L}{dp^2} = -\dfrac{3n_1 + 2n_2 + n_3}{p^2} - \dfrac{n_2 + 2n_3 + 3n_4}{(1 - p)^2}$

whence $\qquad -E\left[\dfrac{d^2\log L}{dp^2}\right] = \dfrac{N}{p^2}[3p^3 + 6p^2(1 - p) + 3p(1 - p^2)] +$

$$+ \dfrac{N}{(1 - p)^2}[3p^2(1 - p) + 6p(1 - p)^2 + 3(1 - p)^3]$$

$$= 3N/\{p(1 - p)\}, \text{ on reduction.}$$

Therefore $\qquad$ $\text{var}(\hat{p}) = p(1 - p)/3N$.

This is otherwise obvious since $\hat{p}$ is the observed proportion of heads in $3N$ independent tosses of the penny.

With the numerical values of n_i and r_i given,

$$\hat{p} = 424/1011 = 0\cdot419\ 387.$$

The unbiased estimate of $\text{var}(\hat{p})$ is

$$\frac{0\cdot419\ 387 \times 0\cdot580\ 613}{1010} = 10^{-3} \times 0\cdot241\ 091,$$

and $\qquad$ $\text{s.e.}\,(\hat{p}) = 10^{-1} \times 0\cdot155\ 271.$

The two per cent point of the unit normal distribution is $2 \cdot 326$. Therefore the required confidence interval for p is

$$\hat{p} \pm 2 \cdot 326 \times \text{s.e.} (\hat{p}), \quad \text{i.e.} \quad 0 \cdot 3833, \ 0 \cdot 4555.$$

Since the point $0 \cdot 5$ is not included in the confidence interval, therefore the null hypothesis is not tenable at the two per cent level of significance.

26 Assuming that the two experiments are independent, the joint sample likelihood is

$$L = \binom{N_1}{n_1} \left[\frac{1}{64}(27 - \theta) \right]^{n_1} \left[\frac{1}{64}(37 + \theta) \right]^{N_1 - n_1} \times \binom{N_2}{n_2} \left[\frac{\theta}{37 + \theta} \right]^{n_2} \left[\frac{37}{37 + \theta} \right]^{N_2 - n_2}$$

Therefore

$$\log L = \text{constant} + n_1 \log(27 - \theta) + (N_1 - N_2 - n_1) \log(37 + \theta) + n_2 \log \theta$$

$$\frac{d \log L}{d\theta} = - \frac{n_1}{27 - \theta} + \frac{N_1 - N_2 - n_1}{37 + \theta} + \frac{n_2}{\theta}$$

and

$$\frac{d^2 \log L}{d\theta^2} = - \frac{n_1}{(27 - \theta)^2} - \frac{N_1 - N_2 - n_1}{(37 + \theta)^2} - \frac{n_2}{\theta^2}.$$

Hence the equation for the maximum-likelihood estimate $\hat{\theta}$ is

$$- n_1 \hat{\theta}(37 + \hat{\theta}) + (N_1 - N_2 - n_1)\hat{\theta}(27 - \hat{\theta}) + n_2(27 - \hat{\theta})(37 + \hat{\theta}) = 0,$$

which can be rewritten in the stated form. Again,

$$\frac{1}{\text{var}(\hat{\theta})} = - E\left[\frac{d^2 \log L}{d\theta^2} \right] = \frac{N_1}{64(27 - \theta)} + \frac{N_1 - N_2}{(37 + \theta)^2} - \frac{N_1(27 - \theta)}{64(37 + \theta)^2} + \frac{N_2}{\theta(37 + \theta)}$$

$$= \frac{N_1 \theta(37 + \theta) + 37 N_2(27 - \theta)}{\theta(27 - \theta)(37 + \theta)^2}, \quad \text{on reduction.}$$

For the numerical values given, the equation for $\hat{\theta}$ is

$$382\hat{\theta}^2 + 175\hat{\theta} - 4995 = 0,$$

whence $\hat{\theta} = 3 \cdot 394 \ 254$. A consistent estimate of $\text{var}(\hat{\theta})$ is

$$\frac{\hat{\theta}(27 - \hat{\theta})(37 + \hat{\theta})^2}{N_1 \hat{\theta}(37 + \hat{\theta}) + 37 N_2(27 - \hat{\theta})} = 1 \cdot 163 \ 965.$$

Therefore $\qquad \qquad \text{s.e.} (\hat{\theta}) = 1 \cdot 078 \ 872.$

The 95 per cent confidence interval for θ is

$$\hat{\theta} \pm 1 \cdot 96 \times \text{s.e.} (\hat{\theta}) \quad \text{i.e.} \quad 1 \cdot 2797, \ 5 \cdot 5088.$$

Since the point $3 \cdot 5$ is contained in the confidence interval, therefore the null hypothesis $H(\theta = 3 \cdot 5)$ is acceptable at the 5 per cent level of significance.

27 Mean $\bar{x}$ = 4·2059.

Variance $s_x^2 = 10^{-1} \times 0·343\ 975 = 0·0344$ to 4 d.p.

The best estimate of var $(\bar{x})$ is

$$\frac{10^{-1} \times 0·343\ 975}{1110} = 10^{-4} \times 0·309\ 887.$$

Therefore   s.e. $(\bar{x}) = 10^{-2} \times 0·556\ 675.$

The one per cent point of the unit normal distribution is 2·576. Hence the 99 per cent confidence interval for the true mean length of the eggs is

$$\bar{x} \pm 2·576 \times \text{s.e.}\,(\bar{x}) \quad \text{i.e.} \quad 4·1916,\ 4·2202.$$

28 The observed frequency distribution is as follows:

No. of cells per square (x)	Frequency	No. of cells per square (x)	Frequency
0	0	7	37
1	20	8	18
2	43	9	10
3	53	10	5
4	86	11	2
5	70	12	2
6	54		
Total			400

Mean $\bar{x}$ = 4·6800;   variance $s_x^2 = 4·468\ 772.$

Therefore $s_x/20 = 0·105\ 697$; $\sqrt{\bar{x}}/20 = 0·108\ 167.$

(i) If we assume that x has a Poisson distribution with mean μ, then the best estimate of μ is $\bar{x}$ and s.e. $(\bar{x}) = 0·108\ 167$. Hence, since the average is based on a large number of observations, $\bar{x}$ is approximately normally distributed with mean μ and variance $\mu/400$. Therefore the 99 per cent confidence interval for μ is

$$4·6800 \pm 2·576 \times 0·108\ 167 \quad \text{i.e.} \quad 4·9586,\ 4·4014.$$

(ii) If we consider that the difference between $\bar{x}$ and s_x^2 is too large, then the Poisson model is inapplicable. If we assume that the distribution of x has mean μ and variance $\sigma^2 (\neq \mu)$, then $\bar{x}$ is an unbiased estimate of μ and s_x^2 that of σ^2. The approximate normality of the distribution of $\bar{x}$ is still assured by the central limit theorem, though s.e. $(\bar{x}) = 0·105\ 697$. Hence the 99 per cent confidence interval for μ is

$$4·6800 \pm 2·576 \times 0·105\ 697 \quad \text{i.e.} \quad 4·9523,\ 4·4077.$$

29 *Men:*   Mean $\bar{x}_1$ = 3·7436;   variance $s_1^2 = 4·232\ 836$;

s.e. $(\bar{x}_1) = 10^{-1} \times 0·621\ 457.$

Women: Mean $\bar{x}_2$ = 3·5301;   variance $s_2^2 = 4·713\ 248$;

s.e. $(\bar{x}_2) = 10^{-1} \times 0·655\ 776.$

The best estimates of θ_1 and θ_2 are $\bar{x}_1$ and $\bar{x}_2$ respectively. Therefore the best estimate of $\theta_1 - \theta_2$ is $\bar{x}_1 - \bar{x}_2 = 0 \cdot 2135$. The unbiased estimate of $\text{var}(\bar{x}_1 - \bar{x}_2)$ is

$$(s_1^2 + s_2^2)/1096 = 10^{-2} \times 0 \cdot 816\ 249,$$

so that $\text{s.e.}(\bar{x}_1 - \bar{x}_2) = 10^{-1} \times 0 \cdot 903\ 465.$

The 4 per cent point of the unit normal distribution is $2 \cdot 054$. Hence the large-sample 96 per cent confidence interval for $\theta_1 - \theta_2$ is

$$\bar{x}_1 - \bar{x}_2 \pm 2 \cdot 054 \times \text{s.e.}(\bar{x}_1 - \bar{x}_2) \quad \text{i.e.} \quad 0 \cdot 3991,\ 0 \cdot 0279.$$

This confidence interval does not include the point zero and so the null hypothesis $H(\theta_1 - \theta_2 = 0)$ may be rejected at the 4 per cent level of significance.

30   *Normal convicts*:

Mean $\bar{x}_1 = 73 \cdot 3299$;   variance $s_1^2 = 115 \cdot 3593$;
s.e.$(\bar{x}_1) = 0 \cdot 396\ 170.$

Weak-minded convicts:

Mean $\bar{x}_2 = 77 \cdot 6204$;   variance $s_2^2 = 142 \cdot 5486$;
s.e.$(\bar{x}_2) = 0 \cdot 863\ 903.$

The best estimates of θ_1 and θ_2 are $\bar{x}_1$ and $\bar{x}_2$ respectively. Hence the 95 per cent confidence interval for θ_1 is

$$\bar{x}_1 \pm 1 \cdot 96 \times \text{s.e.}(\bar{x}_1) \quad \text{i.e.} \quad 74 \cdot 1064,\ 72 \cdot 5534;$$

and that for θ_2 is

$$\bar{x}_2 \pm 1 \cdot 96 \times \text{s.e.}(\bar{x}_2) \quad \text{i.e.} \quad 79 \cdot 3136,\ 75 \cdot 9272.$$

The best estimate of $\theta_2 - \theta_1$ is $\bar{x}_2 - \bar{x}_1 = 4 \cdot 2905$ and the best estimate of $\text{var}(\bar{x}_2 - \bar{x}_1)$ is

$$\frac{s_2^2}{191} + \frac{s_1^2}{735} = 0 \cdot 903\ 279.$$

Therefore $\text{s.e.}(\bar{x}_2 - \bar{x}_1) = 0 \cdot 950\ 410.$
Hence the 95 per cent confidence interval for $\theta_2 - \theta_1$ is

$$\bar{x}_2 - \bar{x}_1 \pm 1 \cdot 96 \times \text{s.e.}(\bar{x}_2 - \bar{x}_1) \quad \text{i.e.} \quad 6 \cdot 1533,\ 2 \cdot 4277.$$

This confidence interval does not include the zero point and so the null hypothesis $H(\theta_2 - \theta_1 = 0)$ is not tenable at the 5 per cent level of significance.

31   *Boys*:   Mean $\bar{x}_1 = 66 \cdot 2677$;   variance $s_1^2 = 60 \cdot 5800.$

Girls: Mean $\bar{x}_2 = 64 \cdot 3755$;   variance $s_2^2 = 69 \cdot 6746.$

The best estimate of $\theta_1 - \theta_2$ is $\bar{x}_1 - \bar{x}_2 = 1\cdot8922$ and the best estimate of $\text{var}(\bar{x}_1 - \bar{x}_2)$ is

$$\frac{s_1^2}{594} + \frac{s_2^2}{498} = 0\cdot241\ 896.$$

Therefore $\text{s.e.}(\bar{x}_1 - \bar{x}_2) = 0\cdot491\ 829.$
Hence the 95 per cent confidence interval for $\theta_1 - \theta_2$ is

$$\bar{x}_1 - \bar{x}_2 \pm 1\cdot96 \times \text{s.e.}(\bar{x}_1 - \bar{x}_2) \quad \text{i.e.} \quad 2\cdot8562,\ 0\cdot9282.$$

(i) Since the zero point is not included in the above confidence interval, the null hypothesis $H(\theta_1 - \theta_2 = 0)$ can be rejected at the 5 per cent level of significance.

(ii) For large samples,

$$z = \frac{\bar{x}_1 - \bar{x}_2 - (\theta_1 - \theta_2)}{\text{s.e.}(\bar{x}_1 - \bar{x}_2)} \quad \text{is} \sim N(0,\ 1).$$

Also, if the null hypothesis $H(\theta_1 - \theta_2 = 2)$ is true, then the observed value of z is $-0\cdot219$. Since the alternatives to the null hypothesis under test are $\theta_1 - \theta_2 > 0$, therefore the relevant probability for testing significance is

$$P(z \geqslant -0\cdot219) \sim 1 - \Phi(-0\cdot219) = 0\cdot586\ 67.$$

This probability is large and so the null hypothesis can be accepted. The data do not suggest that the difference between the true mean weights of boys and girls is greater than 2 lb.

32   Leicester, 1968.

(i) *Males:*   Mean $\bar{x}_1 = 1400\cdot4808$;   variance $s_1^2 = 11\ 541\cdot9375.$

Females: Mean $\bar{x}_2 = 1252\cdot6824$;   variance $s_2^2 = 10\ 404\cdot9500.$

(ii) Suppose θ_1 and θ_2 are the means of the brain-weight populations of the males and females respectively, and let σ^2 be the common population variance. Then the pooled estimate of σ^2 is
$$s^2 = (415s_1^2 + 232s_2^2)/647 = 11\ 134\cdot2387.$$

(iii) The best estimate of $\theta_1 - \theta_2$ is $\bar{x}_1 - \bar{x}_2 = 147\cdot7984$, and

$$\text{s.e.}(\bar{x}_1 - \bar{x}_2) = s\left[\frac{1}{416} + \frac{1}{233}\right]^{\frac{1}{2}} = 8\cdot6343.$$

The 10 per cent point of the unit normal distribution is $1\cdot645$. Therefore the required confidence interval for $\theta_1 - \theta_2$ is

$$\bar{x}_1 - \bar{x}_2 \pm 1\cdot645 \times \text{s.e.}(\bar{x}_1 - \bar{x}_2) \quad \text{i.e.} \quad 162\cdot0018,\ 133\cdot5950.$$

Hence the null hypothesis $H(\theta_1 - \theta_2 = 150)$ is acceptable at the 10 per cent level of significance.

33 (i) *Recruits:*   Mean $\bar{x}_1 = 166 \cdot 3551$; variance $s_1^2 = 30 \cdot 3261$.

   Rejected conscripts: Mean $\bar{x}_2 = 164 \cdot 8883$; variance $s_2^2 = 43 \cdot 9454$.

(ii) Pooled estimate of the common population variance is

$$s^2 = \frac{3809 s_1^2 + 12\ 392 s_2^2}{16\ 201} = 40 \cdot 7434.$$

If θ_1 and θ_2 are the true mean statures of the recruits and rejected conscripts, then the best estimate of $\theta_1 - \theta_2$ is

$$\bar{x}_1 - \bar{x}_2 = 1 \cdot 4688$$

and $\text{s.e.} (\bar{x}_1 - \bar{x}_2) = s \left[\dfrac{1}{3810} + \dfrac{1}{12\ 393} \right]^{\frac{1}{2}} = 0 \cdot 118\ 243.$

(a) To test the null hypothesis $H(\theta_1 - \theta_2 = 2)$ with alternatives $\theta_1 - \theta_2 > 2$, we have

$$z = \frac{\bar{x}_1 - \bar{x}_2 - (\theta_1 - \theta_2)}{\text{s.e.} (\bar{x}_1 - \bar{x}_2)} = -4 \cdot 509 \text{ is } \sim N(0, 1).$$

Hence $P(z > -4 \cdot 509) \sim 1$, and so the null hypothesis is acceptable.

(b) To test the null hypothesis $H(\theta_1 - \theta_2 = 3)$ with alternatives $\theta_1 - \theta_2 < 3$, we have

$$z = -12 \cdot 967 \text{ is } \sim N(0, 1).$$

Hence $P(z < -12 \cdot 967) \sim 0$, and so the null hypothesis is rejected.

(iii) We now have

$$\text{s.e.} (\bar{x}_1 - \bar{x}_2) = \left[\frac{30 \cdot 3261}{3810} + \frac{43 \cdot 9454}{12\ 393} \right]^{\frac{1}{2}} = 0 \cdot 107\ 264.$$

Since this standard error is less than that obtained in (ii), the conclusions of the tests in (a) and (b) remain unaffected.

(iv) The required 95 per cent confidence interval for $\theta_1 - \theta_2$ is

$$1 \cdot 4668 \pm 1 \cdot 96 \times 0 \cdot 107\ 264 \quad \text{i.e.} \quad 1 \cdot 6770, \ 1 \cdot 2566.$$

34 (i) *1906 data:*   Mean $\bar{x}_1 = 11 \cdot 5275$;   variance $s_1^2 = 15 \cdot 4980$.

   1907 data: Mean $\bar{x}_2 = 9 \cdot 7562$;   variance $s_2^2 = 15 \cdot 9893$.

(ii) Pooled estimate of the common population variance is

$$s^2 = \frac{1999 s_1^2 + 799 s_2^2}{2798} = 15 \cdot 6383.$$

If θ_1 and θ_2 are the true mean number of seeds per placenta in 1906 and 1907 respectively, then the best estimate of $\theta_1 - \theta_2$ is

$$\bar{x}_1 - \bar{x}_2 = 1 \cdot 7713$$

and $\qquad$ s.e.$(\bar{x}_1 - \bar{x}_2) = s\left[\dfrac{1}{2000} + \dfrac{1}{800}\right]^{\frac{1}{2}} = 0\cdot165\ 429.$

(a) To test the null hypothesis $H(\theta_1 - \theta_2 = 1\cdot5)$ with alternatives $\theta_1 - \theta_2 > 1\cdot5$, we have

$$z = \frac{\bar{x}_1 - \bar{x}_2 - (\theta_1 - \theta_2)}{\text{s.e.}(\bar{x}_1 - \bar{x}_2)} = 1\cdot640 \text{ is } \sim N(0, 1).$$

Hence $P(z > 1\cdot640) = 0\cdot0505$ and so the null hypothesis is acceptable at the one per cent level of significance.

(b) To test the null hypothesis $H(\theta_1 - \theta_2 = 2)$ with alternatives $\theta_1 - \theta_2 < 2$, we have $z = -1\cdot382$ is $\sim N(0, 1)$. Hence $P(z < -1\cdot382) = 0\cdot083\ 48$ and so the null hypothesis is acceptable.

(c) To test the null hypothesis $H(\theta_1 - \theta_2 = 2)$ with alternatives $\theta_1 - \theta_2 \neq 2$, the relevant probability is $P(|z| \geqslant 1\cdot382) = 0\cdot166\ 96$. Hence the null hypothesis is acceptable.

(iii) We now have

$$\text{s.e.}(\bar{x}_1 - \bar{x}_2) = \left[\frac{15\cdot4980}{2000} + \frac{15\cdot9893}{800}\right]^{\frac{1}{2}} = 0\cdot166\ 540.$$

Since this standard error is larger than the one obtained above on the assumption of equality of the two population variances, it is easily seen that the conclusions based on the tests of significance in (a), (b) and (c) remain unaffected if the latter standard error is used.

(iv) The $0\cdot1$ per cent point of the unit normal distribution is $3\cdot29$. Therefore the $99\cdot9$ per cent confidence interval for $\theta_1 - \theta_2$ is

$$1\cdot7713 \pm 3\cdot29 \times 0\cdot165\ 429 \quad \text{i.e.} \quad 2\cdot3156,\ 1\cdot2270.$$

35 (i) *Deep water males:* $\qquad$ Mean $\bar{x}_1 = 8\cdot5754$; variance $s_1^2 = 2\cdot7933$.

Shallow water males: $\quad$ Mean $\bar{x}_2 = 8\cdot4064$; variance $s_2^2 = 2\cdot2296$.

Deep water females: $\qquad$ Mean $\bar{x}_3 = 7\cdot5404$; variance $s_3^2 = 0\cdot8898$.

Shallow water females: Mean $\bar{x}_4 = 7\cdot1222$; variance $s_4^2 = 0\cdot7464$.

(ii) Let σ_m^2 and σ_f^2 be the common unknown variances of the males and females respectively. Then their best estimates are

$$\sigma_m^{2*} = \frac{482s_1^2 + 549s_2^2}{1031} = 2\cdot4931$$

and $\qquad$ $$\sigma_f^{2*} = \frac{469s_3^2 + 553s_4^2}{1022} = 0\cdot8122.$$

(iii) Let $\theta_1, \theta_2, \theta_3$ and θ_4 be the true means of the populations of deep water males, shallow water males, deep water females and shallow females respectively.

(a) The best estimate of $\theta_1 - \theta_2$ is $\bar{x}_1 - \bar{x}_2 = 0 \cdot 1690$ and

$$\text{s.e.}(\bar{x}_1 - \bar{x}_2) = \sigma_m^* \left[\frac{1}{483} + \frac{1}{550} \right]^{\frac{1}{2}} = 10^{-1} \times 0 \cdot 984\ 611.$$

Therefore the 95 per cent confidence interval for $\theta_1 - \theta_2$ is

$$\bar{x}_1 - \bar{x}_2 \pm 1 \cdot 96 \times \text{s.e.}(\bar{x}_1 - \bar{x}_2) \quad \text{i.e.} \quad 0 \cdot 3620, \ -0 \cdot 0240.$$

(b) The best estimate of $\theta_3 - \theta_4$ is $\bar{x}_3 - \bar{x}_4 = 0 \cdot 4182$ and

$$\text{s.e.}(\bar{x}_3 - \bar{x}_4) = \sigma_f^* \left[\frac{1}{470} + \frac{1}{554} \right]^{\frac{1}{2}} = 10^{-1} \times 0 \cdot 565\ 168.$$

Therefore the 95 per cent confidence interval for $\theta_3 - \theta_4$ is

$$\bar{x}_3 - \bar{x}_4 \pm 1 \cdot 96 \times \text{s.e.}(\bar{x}_3 - \bar{x}_4). \quad \text{i.e.} \quad 0 \cdot 5290, \ 0 \cdot 3074.$$

(c) The best estimate of $\theta_2 - \theta_4$ is $\bar{x}_2 - \bar{x}_4 = 1 \cdot 2842$ and

$$\text{s.e.}(\bar{x}_2 - \bar{x}_4) = \left[\frac{\sigma_m^{2*}}{550} + \frac{\sigma_f^{2*}}{554} \right]^{\frac{1}{2}} = 10^{-1} \times 0 \cdot 774\ 530.$$

Therefore the 95 per cent confidence interval for $\theta_2 - \theta_4$ is

$$\bar{x}_2 - \bar{x}_4 \pm 1 \cdot 96 \times \text{s.e.}(\bar{x}_2 - \bar{x}_4) \quad \text{i.e.} \quad 1 \cdot 4360, \ 1 \cdot 1324.$$

(d) The best estimate of $\theta_1 - \theta_3$ is $\bar{x}_1 - \bar{x}_3 = 1 \cdot 0350$ and

$$\text{s.e.}(\bar{x}_1 - \bar{x}_3) = \left[\frac{\sigma_m^{2*}}{483} + \frac{\sigma_f^{2*}}{470} \right]^{\frac{1}{2}} = 10^{-1} \times 0 \cdot 830\ 048.$$

Therefore the 95 per cent confidence interval for $\theta_1 - \theta_3$ is

$$\bar{x}_1 - \bar{x}_3 \pm 1 \cdot 96 \times \text{s.e.}(\bar{x}_1 - \bar{x}_3) \quad \text{i.e.} \quad 1 \cdot 1977, \ 0 \cdot 8723.$$

(e) Let μ_1 and μ_2 be the true means of the populations of males and females, ignoring water depth. Then the best estimates of μ_1 and μ_2 are

$$\mu_1^* = \frac{483 \bar{x}_1 + 550 \bar{x}_2}{1033} = 8 \cdot 4854 \quad \text{and} \quad \mu_2^* = \frac{470 \bar{x}_3 + 554 \bar{x}_4}{1024} = 7 \cdot 3141.$$

Therefore the best estimate of $\mu_1 - \mu_2$ is $\mu_1^* - \mu_2^* = 1 \cdot 1713$ and

$$\text{s.e.}(\mu_1^* - \mu_2^*) = \left[\frac{\sigma_m^{2*}}{1033} + \frac{\sigma_f^{2*}}{1024} \right]^{\frac{1}{2}} = 10^{-1} \times 0 \cdot 566\ 270.$$

Therefore the 95 per cent confidence interval for $\mu_1 - \mu_2$ is

$$\mu_1^* - \mu_2^* \pm 1 \cdot 96 \times \text{s.e.}(\mu_1^* - \mu_2^*) \quad \text{i.e.} \quad 1 \cdot 2823, \ 1 \cdot 0603.$$

(iv) We now have

$$\text{s.e.}(\bar{x}_1 - \bar{x}_2) = \left[\frac{s_1^2}{483} + \frac{s_2^2}{550} \right]^{\frac{1}{2}} = 10^{-1} \times 0 \cdot 991\ 819;$$

$$\text{s.e.}(\bar{x}_3 - \bar{x}_4) = \left[\frac{s_3^2}{470} + \frac{s_4^2}{554}\right]^{\frac{1}{2}} = 10^{-1} \times 0 \cdot 569\ 252;$$

$$\text{s.e.}(\bar{x}_2 - \bar{x}_4) = \left[\frac{s_2^2}{550} + \frac{s_4^2}{554}\right]^{\frac{1}{2}} = 10^{-1} \times 0 \cdot 734\ 922;$$

$$\text{s.e.}(\bar{x}_1 - \bar{x}_3) = \left[\frac{s_1^2}{483} + \frac{s_3^2}{470}\right]^{\frac{1}{2}} = 10^{-1} \times 0 \cdot 876\ 152.$$

Hence the corresponding 95 per cent confidence intervals for the parametric differences are

$0 \cdot 3634, -0 \cdot 0254;\quad 0 \cdot 5298, 0 \cdot 3066;\quad 1 \cdot 4282, 1 \cdot 1402;\quad 1 \cdot 2067, 0 \cdot 8633.$

36 Suppose x is a random variable denoting the difference in mean elongation obtained in any experiment. We assume that x is $N(\mu, \sigma^2)$. The observed values (x_i) of x are 10 in all.

$$\sum x_i = 42 \cdot 9;\quad \sum x_i^2 = 1121 \cdot 23;\quad \text{Mean } \bar{x} = 4 \cdot 29;\quad \text{C.F.}(x) = 184 \cdot 041;$$

Total S.S. $= 937 \cdot 189$ with 9 d.f.

Therefore   s.e.$(\bar{x}) = (937 \cdot 189/90)^{\frac{1}{2}} = 3 \cdot 226\ 95.$

Also, the 5 per cent point of Student's distribution with 9 d.f. is $2 \cdot 262$. Hence the 95 per cent confidence interval for μ is

$$\bar{x} \pm 2 \cdot 262 \times \text{s.e.}(\bar{x})\quad \text{i.e.}\quad 11 \cdot 589, -3 \cdot 009.$$

Since the zero point is included in this interval, it follows that the null hypothesis $H(\mu = 0)$ is acceptable at the 5 per cent level of significance. Therefore the sample data are consistent with the hypothesis that electrical treatment does not increase the rate of growth of the maize seedlings.

37 *Percentage loss of weight analysis*

$$\sum x_i = 291 \cdot 3;\quad \sum x_i^2 = 8749 \cdot 43;\quad \text{Mean } \bar{x} = 29 \cdot 13;\quad \text{C.F.}(x) = 8485 \cdot 569;$$

Total S.S. $= 263 \cdot 861$ with 9 d.f.

Therefore   s.e.$(\bar{x}) = (263 \cdot 861/90)^{\frac{1}{2}} = 1 \cdot 712\ 25.$

Also, the 2 per cent point of Student's distribution with 9 d.f. is $2 \cdot 821$. Therefore the 98 per cent confidence interval for the true mean percentage loss in weight is

$$\bar{x} \pm 2 \cdot 821 \times \text{s.e.}(\bar{x})\quad \text{i.e.}\quad 33 \cdot 960, 24 \cdot 300.$$

Acidity analysis

It is convenient to work with the transformed values $u_i = 1000 y_i$.

$$\sum u_i = 813;\quad \sum u_i^2 = 71\ 755;\quad \text{Mean } \bar{u} = 81 \cdot 3;\quad \text{C.F.}(u) = 66\ 096 \cdot 9;$$

Total S.S. $= 5658 \cdot 1$ with 9 d.f.

Therefore   s.e. $(\bar{u}) = (5658 \cdot 1/90)^{\frac{1}{2}} = 7 \cdot 9289$.

Hence the 98 per cent confidence interval for $E(u)$ is

$$\bar{u} \pm 2 \cdot 821 \times \text{s.e.} (\bar{u}) \quad \text{i.e.} \quad 103 \cdot 67, \; 58 \cdot 93.$$

Therefore the corresponding confidence interval for the true mean acidity is $0 \cdot 1037, \; 0 \cdot 0589$.

Specific gravity analysis

It is convenient to use the transformation $v_i = 1000 \, (z_i - 1 \cdot 000)$.

$$\sum v_i = 616; \quad \sum v_i^2 = 38 \, 034; \quad \text{Mean } \bar{v} = 61 \cdot 6; \quad \text{C.F.} (v) = 37 \, 945 \cdot 6;$$

Total S.S. $= 88 \cdot 4$ with 9 d.f.

Therefore   s.e. $(\bar{v}) = (88 \cdot 4/90)^{\frac{1}{2}} = 0 \cdot 991 \, 071$.

Hence the 98 per cent confidence interval for $E(v)$ is

$$\bar{v} \pm 2 \cdot 821 \times \text{s.e.} (\bar{v}) \quad \text{i.e.} \quad 64 \cdot 40, \; 58 \cdot 80.$$

Therefore the corresponding confidence interval for the true mean specific gravity is $1 \cdot 0644, \; 1 \cdot 0588$.

38   Leicester, 1960.

(i)

$$\sum x_i = 1415; \; \sum x_i^2 = 79 \, 349; \; \text{C.F.} (x) = 71 \, 508 \cdot 04; \; \sum (x_i - \bar{x})^2 = 7840 \cdot 96.$$

$$\sum y_i = 1293; \; \sum y_i^2 = 65 \, 647; \; \text{C.F.} (y) = 59 \, 708 \cdot 89; \; \sum (y_i - \bar{y})^2 = 5938 \cdot 11.$$

Grand total $= 2708$;   Raw S.S. $= 144 \, 996$;   C.F. (total) $= 130 \, 951 \cdot 14$.

Therefore   total S.S. $= 14 \, 044 \cdot 86$ with 55 d.f.

The analysis of variance is as follows:

Source of variation	d.f.	S.S.	M.S.	F.R.
Between directions	1	265·79	265·79	1·042
Error	54	13 779·07	255·17	
Total	55	14 044·86		

The 5 per cent table value of F with 1, 54 d.f. lies between $4 \cdot 02$ and $4 \cdot 03$, and so the observed F ratio is not significant.

(ii) Let $z_i = x_i - y_i$. Then for the 28 z observations, we have

$$\sum z_i = 122; \quad \sum z_i^2 = 2228; \quad \text{C.F.} (z) = 531 \cdot 57;$$

$$\sum (z_i - \bar{z})^2 = 1696 \cdot 43 \text{ with 27 d.f.}$$

The analysis of variance is as follows:

Source of variation	d.f.	S.S.	M.S.	F.R.
Hypothesis $E(z) = 0$	1	531·57	531·57	8·460
Error	27	1696·43	62·83	
Total	28	2228·00		

The one per cent table value of F with 1, 27 d.f. is 7·68, and so the observed ratio is highly significant.

The two tests of significance give opposite conclusions, but the second analysis is preferable. The x and y measurements are correlated and the first analysis neglects this fact. Consequently, in (i)

$$\text{var } (x - y) = \text{var } (x) + \text{var } (y),$$

whereas in (ii),

$$\text{var } (x - y) = \text{var } (x) + \text{var } (y) - 2 \text{ cov } (x, y)$$
$$< \text{var } (x) + \text{var } (y) \text{ if cov } (x, y) > 0.$$

The second analysis is more sensitive than the first, despite the fact that the error d.f. are halved.

39   Leicester, 1962.

Let x_i and y_i denote the observations of Samples I and II. Assume that $E(x_i) = \theta_1$, $E(y_i) = \theta_2$, $\text{var}(x_i) = \sigma_1^2$ and $\text{var}(y_i) = \sigma_2^2$.

$\sum x_i = 1146 \cdot 1$; $\sum x_i^2 = 87\ 669 \cdot 89$; Mean $\bar{x} = 76 \cdot 41$; C.F. $(x) = 87\ 569 \cdot 68$; $\sum (x_i - \bar{x})^2 = 100 \cdot 21$ with 14 d.f.

(i) To test the null hypothesis $H(\sigma_1^2 = 9)$ we use the fact that

$$\chi^2 = 100 \cdot 21/9 = 11 \cdot 134 \text{ has the } \chi^2 \text{ distribution with 14 d.f.}$$

For 14 d.f. the upper and lower 2·5 per cent points of the χ^2 distribution are 26·12 and 5·63 respectively, where

$$P(\chi^2 \geqslant 26 \cdot 12) = P(\chi^2 \leqslant 5 \cdot 63) = 0 \cdot 025.$$

Therefore, if the alternatives to the null hypothesis under test are $\sigma_1^2 \neq 9$, the two-sided test of significance shows that the observed value of χ^2 is not significant at the 5 per cent level. Thus Sample I could have arisen from a normal population with $\sigma_1 = 3$.

$\sum y_i = 958 \cdot 2$; $\sum y_i^2 = 70\ 716 \cdot 72$; Mean $\bar{y} = 73 \cdot 71$; C.F. $(y) = 70\ 626 \cdot 71$; $\sum (y_i - \bar{y})^2 = 90 \cdot 01$ with 12 d.f.

Hence   s.e. $(\bar{y}) = \{90 \cdot 01/(13 \times 12)\}^{\frac{1}{2}} = 0 \cdot 7596$.

(ii) To test the null hypothesis $H(\theta_2 = 72 \cdot 0)$ with the alternatives $\theta_2 \neq 72 \cdot 0$, we use the fact that

$$t = \frac{\bar{y} - 72 \cdot 0}{\text{s.e.} (\bar{y})} = 2 \cdot 251 \text{ has Student's distribution with 12 d.f.}$$

For 12 d.f., the 5 per cent table value of t is 2·179. Hence the observed result is significant and we conclude that Sample II is unlikely to have arisen from a population with mean cephalic index of 72·0.

(iii) If $\sigma_1^2 = \sigma_2^2$, then the pooled estimate of the error variance is

$$190 \cdot 22/26 = 7 \cdot 3162 \text{ with 26 d.f.}$$

and $\qquad$ s.e. $(\bar{y}) = (7\cdot3162/13)^{\frac{1}{2}} = 0\cdot7502.$

The test for the null hypothesis $H(\theta_2 = 72\cdot0)$ with the alternatives $\theta_2 \neq 72\cdot0$ is now obtained by using the fact that

$t = 1\cdot71/0\cdot7502 = 2\cdot279$ has Student's distribution with 26 d.f.

The 5 per cent table value of t with 26 d.f. is $2\cdot056$, and so the significance of the observed result is reinforced.

(iv) If we set $\lambda = \sigma_2^2/\sigma_1^2$, then the ratio

$$F = \frac{90\cdot01}{12\sigma_2^2} \Big/ \frac{100\cdot21}{14\sigma_1^2} = 1\cdot0479\lambda^{-1}$$

has the F distribution with 12, 14 d.f. The upper 5 per cent table value of the F distribution with 12, 14 d.f. is $2\cdot53$, and the lower 5 per cent table value is $1/2\cdot64$. Hence the required 90 per cent confidence interval for λ is

$$1/2\cdot64 \leqslant 1\cdot0479\lambda^{-1} \leqslant 2\cdot53 \quad \text{i.e.} \quad 2\cdot7663,\ 0\cdot4142.$$

This interval includes unity and so the null hypothesis $H(\lambda = 1)$ is acceptable at the 10 per cent level against the alternatives $\lambda \neq 1$.

40 Leicester, 1970.

Let μ and σ^2 be the mean and variance of the population of percentage ash contents of the coal.

For the 100 observations given, we have

Grand total $= 1453\cdot09$; Raw S.S. $= 21\ 843\cdot2577$; Mean $\bar{x} = 14\cdot5309$;

Sample variance $s^2 = 7\cdot359\ 113$.

Therefore $\qquad$ s.e. $(\bar{x}) = (10^{-1} \times 0\cdot735\ 911)^{\frac{1}{2}} = 0\cdot271\ 277.$

The 5 per cent table value of Student's distribution with 99 d.f. is $1\cdot987$. Therefore the 95 per cent confidence interval for μ is

$$\bar{x} \pm 1\cdot987 \times \text{s.e.}\,(\bar{x}) \quad \text{i.e.} \quad 15\cdot0699,\ 13\cdot9919.$$

If χ^2 has the χ^2 distribution with ν d.f. and ν is large, then

$$\sqrt{2\chi^2} - \sqrt{2\nu - 1} \text{ is } N(0,\ 1).$$

If we use this approximation for $\nu = 99$, then the upper and lower one per cent points of the χ^2 distribution, denoted by χ_1^2 and χ_2^2, are obtained from the equations

$$\sqrt{2\chi_1^2} - \sqrt{197} = 2\cdot326; \quad \sqrt{2\chi_2^2} - \sqrt{197} = -2\cdot326.$$

Hence $\chi_1^2 = 133\cdot8575$ and $\chi_2^2 = 68\cdot5621$. Therefore the 98 per cent symmetrical confidence interval for σ^2 is

$$68\cdot5621 \leqslant 99s^2/\sigma^2 \leqslant 133\cdot8575 \quad \text{i.e.} \quad 10\cdot6262,\ 5\cdot4427.$$

41   The engine variation is ignored if the ratings of the two fuels are treated as two independent samples. We then have the following basic calculations.

Grand total = 2016·0;   Raw S.S. = 203 250·16;   C.F. = 203 212·80.

Between-samples S.S. $= \dfrac{(1012 \cdot 5)^2}{10} + \dfrac{(1003 \cdot 5)^2}{10}$ – C.F. = 4·050 with 1 d.f.

The analysis of variance for testing the equality of the two mean ratings is as follows:

Source of variation	d.f.	S.S.	M.S.	F.R.
Between samples	1	4·05	4·05	2·19
Error	18	33·31	1·851	
Total	19	37·36		

The 5 per cent table value of F with 1, 18 d.f. is 4·41 and so the observed F ratio is not significant.

The engine variation is eliminated if the differences of the A and B ratings are considered. If the observed differences are denoted by z_i, then the null hypothesis is now equivalent to $E(z) = 0$. For the observed z_i we have

Raw S.S. = 16·64;   C.F. = 8·10;   Mean $\bar{z}$ = 0·9;   Total S.S. = 8·54;

Error M.S. = 0·9489 with 9 d.f.

Hence   s.e. $(\bar{z}) = \sqrt{(0 \cdot 094\ 89)} = 0 \cdot 3080$.

Therefore if the null hypothesis $H[E(z) = 0]$ is true, then

$$t = \bar{z}/\{\text{s.e.}(\bar{z})\} = 3 \cdot 081$$

has Student's distribution with 9 d.f. For 9 d.f. the 5 per cent table value of t is 2·262. Therefore the observed t is significant and the null hypothesis can be rejected at the 5 per cent level.

(ii) The 5 per cent table value of Student's distribution with 9 d.f. is 2·262. Hence the required 95 per cent confidence interval is

$$\bar{z} \pm 2 \cdot 262 \times \text{s.e.}(\bar{z}) \quad \text{i.e.} \quad 1 \cdot 6456,\ 0 \cdot 2522.$$

42   *Thiamin analysis*

(i) *Hard wheat*

Grand total = 76·57;   Raw S.S. = 401·2741;   C.F. = 390·8643.
Mean = 5·1047;   Total S.S. = 10·4098 with 14 d.f.

Soft wheat

Grand total = 56·46;   Raw S.S. = 203·5342;   C.F. = 199·2332.
Mean = 3·5288;   Total S.S. = 4·3010 with 15 d.f.
Hence
Pooled error M.S. (s_1^2) = (10·4098 + 4·3010)/29 = 0·507 269 with 29 d.f.

If θ_1 and θ_2 are the true mean thiamin contents of hard and soft wheat, then the best estimate of $\theta_1 - \theta_2$ is

$$\theta_1^* - \theta_2^* = 1\cdot5759,$$

and   s.e. $(\theta_1^* - \theta_2^*) = (31s_1^2/240)^{\frac{1}{2}} = 0\cdot255\ 973.$

The 5 per cent table value of Student's distribution with 29 d.f. is 2·045. Therefore the required 95 per cent confidence interval for $\theta_1 - \theta_2$ is

$$\theta_1^* - \theta_2^* \pm 2\cdot045 \times \text{s.e.}\,(\theta_1^* - \theta_2^*)\quad \text{i.e.}\quad 2\cdot0994,\ 1\cdot0524.$$

Hence the null hypothesis $H(\theta_1 - \theta_2 = 0)$ is rejected at the 5 per cent level of significance.

(ii) To test the equality of the variances of the thiamin measurements of hard and soft wheat, we have

$$F = \frac{10\cdot4098 \times 15}{4\cdot3010 \times 14} = 2\cdot59 \text{ with } 14,\ 15 \text{ d.f.}$$

For 14, 15 d.f. the one per cent table value of F is 3·56, and so the observed F ratio is not significant.

Riboflavin analysis

(i) *Hard wheat*

Grand total = 17·77;   Raw S.S. = 21·9245;   C.F. = 21·0515.
Mean = 1·1847;   Total S.S. = 0·8730 with 14 d.f.

Soft wheat

Grand total = 17·18;   Raw. S.S. = 19·0456;   C.F. = 18·4470.
Mean = 1·0738;   Total S.S. = 0·5986 with 15 d.f.
Hence
Pooled error M.S. $(s_2^2) = (0\cdot8730 + 0\cdot5986)/29 = 10^{-1} \times 0\cdot507\ 448$ with 29 d.f.

If δ_1 and δ_2 are the true mean riboflavin contents of hard and soft wheat, then the best estimate of $\delta_1 - \delta_2$ is

$$\delta_1^* - \delta_2^* = 0\cdot1109,$$

and   s.e. $(\delta_1^* - \delta_2^*) = (31s_2^2/240)^{\frac{1}{2}} = 10^{-1} \times 0\cdot809\ 600.$

Therefore the required 95 per cent confidence interval for $\delta_1 - \delta_2$ is

$$\delta_1^* - \delta_2^* \pm 2\cdot045 \times \text{s.e.}\,(\delta_1^* - \delta_2^*)\quad \text{i.e.}\quad 0\cdot2765,\ -0\cdot0547.$$

Hence the null hypothesis $H(\delta_1 - \delta_2 = 0)$ is acceptable at the 5 per cent level of significance.

If δ is the common value of the true riboflavin content of hard and soft wheat, then the best estimate of δ is

$$\delta^* = (15\delta_1^* + 16\delta_2^*)/31 = 1\cdot1274,$$

and   s.e. $(\delta^*) = s_2/\sqrt{31} = 10^{-1} \times 0\cdot404\ 590.$

The 2 per cent point of Student's distribution with 29 d.f. is 2·462. There-fore the 98 per cent confidence interval for δ is

$$\delta^* \pm 2\text{·}462 \times \text{s.e.}(\delta^*) \quad \text{i.e.} \quad 1\text{·}2270, \ 1\text{·}0278.$$

(ii) To test the equality of the variances of the riboflavin measure-ments of hard and soft wheat, we have

$$F = \frac{0\text{·}8730 \times 15}{0\text{·}5986 \times 14} = 1\text{·}56 \text{ with } 14, 15 \text{ d.f.}$$

For 14, 15 d.f. the 5 per cent table value of F is 2·37, and so the ob-served F is not significant.

43 (i) *Protein analysis*

It is convenient to work in terms of protein percentage – 10.
Grand total = 124·05; Raw S.S. = 627·0519; C.F. = 615·5361.
Mean = 4·962; Total S.S. = 11·5158 with 24 d.f.

The estimated difference between the Chicago and Connecticut means is

$$4\text{·}962 - 4\text{·}69 = 0\text{·}272.$$

(a) If the Connecticut average is sensibly without error, then the stan-dard error of the estimated difference is

$$\{11\text{·}5158/(24 \times 25)\}^{\frac{1}{2}} = 0\text{·}138 \ 54.$$

Hence, to test the equality of the two means, we have

$$t = 0\text{·}272/0\text{·}138 \ 54 = 1\text{·}963 \text{ with } 24 \text{ d.f.}$$

The one per cent table value of Student's distribution with 24 d.f. is 2·797. Therefore the null hypothesis is acceptable.

(b) If the Connecticut average is based on observations with the same intrinsic variability as the Chicago one, then the standard error of the estimated difference is

$$\left[\frac{11\text{·}5158 \times 226}{24 \times 25 \times 201} \right]^{\frac{1}{2}} = 0\text{·}146 \ 90.$$

Hence the observed value of t is

$$0\text{·}272/0\text{·}146 \ 90 = 1\text{·}852 \text{ with } 24 \text{ d.f.}$$

The null hypothesis is still acceptable at the one per cent level.

(ii) *N.F. extract + crude fibre analysis*

It is convenient to work in terms of N.F. extract + crude fibre – 70.

Grand total = 187·26; Raw S.S. = 1418·1526; C.F. = 1402·6523.
Mean = 7·4904; Total S.S. = 15·5003 with 24 d.f.

The estimated difference between the Chicago and Connecticut means is
$$11\text{·}33 - 7\text{·}4904 = 3\text{·}8396.$$

(a) If the Connecticut average is sensibly without error, then the standard error of the estimated difference is

$$\{15 \cdot 5003/(24 \times 25)\}^{\frac{1}{2}} = 0 \cdot 160\ 73.$$

Hence, to test the equality of the two means, we have

$$t = 3 \cdot 8396/0 \cdot 160\ 73 = 23 \cdot 89 \text{ with 24 d.f.}$$

The observed t is obviously highly significant and so the null hypothesis can be rejected quite confidently.

(b) If the Connecticut average is based on observations with the same intrinsic variability as the Chicago one, then the standard error of the estimated difference is

$$\left[\frac{15 \cdot 5003 \times 226}{24 \times 25 \times 201}\right]^{\frac{1}{2}} = 0 \cdot 170\ 43.$$

Hence the observed value of t is

$$3 \cdot 8396/0 \cdot 170\ 43 = 22 \cdot 53 \text{ with 24 d.f.}$$

The null hypothesis is again rejected.

(iii)   *Ether extract analysis*

Grand total = 118·39;   Raw S.S. = 563·8535;   C.F. = 560·6477.
Mean = 4·7356;   Total S.S. = 3·2058 with 24 d.f.

The estimated difference between the Chicago and Connecticut means is

$$4 \cdot 7356 - 1 \cdot 83 = 2 \cdot 9056.$$

(a) If the Connecticut average is sensibly without error then the standard error of the estimated difference is

$$\{3 \cdot 2058/(24 \times 25)\}^{\frac{1}{2}} = 10^{-1} \times 0 \cdot 730\ 96.$$

Hence, to test the equality of the two means, we have

$$t = 2 \cdot 9056/(10^{-1} \times 0 \cdot 730\ 96) = 39 \cdot 75 \text{ with 24 d.f.}$$

The observed t is highly significant and so the null hypothesis can be rejected with confidence.

(b) If the Connecticut average is based on observations with the same intrinsic variability as the Chicago one, then the standard error of the estimated difference is

$$\left[\frac{3 \cdot 2058 \times 226}{24 \times 25 \times 201}\right]^{\frac{1}{2}} = 10^{-1} \times 0 \cdot 775\ 08.$$

Hence the observed value of t is

$$2 \cdot 9056/(10^{-1} \times 0 \cdot 775\ 08) = 37 \cdot 49 \text{ with 24 d.f.}$$

The null hypothesis is again rejected.

(iv)   *Ash analysis*

Grand total = 70·30;   Raw S.S. = 198·5898;   C.F. = 197·6836.
Mean = 2·812;   Total S.S. = 0·9062 with 24 d.f.

The estimated difference between the Chicago and Connecticut means is

$$2·812 - 2·15 = 0·662.$$

(a) If the Connecticut average is sensibly without error, then the standard error of the estimated difference is

$$\{0·9062/(24 \times 25)\}^{\frac{1}{2}} = 10^{-1} \times 0·388\ 63.$$

Hence, to test the equality of the two means, we have

$$t = 0·662/(10^{-1} \times 0·388\ 63) = 17·03 \text{ with } 24 \text{ d.f.}$$

The observed t is highly significant and so the null hypothesis can be rejected with confidence.

(b) If the Connecticut average is based on observations with the same intrinsic variability as the Chicago one, then the standard error of the estimated difference is

$$\left[\frac{0·9062 \times 226}{24 \times 25 \times 201}\right]^{\frac{1}{2}} = 10^{-1} \times 0·412\ 09.$$

Hence the observed value of t is

$$0·662/(10^{-1} \times 0·412\ 09) = 16·06 \text{ with } 24 \text{ d.f.}$$

The null hypothesis is again rejected.

44   Grand total = 6706;   Raw S.S. = 1 689 052;   C.F. = 1 606 087·00.
Mean $\bar{x}$ = 239·50;   Total S.S. = 82 965·00 with 27 d.f.

Therefore   s.e. $(\bar{x})$ = $\{82\ 965·00/(27 \times 28)\}^{\frac{1}{2}}$ = 10·4758.

The 5 per cent table value of Student's distribution with 27 d.f. is 2·052. Therefore the 95 per cent confidence interval for the true mean is

$$\bar{x} \pm 2·052 \times \text{s.e.} (\bar{x}) \quad \text{i.e.} \quad 261·00,\ 218·00.$$

45   Grand total = 65 713;   Raw S.S. = 36 045 275;   C.F. = 35 984 986·5;
Mean $\bar{x}$ = 547·61;   Total S.S. = 60 288·5 with 119 d.f.
Between-samples S.S. = 215 952 881/6 – C.F. = 7160·3 with 19 d.f.

(i) The analysis of variance is as follows:

Source of variation	d.f.	S.S.	M.S.	F.R.
Between samples	19	7 160·3	376·86	< 1
Error	100	53 128·2	531·28	
Total	119	60 288·5		

The F ratio is obviously not significant and so, on the average, the tensile strengths of the briquettes produced by the different mixings are the same.

Also,   s.e. $(\bar{x}) = (531 \cdot 28/120)^{\frac{1}{2}} = 2 \cdot 104\ 123$.

The 2 per cent table value of Student's distribution with 100 d.f. is $2 \cdot 364$. Therefore the 98 per cent confidence interval for the true mean strength is

$$\bar{x} \pm 2 \cdot 364 \times \text{s.e.} (\bar{x}) \quad \text{i.e.} \quad 552 \cdot 58,\ 542 \cdot 64.$$

(ii) Let σ^2 denote the population variance of tensile strength. For 100 d.f.,

$$P(\chi^2 \geqslant 129 \cdot 6) = P(\chi^2 \leqslant 74 \cdot 22) = 0 \cdot 025.$$

Hence the symmetrical 95 per cent confidence interval for σ^2 is
$$74 \cdot 22 \leqslant 53\ 128 \cdot 2/\sigma^2 \leqslant 129 \cdot 6,$$
whence the limits are $715 \cdot 82,\ 409 \cdot 94$.

46   Leicester, 1964.

It is convenient to subtract 1400 from the individual yield measurements. The analysis in terms of the transformed data follows.

Grand total $= 3825$;   Raw S.S. $= 602\ 875$;   C.F. $= 487\ 687 \cdot 5$;
Total S.S. $= 115\ 187 \cdot 5$ with 29 d.f.
Between-samples S.S. $= 2\ 720\ 225/5 - $ C.F. $= 56\ 357 \cdot 5$ with 5 d.f.

(i) The analysis of variance is as follows:

Source of variation	d.f.	S.S.	M.S.	F.R.
Between samples	5	56 357·5	11 271·50	4·60
Error	24	58 830·0	2 451·25	
Total	29	115 187·5		

The one per cent table value of the F distribution with 5, 24 d.f. is $3 \cdot 90$. Therefore the observed F ratio is highly significant and we infer that possibly real differences exist in the quality of H-acid from batch to batch.

(ii) If σ^2 is the true error variance, then under usual assumptions

$$\chi^2 = 58\ 830 \cdot 0/\sigma^2 \text{ has the } \chi^2 \text{ distribution with 24 d.f.}$$

But for 24 d.f., $P(\chi^2 \leqslant 12 \cdot 40) = P(\chi^2 \geqslant 39 \cdot 36) = 0 \cdot 025$.
Therefore the 95 per cent confidence interval for σ^2 is

$$12 \cdot 40 \leqslant 58\ 830 \cdot 0/\sigma^2 \leqslant 39 \cdot 36,$$

whence the required confidence limits for σ are $68 \cdot 88,\ 38 \cdot 66$.

47   Leicester, 1966.
Grand total $= 5298$;   Raw S.S. $= 949\ 898$;   C.F. $= 935\ 626 \cdot 80$;
Total S.S. $= 14\ 271 \cdot 20$ with 29 d.f.
Between-processes S.S. $= 5\ 644\ 326/6 - $ C.F. $= 5094 \cdot 20$ with 4 d.f.

The analysis of variance is as follows:

Source of variation	d.f.	S.S.	M.S.	F.R.
Between processes	4	5 094·20	1273·55	3·47
Error	25	9 177·00	367·08	
Total	29	14 271·20		

The 5 per cent table value of F with 4, 25 d.f. is 2·76. Therefore the observed ratio is significant and we conclude that possibly there are real differences in the tensile strength of the rubber produced by the five processes.

The least-squares estimates of α and β are

$$\alpha^* = \sum_{r=1}^{5} \bar{y}_r/5 = 5298/30 = 176\cdot6;$$

and $$\beta^* = \sum_{r=1}^{5} (r-3)\bar{y}_r \Big/ \sum_{r=1}^{5} (r-3)^2 = 505/60 = 8\cdot4167.$$

Also, since $\text{var}(\bar{y}_r) = \sigma^2/6$, we have

$$\text{var}(\beta^*) = \sigma^2/60, \text{ where } \sigma^2 \text{ is the error variance.}$$

Hence $\text{s.e.}(\beta^*) = (367\cdot08/60)^{\frac{1}{2}} = 2\cdot4735.$

The 5 per cent table value of Student's distribution with 25 d.f. is 2·060. Therefore the 95 per cent confidence interval for β is

$$\beta^* \pm 2\cdot060 \times \text{s.e.}(\beta^*) \quad \text{i.e.} \quad 13\cdot5121,\ 3\cdot3213.$$

48   Leicester, 1963.

Let x_i, y_i and z_i denote the normal, offtype and aberrant measurements respectively.

$\Sigma\, x_i = 183\cdot88;$ Mean $\bar{x} = 6\cdot8104;$ C.F. $= 33\ 811\cdot8544/27 = 1252\cdot2909;$

$\Sigma\, y_i = 84\cdot25;$ Mean $\bar{y} = 5\cdot6167;$ C.F. $= 7098\cdot0625/15 = 473\cdot2042;$

$\Sigma\, z_i = 80\cdot92;$ Mean $\bar{z} = 6\cdot7433;$ C.F. $= 6548\cdot0464/12 = 545\cdot6705.$

Grand total $= 349\cdot05;$ Raw S.S. $= 2298\cdot9791;$ C.F. $= 2256\cdot2204;$
Total S.S. $= 42\cdot7587$ with 53 d.f.

(i) The analysis of variance is as follows:

Source of variation	d.f.	S.S.	M.S.	F.R.
$H(\theta_n = \theta_o = \theta_a)$	2	14·9452	7·472 6	13·70
Error	51	27·8135	0·545 363	
Total	53	42·7587		

The 0·1 per cent table value of F with 2, 51 d.f. lies between 7·76 and 8·25 and so the observed F ratio is highly significant. The rubber content of the three types of plants is almost certainly different.

(ii) The best estimate of $\theta_n - \theta_a$ is $\bar{x} - \bar{z} = 0\cdot0671$,

and $\qquad$ s.e.$(\bar{x} - \bar{z}) = \left[\dfrac{39 \times 0\cdot545\ 363}{27 \times 12}\right]^{\frac{1}{2}} = 0\cdot2562.$

To test the null hypothesis $H(\theta_n - \theta_a = 0)$ we have

$$t = 0\cdot0671/0\cdot2562 = 0\cdot26 \text{ with } 51 \text{ d.f.}$$

The observed t is clearly not significant and so the null hypothesis is acceptable.

The best estimate of $2\theta_o - \theta_n - \theta_a$ is $2\bar{y} - \bar{x} - \bar{z} = -2\cdot3203$, and

$$\text{s.e.} (2\bar{y} - \bar{x} - \bar{z}) = \left[0\cdot545\ 363 \times \left(\frac{4}{15} + \frac{1}{27} + \frac{1}{12}\right)\right]^{\frac{1}{2}} = 0\cdot4594.$$

To test the null hypothesis $H(2\theta_o - \theta_n - \theta_a = 0)$ we have

$$|t| = 2\cdot3203/0\cdot4594 = 5\cdot05 \text{ with } 51 \text{ d.f.}$$

The $0\cdot1$ per cent table value of Student's distribution with 51 d.f. is $< 3\cdot551$ and so the observed ratio is highly significant. The null hypothesis can be rejected confidently.

(iii) The analysis of variance shows that the overall null hypothesis $H(\theta_n = \theta_o = \theta_a)$ is not tenable. The t tests indicate that the composite null hypothesis can be considered separately in terms of the null hypotheses $H(\theta_n - \theta_a = 0)$ and $H(2\theta_o - \theta_n - \theta_a = 0)$. These indicate that significance is due to a possible real difference between the average rubber content of normal and aberrant plants, and offtype plants, though the two t tests are not independent.

49 Let $x_{1i}, x_{2i}, x_{3i}, x_{4i}$ denote the observations in the groups C_1, C_2, NC_1 and NC_2 respectively.

$\Sigma\ x_{1i} = 35\cdot860$; Mean $\bar{x}_1 = 2\cdot5614$; C.F. $= \dfrac{1285\cdot939\ 600}{14} = 91\cdot852\ 829$;

$\Sigma\ x_{2i} = 18\cdot029$; Mean $\bar{x}_2 = 2\cdot0032$; C.F. $= \dfrac{325\cdot044\ 841}{9} = 36\cdot116\ 093$;

$\Sigma\ x_{3i} = 12\cdot173$; Mean $\bar{x}_3 = 2\cdot0288$; C.F. $= \dfrac{148\cdot181\ 929}{6} = 24\cdot696\ 988$;

$\Sigma\ x_{4i} = 9\cdot798$; Mean $\bar{x}_4 = 1\cdot6330$; C.F. $= \dfrac{96\cdot000\ 804}{6} = 16\cdot000\ 134.$

Grand total $= 75\cdot860$; Raw S.S. $= 170\cdot980\ 696$; C.F. $= 164\cdot421\ 132$; Total S.S. $= 6\cdot5596$ with 34 d.f.

(i) The analysis of variance is as follows:

Source of variation	d.f.	S.S.	M.S.	F.R.
Between groups	3	4·2449	1·415 0	18·95
Error	31	2·3147	0·074 668	
Total	34	6·5596		

The one per cent table value of F with 3, 31 d.f. lies between $4\cdot46$ and $4\cdot51$ and so the observed ratio is highly significant.

(ii) If μ_1, μ_2, μ_3, μ_4 are the true mean calcium contents of eggs laid by the four pullets, then set $\delta = \mu_1 + \mu_2 - \mu_3 - \mu_4$. The best estimate of δ is

$$\delta^* = \bar{x}_1 + \bar{x}_2 - \bar{x}_3 - \bar{x}_4 = 0 \cdot 9028,$$

and $$\text{s.e.}(\delta^*) = \left[\frac{2 \cdot 3147}{31} \times \left(\frac{1}{14} + \frac{1}{9} + \frac{1}{6} + \frac{1}{6} \right) \right]^{\frac{1}{2}} = 0 \cdot 1963.$$

The 5 per cent table value of Student's distribution with 31 d.f. is $2 \cdot 040$. Hence the 95 per cent confidence interval for δ is

$$\delta^* \pm 2 \cdot 040 \times \text{s.e.}(\delta^*) \quad \text{i.e.} \quad 1 \cdot 3033, \ 0 \cdot 5023.$$

50

$\Sigma \, x_{1i} = 173 \cdot 8$; Mean $\bar{x}_1 = 5 \cdot 6065$; C.F. $= 30 \ 206 \cdot 44/31 = 974 \cdot 4013$;

$\Sigma \, x_{2i} = 226 \cdot 6$; Mean $\bar{x}_2 = 6 \cdot 1243$; C.F. $= 51 \ 347 \cdot 56/37 = 1387 \cdot 7719$;

$\Sigma \, x_{3i} = 122 \cdot 8$; Mean $\bar{x}_3 = 5 \cdot 3391$; C.F. $= 15 \ 079 \cdot 84/23 = 655 \cdot 6452$;

$\Sigma \, x_{4i} = 151 \cdot 2$; Mean $\bar{x}_4 = 7 \cdot 2000$; C.F. $= 22 \ 861 \cdot 44/21 = 1088 \cdot 6400$;

$\Sigma \, x_{5i} = \ \ 48 \cdot 2$; Mean $\bar{x}_5 = 4 \cdot 8200$; C.F. $= 2323 \cdot 24/10 = 232 \cdot 3240$;

$\Sigma \, x_{6i} = \ \ 29 \cdot 9$; Mean $\bar{x}_6 = 5 \cdot 9800$; C.F. $= 894 \cdot 01/5 = 178 \cdot 8020$;

$\Sigma \, x_{7i} = \ \ 21 \cdot 7$; Mean $\bar{x}_7 = 7 \cdot 2333$; C.F. $= 470 \cdot 89/3 = 156 \cdot 9633$.

Grand total $= 774 \cdot 2$;   Raw S.S. $= 4844 \cdot 58$;   C.F. $= 4610 \cdot 66$; Total S.S. $= 233 \cdot 92$ with 129 d.f.

The analysis of variance is as follows:

Source of variation	d.f.	S.S.	M.S.	F.R.
Between cereals	6	63·89	10·648 3	7·70
Error	123	170·03	1·382 358	
Total	129	233·92		

The one per cent table value of F with 6, 123 d.f. is $< 2 \cdot 95$, and so the observed F ratio is highly significant. There are real differences in the thiamin content of the seven cereals.

$$\text{s.e.}(\bar{x}_1) = (1 \cdot 382 \ 358/31)^{\frac{1}{2}} = 0 \cdot 211 \ 17;$$
$$\text{s.e.}(\bar{x}_2) = (1 \cdot 382 \ 358/37)^{\frac{1}{2}} = 0 \cdot 193 \ 29;$$
$$\text{s.e.}(\bar{x}_3) = (1 \cdot 382 \ 358/23)^{\frac{1}{2}} = 0 \cdot 245 \ 16;$$
$$\text{s.e.}(\bar{x}_4) = (1 \cdot 382 \ 358/21)^{\frac{1}{2}} = 0 \cdot 256 \ 57;$$
$$\text{s.e.}(\bar{x}_5) = (1 \cdot 382 \ 358/10)^{\frac{1}{2}} = 0 \cdot 371 \ 80;$$
$$\text{s.e.}(\bar{x}_6) = (1 \cdot 382 \ 358/5)^{\frac{1}{2}} = 0 \cdot 525 \ 80;$$
$$\text{s.e.}(\bar{x}_7) = (1 \cdot 382 \ 358/3)^{\frac{1}{2}} = 0 \cdot 678 \ 81.$$

The 5 per cent table value of Student's distribution with 123 d.f. is $1 \cdot 980$. Hence the required confidence intervals are:

Wheat: $\bar{x}_1 \pm 1 \cdot 980 \times \text{s.e.}(\bar{x}_1)$ i.e. $6 \cdot 0246, \ 5 \cdot 1884$;

Barley: $\bar{x}_2 \pm 1 \cdot 980 \times \text{s.e.}(\bar{x}_2)$ i.e. $6 \cdot 5070, \ 5 \cdot 7416$;

Maize:	$\bar{x}_3 \pm 1\cdot980 \times$ s.e.$(\bar{x}_3)$	i.e.	5·8245, 4·8537;
Oats:	$\bar{x}_4 \pm 1\cdot980 \times$ s.e.$(\bar{x}_4)$	i.e.	7·7080, 6·6920;
Rye:	$\bar{x}_5 \pm 1\cdot980 \times$ s.e.$(\bar{x}_5)$	i.e.	5·5562, 4·0838;
Buckwheat:	$\bar{x}_6 \pm 1\cdot980 \times$ s.e.$(\bar{x}_6)$	i.e.	7·0211, 4·9389;
Millet:	$\bar{x}_7 \pm 1\cdot980 \times$ s.e.$(\bar{x}_7)$	i.e.	8·5773, 5·8893.

The 95 per cent confidence interval for the error variance σ^2 is

$$\chi_1^2 \leqslant \{170\cdot03/\sigma^2\} \leqslant \chi_2^2$$

where, for 123 d.f., $P(\chi^2 \leqslant \chi_1^2) = P(\chi^2 \geqslant \chi_2^2) = 0\cdot025$. But for $\nu = 123$,

$$z = \sqrt{2\chi^2} - \sqrt{2\nu - 1} \text{ is } \sim N(0, 1).$$

Therefore $-1\cdot96 = \sqrt{2\chi_1^2} - \sqrt{245}$ and $1\cdot96 = \sqrt{2\chi_2^2} - \sqrt{245}$,

whence $\chi_1^2 = 93\cdot71$ and $\chi_2^2 = 155\cdot06$.

Therefore $93\cdot71 \leqslant \{170\cdot03/\sigma^2\} \leqslant 155\cdot06$,

so that the limits for σ^2 are 1·8144, 1·0965.

51 Leicester, 1969.

It is simpler to work with the transformed observations $y_{ij} = 100(x_{ij} - 14)$.

Grand total = 4383;   Raw S.S. = 536 191;   C.F. = 519 207·81; Total S.S. = 16 983·19 with 36 d.f.

Between-pens S.S. $= \dfrac{1\ 100\ 401}{9} + \dfrac{2\ 781\ 926}{7} - $ C.F. $= 476\cdot97$ with 4 d.f.

(i) The analysis of variance is as follows:

Source of variation	d.f.	S.S.	M.S.	F.R.
Between pens	4	476·97	119·24	< 1
Error	32	16 506·22	515·82	
Total	36	16 983·19		

Obviously the observed F ratio is not significant. Therefore, on the average, there is no difference in nitrogen content due to the diets.

(ii) Let $\theta_1, \theta_2, \theta_3, \theta_4, \theta_5$ be the true group means in terms of the y values. Then their best estimates are the corresponding pen means:

$$\bar{y}_1 = 120\cdot43; \quad \bar{y}_2 = 124\cdot57; \quad \bar{y}_3 = 116\cdot56; \quad \bar{y}_4 = 113\cdot86; \quad \bar{y}_5 = 117\cdot43.$$

The best estimate of $\delta \equiv \theta_1 - (\theta_2 + \theta_3 + \theta_4 + \theta_5)/4$ is

$$\delta^* = \bar{y}_1 - (\bar{y}_2 + \bar{y}_3 + \bar{y}_4 + \bar{y}_5)/4 = 2\cdot33.$$

Also, $$\text{var}(\delta^*) = \sigma^2\left[\frac{1}{7} + \frac{1}{16}\left(\frac{3}{7} + \frac{1}{9}\right)\right] = \frac{89\sigma^2}{504},$$

where σ^2 is the variance of the individual y observations. Hence

$$\text{s.e.}(\delta^*) = \{(89 \times 515\cdot82)/504\}^{\frac{1}{2}} = 9\cdot5440.$$

The 5 per cent table value of Student's distribution with 32 d.f. is $2\cdot037$. Therefore the required 95 per cent confidence interval for δ is

$$\delta^* \pm 2\cdot037 \times \text{s.e.}(\delta^*) \quad \text{i.e.} \quad 21\cdot77, \; -17\cdot11.$$

Therefore, in terms of the original x units the confidence limits are $0\cdot2177$, $-0\cdot1711$.

52 *Ailsa Craig analysis*

Grand total = 28 300;   Raw S.S. = 2 022 348;
C.F. = 800 890 000/407 = 1 967 788·70.
Total S.S. = 54 559·30 with 406 d.f.   Grand mean = 69·53.

$$\text{Between-plants S.S.} = \frac{(4008)^2}{58} + \frac{(4121)^2}{60} + \frac{(2999)^2}{42} + \frac{(3232)^2}{46} + \frac{(3552)^2}{53} +$$

$$+ \frac{(3299)^2}{48} + \frac{(3654)^2}{52} + \frac{(3435)^2}{48} - \text{C.F.}$$

$$= 817\cdot45 \text{ with 7 d.f.}$$

Hence error S.S. = 53 741·85 with 399 d.f.

ES1 analysis

Grand total = 24 161;   Raw S.S. = 1 821 143;
C.F. = 583 753 921/347 = 1 682 287·96.
Total S.S. = 138 855·04 with 346 d.f.   Grand mean = 69·63.

$$\text{Between-plants S.S.} = \frac{(3693)^2}{57} + \frac{(4264)^2}{61} + \frac{(3864)^2}{58} + \frac{(3584)^2}{51} + \frac{(3717)^2}{55} +$$

$$+ \frac{(5039)^2}{65} - \text{C.F.}$$

$$= 6166\cdot74 \text{ with 5 d.f.}$$

Hence error S.S. = 132 688·30 with 341 d.f.

(i) The best estimate of the error variance of Ailsa Craig is

$$53\ 741\cdot85/399 = 134\cdot69 \text{ with 399 d.f.}$$

and that of ES1 is

$$132\ 688\cdot30/341 = 389\cdot12 \text{ with 341 d.f.}$$

(ii) To test the equality of the error variances of the two varieties, we have

$$F = 389\cdot12/134\cdot69 = 2\cdot89 \text{ with 341, 399 d.f.}$$

The 0·1 per cent table value of F with 341, 399 d.f. is < 2·40. Therefore the observed result is highly significant. The error variance of the two varieties is almost certainly not the same.

(iii) If θ_1 and θ_2 are the true mean maturation periods of Ailsa Craig and ES1, then the best estimate of $\theta_2 - \theta_1$ is

$$69 \cdot 63 - 69 \cdot 53 = 0 \cdot 10,$$

and the standard error of the estimated difference is

$$\left[\frac{389 \cdot 12}{347} + \frac{134 \cdot 69}{407} \right]^{\frac{1}{2}} = 1 \cdot 2051.$$

Hence the required 95 per cent confidence interval for $\theta_2 - \theta_1$ is

$$0 \cdot 10 \pm 1 \cdot 96 \times 1 \cdot 2051 \quad \text{i.e.} \quad 2 \cdot 46, \; -2 \cdot 26.$$

53 This is an example of a weighted analysis of variance. There are 15 independent frequency distributions of the percentage fat content, each distribution corresponding to one herd. In the jth herd distribution, let f_{ij} be the observed frequency in the class-interval with midpoint x_i. Make the transformation $x_i = 0 \cdot 2z_i + 4 \cdot 1$, and then the following calculations are obtained.

Herd no.	$\sum_i f_{ij} z_i$	$\sum_i f_{ij} z_i^2$	C.F.	Error S.S.	d.f.
1	− 107	559	190·82	368·18	59
2	− 83	1079	104·38	974·62	65
3	− 92	464	132·25	331·75	63
4	− 79	479	113·47	365·53	54
5	− 243	1043	602·54	440·46	97
6	− 274	1356	766·08	589·92	97
7	− 95	869	205·11	663·89	43
8	− 49	177	80·03	96·97	29
9	− 60	206	128·57	77·43	27
10	12	178	5·14	172·86	27
11	− 5	323	0·86	322·14	28
12	− 55	451	100·83	350·17	29
13	− 71	263	219·17	43·83	22
14	− 219	935	715·84	219·16	66
15	− 1	91	0·08	90·92	11
Total	−1421	8473	3365·17	5107·83	717

Grand C.F. $= (-1421)^2/732 = 2758 \cdot 53$;   Grand mean $\bar{z} = -1 \cdot 9413$.

The best estimate of the error variance is

$$s_z^2 = 5107 \cdot 83/717 = 7 \cdot 1239 \text{ with } 717 \text{ d.f.}$$

(i) In original x units, we have

$$\text{grand mean } \bar{x} = 0 \cdot 2\bar{z} + 4 \cdot 1 = 3 \cdot 7117$$

and estimate of error variance $s_x^2 = (0 \cdot 2)^2 s_z^2 = 0 \cdot 284\,956$.

Therefore   s.e. $(\bar{x}) = s_x/\sqrt{732} = 10^{-1} \times 0 \cdot 197\,30.$

Hence the 99 per cent confidence interval for the overall true mean percentage fat content is

$$\bar{x} \pm 2 \cdot 576 \times \text{s.e.}(\bar{x}) \quad \text{i.e.} \quad 3 \cdot 7625, \ 3 \cdot 6609.$$

(ii) The analysis of variance in terms of the transformed z values is as follows:

Source of variation	d.f.	S.S.	M.S.	F.R.
Between herds	14	606·64	43·3314	6·08
Error	717	5107·83	7·1239	
Total	731	5714·47		

The one per cent table value of F with 14, 717 d.f. is $< 2 \cdot 12$. Therefore the observed F ratio is highly significant, and we conclude that possibly there are real differences in the true mean percentage fat content of the milk of the 15 herds.

54 Let the time variable be $x = \text{Year} - 1956$.

Year	1954	1955	1956	1957	1958
Babies (y_i)	5	7	6	10	13
x_i	−2	−1	0	1	2

$$\Sigma \, x_i = 0; \quad \Sigma \, x_i^2 = 10; \quad \Sigma \, y_i = 41; \quad \Sigma \, y_i^2 = 379; \quad \Sigma \, x_i y_i = 19;$$

$$\text{C.F.}(x) = 0; \quad \text{C.F.}(y) = 336 \cdot 2; \quad \text{C.F.}(x, y) = 0.$$

Hence S.S. $(x) \equiv X = 10$;   S.S. $(y) \equiv Y = 42 \cdot 8$;   S.P. $(x, y) \equiv Z = 19$.

If the linear regression equation is $E(y_i) = \alpha + \beta (x_i - \bar{x})$, then the least-squares estimate of β is $\beta^* = Z/X = 1 \cdot 9$.

S.S. due to regression $= 36 \cdot 10$ with 1 d.f.

Error S.S. $= Y - Z^2/X = 6 \cdot 7$ with 3 d.f.

Hence the F ratio for testing the null hypothesis $H(\beta = 0)$ is

$$3 \times 36 \cdot 10 / 6 \cdot 7 = 16 \cdot 16 \text{ with 1, 3 d.f.}$$

The 5 per cent table value of F with 1, 3 d.f. is $10 \cdot 13$ and the one per cent value is $34 \cdot 12$. Therefore the observed result is significant at the 5 per cent level and not at the one per cent level.

(i) Replace 13 by m as the y value for 1958. Then

$$\Sigma \, y_i = 28 + m; \quad \Sigma \, y_i^2 = 210 + m^2; \quad \Sigma \, x_i y_i = 2m - 7; \quad \text{C.F.}(y) = (28 + m)^2/5;$$

$$Y = 0 \cdot 8 m^2 - 11 \cdot 2 m + 53 \cdot 2; \quad Z = 2m - 7; \quad \beta^* = 0 \cdot 2 m - 0 \cdot 7.$$

Hence   S.S. due to regression $= 10(0 \cdot 2 m - 0 \cdot 7)^2$ with 1 d.f.

Error S.S. $= 0 \cdot 4 m^2 - 8 \cdot 4 m + 48 \cdot 3$ with 3 d.f.

Therefore the F ratio for testing the null hypothesis $H(\beta = 0)$ is

$$\frac{3(0 \cdot 4 m^2 - 2 \cdot 8 m + 4 \cdot 9)}{0 \cdot 4 m^2 - 8 \cdot 4 m + 48 \cdot 3} \text{ with 1, 3 d.f.}$$

Since the 5 per cent value of F with 1, 3 d.f. is $10 \cdot 13$, the minimum value of m is determined from the equation

$$\frac{3(0 \cdot 4m^2 - 2 \cdot 8m + 4 \cdot 9)}{0 \cdot 4m^2 - 8 \cdot 4m + 48 \cdot 3} = 10 \cdot 13,$$

which reduces to the quadratic

$$2852m^2 - 76\,692m + 474\,579 = 0.$$

The two roots of this equation are $17 \cdot 24$ and $9 \cdot 65$, so the least value of m is $9 \cdot 65$, i.e. 10.

(ii) Since the observed F ratio is not significant at the one per cent level, let m be the least number of births in 1959. Then

x	-2	-1	0	1	2	3
y	5	7	6	10	13	m

$\Sigma\, x_i = 3;\ \Sigma\, x_i^2 = 19;\ \Sigma\, y_i = 41 + m;\ \Sigma\, y_i^2 = 379 + m^2;\ \Sigma\, x_i y_i = 19 + 3m;$

C.F. $(x) = 1 \cdot 5;$ C.F. $(y) = \frac{1}{6}(41 + m)^2;$ C.F. $(x, y) = \frac{1}{2}(41 + m).$

Hence $X = 17 \cdot 5;$ $Y = 379 + m^2 - \frac{1}{6}(41 + m)^2;$ $Z = \frac{1}{2}(5m - 3);$
$\beta^* = \frac{1}{35}(5m - 3).$

S.S. due to regression $= \frac{1}{70}(5m - 3)^2$ with 1 d.f.

Error S.S. $= 379 + m^2 - \frac{1}{6}(41 + m)^2 - \frac{1}{70}(5m - 3)^2$ with 4 d.f.

Therefore the F ratio for testing the null hypothesis $H(\beta = 0)$ is

$$\frac{4(5m - 3)^2 / 70}{379 + m^2 - \frac{1}{6}(41 + m)^2 - \frac{1}{70}(5m - 3)^2}$$

$$= \frac{300m^2 - 360m + 108}{100m^2 - 2780m + 20\,728} \text{ with 1, 4 d.f.}$$

Since the one per cent point of the F distribution with 1, 4 d.f. is $21 \cdot 20$, the equation for m is

$$300m^2 - 360m + 108 = 21 \cdot 20(100m^2 - 2780m + 20\,728)$$

which reduces to

$$18\,200m^2 - 585\,760m + 4\,393\,256 = 0.$$

The two roots of the quadratic are $20 \cdot 28$ and $11 \cdot 90$, so that the least value of m is $11 \cdot 90$, i.e., 12.

55 Leicester, 1963.

Let $E(y_i) = \gamma + \beta(x_i - \bar{x})$, so that $\alpha \equiv \gamma - \beta\bar{x}.$

$\Sigma\, x_i = 69 \cdot 25;$ $\Sigma\, x_i^2 = 354 \cdot 0245;$ Mean $\bar{x} = 1 \cdot 923\,611;$

C.F. $(x) = 133 \cdot 2101;$ S.S. $(x) \equiv X = 220 \cdot 8144;$

$$\Sigma \; y_i = 102 \cdot 71; \quad \Sigma \; y_i^2 = 743 \cdot 5925; \quad \text{Mean } \bar{y} = 2 \cdot 853 \; 056;$$

$$\text{C.F.}(y) = 293 \cdot 0373; \qquad \text{S.S.}(y) \equiv Y = 450 \cdot 5552;$$

$$\Sigma \; x_i y_i = 510 \cdot 8425; \quad \text{C.F.}(x, y) = 197 \cdot 5741;$$

$$\text{S.P.}(x, y) \equiv Z = 313 \cdot 2684.$$

Therefore the least-squares estimates of γ and β are

$$\gamma^* = 2 \cdot 853 \; 056; \qquad \beta^* = Z/X = 1 \cdot 418 \; 696.$$

Hence the least-squares estimate of α is

$$a^* = \gamma^* - \bar{x}\beta^* = 0 \cdot 124 \; 037.$$

Error S.S. $= Y - Z^2/X = 6 \cdot 1225$ with 34 d.f.

Therefore $\text{s.e.}(a^*) = \left[\dfrac{6 \cdot 1225}{34} \times \left(\dfrac{1}{36} + \dfrac{\bar{x}^2}{X} \right) \right]^{\frac{1}{2}} = 10^{-1} \times 0 \cdot 895 \; 525.$

The 5 per cent table value of Student's distribution with 34 d.f. is 2·0323. Therefore the 95 per cent confidence interval for α is

$$a^* \pm 2 \cdot 0323 \times \text{s.e.}(a^*) \quad \text{i.e.} \quad 0 \cdot 3060, \; -0 \cdot 0580.$$

56   Leicester, 1962.
Swift-water analysis

Let $E(y) = \alpha_1 + \beta_1(x - \bar{x}_1)$.

$$\Sigma \; x_\nu = 67 \cdot 1; \quad \Sigma \; x_\nu^2 = 466 \cdot 45; \quad \text{Mean } \bar{x}_1 = 6 \cdot 71;$$

$$\text{C.F.}(x) = 450 \cdot 241; \qquad \text{S.S.}(x) \equiv X_1 = 16 \cdot 209;$$

$$\Sigma \; y_\nu = 7867 \cdot 5; \quad \Sigma \; y_\nu^2 = 6 \; 630 \; 212 \cdot 15; \quad \text{Mean } \bar{y}_1 = 786 \cdot 75;$$

$$\text{C.F.}(y) = 6 \; 189 \; 755 \cdot 625; \qquad \text{S.S.}(y) \equiv Y_1 = 440 \; 456 \cdot 525;$$

$$\Sigma \; x_\nu y_\nu = 55 \; 236 \cdot 87; \; \text{C.F.}(x, y) = 52 \; 790 \cdot 925; \; \text{S.P.}(x, y) \equiv Z_1 = 2445 \cdot 945$$

Therefore the least-squares estimates of α_1 and β_1 are

$$a_1^* = 786 \cdot 75; \qquad \beta_1^* = Z_1/X_1 = 150 \cdot 9004;$$

and the estimated regression equation is $y_{1e}^* = -225 \cdot 79 + 150 \cdot 9004x$.

Error S.S. $= Y_1 - X_1 \beta_1^{*2} = 71 \; 362 \cdot 509$ with 8 d.f.

Slow-water analysis

Let $E(y) = \alpha_2 + \beta_2(x - \bar{x}_2)$.

$$\Sigma \; x_\nu = 58 \cdot 3; \quad \Sigma \; x_\nu^2 = 353 \cdot 19; \quad \text{Mean } \bar{x}_2 = 5 \cdot 83;$$

$$\text{C.F.}(x) = 339 \cdot 889; \qquad \text{S.S.}(x) \equiv X_2 = 13 \cdot 301;$$

$$\Sigma \; y_\nu = 5413 \cdot 9; \quad \Sigma \; y_\nu^2 = 3 \; 290 \; 638 \cdot 57; \quad \text{Mean } \bar{y}_2 = 541 \cdot 39;$$

$$\text{C.F.}(y) = 2 \; 931 \; 031 \cdot 321; \qquad \text{S.S.}(y) \equiv Y_2 = 359 \; 607 \cdot 249;$$

$$\Sigma \; x_\nu y_\nu = 33 \; 222 \cdot 75; \; \text{C.F.}(x, y) = 31 \; 563 \cdot 037; \; \text{S.P.}(x, y) \equiv Z_2 = 1659 \cdot 713.$$

Therefore the least-squares estimates of α_2 and β_2 are

$$a_2^* = 541 \cdot 39; \qquad \beta_2^* = Z_2/X_2 = 124 \cdot 7811;$$

and the estimated regression equation is $y_{2e}^* = -186{\cdot}08 + 124{\cdot}7811x$.

$\qquad$ Error S.S. $= Y_2 - X_2\beta_2^{*2} = 152\,506{\cdot}384$ with 8 d.f.

Pooled error S.S. $= 223\,868{\cdot}893$ with 16 d.f.

Hence

$$\text{s.e.}\,(\beta_1^* - \beta_2^*) = \left[\frac{223\,868{\cdot}893}{16} \times \left(\frac{1}{X_1} + \frac{1}{X_2}\right)\right]^{\frac{1}{2}} = 43{\cdot}7624.$$

The 5 per cent table value of Student's distribution with 16 d.f. is $2{\cdot}120$. Therefore the required confidence interval for $\beta_1 - \beta_2$ is

$$\beta_1^* - \beta_2^* \pm 2{\cdot}120 \times \text{s.e.}\,(\beta_1^* - \beta_2^*) \quad \text{i.e.} \quad 118{\cdot}8956,\; -66{\cdot}6570.$$

Since this interval includes the zero point, the null hypothesis of the equality of the two regression coefficients is acceptable at the 5 per cent level.

57   Leicester, 1961.

$\qquad$ (i)   *Cat 1 analysis*

$$\text{Let } E(y) = a_1 + \beta_1(x - \bar{x}_1).$$

$\Sigma\,x_\nu = 89{\cdot}7;\quad \Sigma\,x_\nu^2 = 730{\cdot}15;\quad \text{Mean } \bar{x}_1 = 7{\cdot}4750;$

C.F.$(x) = 670{\cdot}5075;\qquad$ S.S.$(x) \equiv X_1 = 59{\cdot}6425;$

$\Sigma\,y_\nu = 76;\quad \Sigma\,y_\nu^2 = 674;\quad \text{Mean } \bar{y}_1 = 6{\cdot}3333;$

C.F.$(y) = 481{\cdot}3333;\qquad$ S.S.$(y) \equiv Y_1 = 192{\cdot}6667;$

$\Sigma x_\nu y_\nu = 654{\cdot}0;\quad$ C.F.$(x, y) = 568{\cdot}1000;\quad$ S.P.$(x, y) \equiv Z_1 = 85{\cdot}9000.$

Therefore the least-squares estimates of a_1 and β_1 are

$$a_1^* = 6{\cdot}3333;\qquad \beta_1^* = Z_1/X_1 = 1{\cdot}440\,248;$$

and the estimated regression equation is $y_{1e}^* = -4{\cdot}4326 + 1{\cdot}440\,248x$.

$\qquad$ Error S.S. $= Y_1 - X_1\beta_1^{*2} = 68{\cdot}9494$ with 10 d.f.

$\qquad$ *Cat 2 analysis*

$$\text{Let } E(y) = a_2 + \beta_2(x - \bar{x}_2).$$

$\Sigma\,x_\nu = 161{\cdot}6;\quad \Sigma\,x_\nu^2 = 1810{\cdot}04;\quad \text{Mean } \bar{x}_2 = 10{\cdot}7733;$

C.F.$(x) = 1740{\cdot}9707;\qquad$ S.S.$(x) \equiv X_2 = 69{\cdot}0693;$

$\Sigma\,y_\nu = 153;\quad \Sigma\,y_\nu^2 = 1875;\quad \text{Mean } \bar{y}_2 = 10{\cdot}2000;$

C.F.$(y) = 1560{\cdot}6000;\qquad$ S.S.$(y) \equiv Y_2 = 314{\cdot}4000;$

$\Sigma\,x_\nu y_\nu = 1771{\cdot}7;\;$ C.F.$(x, y) = 1648{\cdot}3200;\;$ S.P.$(x, y) = Z_2 = 123{\cdot}3800.$

Therefore the least-squares estimates of a_2 and β_2 are

$$a_2^* = 10{\cdot}2000;\qquad \beta_2^* = Z_2/X_2 = 1{\cdot}786\,322;$$

and the estimated regression equation is $y_{2e}^* = -9{\cdot}0446 + 1{\cdot}786\,322x$.

$\qquad$ Error S.S. $= Y_2 - X_2\beta_2^{*2} = 94{\cdot}0036$ with 13 d.f.

(ii) If σ_1^2 and σ_2^2 are the true error variances for coronary flow of the two cats, then to test the null hypothesis $H(\sigma_1^2 = \sigma_2^2)$ we have

$$F = \frac{13 \times 68 \cdot 9494}{10 \times 94 \cdot 0036} = 0 \cdot 95 \text{ with } 10, 13 \text{ d.f.}$$

For 10, 13 d.f. the upper and lower one per cent points of the F distribution are $4 \cdot 10$ and $< 1/4 \cdot 60$. Hence the observed ratio is not significant at the 2 per cent level (two-sided test).

Also,   s.e. $(\beta_1^* - \beta_2^*) = \left[\frac{162 \cdot 9530}{23} \times \left(\frac{1}{X_1} + \frac{1}{X_2} \right) \right]^{\frac{1}{2}} = 0 \cdot 470 \ 496.$

Hence, to test the null hypothesis $H(\beta_1 = \beta_2)$, we have

$$|t| = 0 \cdot 346 \ 074 / 0 \cdot 470 \ 496 = 0 \cdot 74 \text{ with } 23 \text{ d.f.}$$

The observed value of t is obviously not significant and so the null hypothesis is acceptable.

58   Let x denote initial weight and y weight after one week.

(i)   *Control group analysis*

$$\text{Let } E(y) = \alpha_1 + \beta_1(x - \bar{x}_1).$$

$\Sigma \ x_\nu = 540; \quad \Sigma \ x_\nu^2 = 29 \ 426; \quad \text{Mean } \bar{x}_1 = 54 \cdot 0;$

$\text{C.F.}(x) = 29 \ 160 \cdot 00; \qquad \text{S.S.}(x) \equiv X_1 = 266 \cdot 00;$

$\Sigma \ y_\nu = 785; \quad \Sigma \ y_\nu^2 = 62 \ 459; \quad \text{Mean } \bar{y}_1 = 78 \cdot 5;$

$\text{C.F.}(y) = 61 \ 622 \cdot 50; \qquad \text{S.S.}(y) \equiv Y_1 = 836 \cdot 50;$

$\Sigma \ x_\nu y_\nu = 42 \ 836; \quad \text{C.F.}(x, y) = 42 \ 390 \cdot 00; \quad \text{S.P.}(x, y) \equiv Z_1 = 446 \cdot 00.$

Therefore the least-squares estimates of α_1 and β_1 are

$$\alpha_1^* = 78 \cdot 5; \qquad \beta_1^* = Z_1/X_1 = 1 \cdot 676 \ 692;$$

and the estimated regression equation is $y_{1e}^* = -12 \cdot 04 + 1 \cdot 676 \ 692x.$

$$\text{Error S.S.} = Y_1 - X_1 \beta_1^{*2} = 88 \cdot 70 \text{ with } 8 \text{ d.f.}$$

Thyroxin group analysis

$$\text{Let } E(y) = \alpha_2 + \beta_2(x - \bar{x}_2).$$

$\Sigma \ x_\nu = 389; \quad \Sigma \ x_\nu^2 = 21 \ 671; \quad \text{Mean } \bar{x}_2 = 55 \cdot 57;$

$\text{C.F.}(x) = 21 \ 617 \cdot 29; \qquad \text{S.S.}(x) \equiv X_2 = 53 \cdot 71;$

$\Sigma \ y_\nu = 531; \quad \Sigma \ y_\nu^2 = 40 \ 529; \quad \text{Mean } \bar{y}_2 = 75 \cdot 86;$

$\text{C.F.}(y) = 40 \ 280 \cdot 14; \qquad \text{S.S.}(y) \equiv Y_2 = 248 \cdot 86;$

$\Sigma \ x_\nu y_\nu = 29 \ 604; \quad \text{C.F.}(x, y) = 29 \ 508 \cdot 43; \quad \text{S.P.}(x, y) \equiv Z_2 = 95 \cdot 57.$

Therefore the least-squares estimates of α_2 and β_2 are

$$\alpha_2^* = 75 \cdot 86; \qquad \beta_2^* = Z_2/X_2 = 1 \cdot 779 \ 371;$$

and the estimated regression equation is $y_{2e}^* = -23 \cdot 02 + 1 \cdot 779 \ 371x.$

$$\text{Error S.S.} = Y_2 - X_2 \beta_2^{*2} = 78 \cdot 81 \text{ with } 5 \text{ d.f.}$$

Thiouracil group analysis

$$\text{Let } E(y) = \alpha_3 + \beta_3(x - \bar{x}_3).$$

$\Sigma\, x_\nu = 547;$ $\Sigma\, x_\nu^2 = 30\ 119;$ Mean $\bar{x}_3 = 54\cdot7;$

C.F. $(x) = 29\ 920\cdot90;$ S.S. $(x) \equiv X_3 = 198\cdot10;$

$\Sigma\, y_\nu = 763;$ $\Sigma\, y_\nu^2 = 58\ 781;$ Mean $\bar{y}_3 = 76\cdot3;$

C.F. $(y) = 58\ 216\cdot90;$ S.S. $(y) \equiv Y_3 = 564\cdot10;$

$\Sigma\, x_\nu y_\nu = 41\ 987;$ C.F. $(x, y) = 41\ 736\cdot10;$ S.P. $(x, y) \equiv Z_3 = 250\cdot90.$

Therefore the least-squares estimates of α_3 and β_3 are

$$\alpha_3^* = 76\cdot3; \qquad \beta_3^* = Z_3/X_3 = 1\cdot266\ 532;$$

and the estimated regression equation is $y_{3e}^* = 7\cdot02 + 1\cdot266\ 532x.$

$$\text{Error S.S.} = Y_3 - X_3\beta_3^{*2} = 246\cdot33 \text{ with 8 d.f.}$$

Comparison of regressions

Hence pooled error S.S. $= 413\cdot84$ with 21 d.f.

Also, $X \equiv \Sigma\, X_i = 517\cdot81$ and $Z \equiv \Sigma\, Z_i = 792\cdot47.$

If β denotes the common regression parameter, then the least-squares estimates of β is

$$\beta^* = Z/X = 1\cdot530\ 426.$$

Hence the S.S. due to regression is

$$\sum_{i=1}^{3} X_i\beta_i^{*2} - X\beta^{*2} = 22\cdot80 \text{ with 2 d.f.}$$

The analysis of variance is as follows:

Source of variation	d.f.	S.S.	M.S.	F.R.
$H(\beta_1 = \beta_2 = \beta_3)$	2	22·80	11·40	< 1
Error	21	413·84	19·7067	
Total	23	436·64		

The F ratio is evidently not significant and so it is reasonable to accept the hypothesis of the equality of the regression coefficients in the three groups.

(ii) We have

$$\text{s.e.}\,(\beta_2^* - \beta_3^*) = \left[\frac{413\cdot84}{21} \times \left(\frac{1}{X_2} + \frac{1}{X_3}\right)\right]^{\frac{1}{2}} = 0\cdot682\ 926.$$

The 5 per cent table value of Student's distribution with 21 d.f. is 2·080. Hence the 95 per cent confidence interval for $\beta_2 - \beta_3$ is

$$\beta_2^* - \beta_3^* \pm 2\cdot080 \times \text{s.e.}\,(\beta_2^* - \beta_3^*) \quad \text{i.e.} \quad 1\cdot9333,\ -0\cdot9076.$$

(iii) Let $z = y - x$ denote the gain in weight. Then the analysis of the z observations is as follows:

$\Sigma\, z_\nu = 603;$ $\Sigma\, z_\nu^2 = 14\ 131;$ C.F. $= (603)^2/27 = 13\ 467\cdot00.$

Therefore   total S.S. $= 664 \cdot 00$ with 26 d.f.

Between-groups S.S. $= \dfrac{(245)^2}{10} + \dfrac{(142)^2}{7} + \dfrac{(216)^2}{10} -$ C.F. $= 81 \cdot 67$ with 2 d.f.

Hence the following analysis of variance:

Source of variation	d.f.	S.S.	M.S.	F.R.
Between groups	2	81·67	40·835	1·68
Error	24	582·33	24·264	
Total	26	664·00		

The 5 per cent table value of the F distribution with 2, 24 d.f. is $3 \cdot 40$. Therefore the observed F ratio is not significant and we may accept the hypothesis that there are, on the average, no differences in the populations of animals treated in the three ways.

(iv) If σ_2^2 and σ_3^2 are the true variances of the final weights of the animals in the thyroxin and thiouracil groups respectively, then from the analysis in (i) the best estimates of the parameters are

$78 \cdot 81/5 = 15 \cdot 762$ with 5 d.f.   and $246 \cdot 33/8 = 30 \cdot 791$ with 8 d.f.

Therefore, to test the null hypothesis $H(\sigma_2^2 = \sigma_3^2)$, we have

$$F = 15 \cdot 762/30 \cdot 791 = 0 \cdot 512 \text{ with } 5, 8 \text{ d.f.}$$

For a two-sided test at the 5 per cent level, the upper and lower percentage points for F with 5, 8 d.f. are $4 \cdot 82$ and $1/6 \cdot 76$. The observed ratio is not significant and the null hypothesis is acceptable.

59   Leicester, 1961.

This is an example of a weighted linear regression analysis. Let y denote yield rate and x the variable $32\times$ intensity. Also, let f_ν denote the frequency corresponding to the x and y values x_ν and y_ν respectively.

Shahabad analysis

$$\text{Let } E(y) = \alpha_1 + \beta_1(x - \bar{x}_1).$$

$\Sigma f_\nu x_\nu = 3475$; $\Sigma f_\nu y_\nu = 1196 \cdot 4$; Mean $\bar{x}_1 = 18 \cdot 3862$; Mean $\bar{y}_1 = 6 \cdot 3302$;

$\Sigma f_\nu x_\nu^2 = 77\,277$; $\Sigma f_\nu y_\nu^2 = 7770 \cdot 60$; $\Sigma f_\nu x_\nu y_\nu = 23\,461 \cdot 6$;

C.F. $(x) = 63\,892 \cdot 20$; C.F. $(y) = 7573 \cdot 40$; C.F. $(x, y) = 21\,997 \cdot 30$;

S.S. $(x) \equiv X_1 = 13\,384 \cdot 80$; S.S. $(y) \equiv Y_1 = 197 \cdot 20$;

S.P. $(x, y) \equiv Z_1 = 1464 \cdot 30$.

Therefore the least-squares estimates of α_1 and β_1 are

$$\alpha_1^* = 6 \cdot 3302; \qquad \beta_1^* = Z_1/X_1 = 0 \cdot 109\,400;$$

and the estimated regression equation is $y_{1e}^* = 4 \cdot 3187 + 0 \cdot 109\,400x$.

$$\text{Error S.S.} = Y_1 - X_1\beta_1^{*2} = 37 \cdot 01 \text{ with } 187 \text{ d.f.}$$

Monghyr analysis

$$\text{Let } E(y) = \alpha_2 + \beta_2(x - \bar{x}_2).$$

$\Sigma f_\nu x_\nu = 4681;$ $\Sigma f_\nu y_\nu = 1557\cdot3;$ Mean $\bar{x}_2 = 18\cdot9514;$ Mean $\bar{y}_2 = 6\cdot3049;$

$\Sigma f_\nu x_\nu^2 = 105\ 911;$ $\Sigma f_\nu y_\nu^2 = 10\ 366\cdot81;$ $\Sigma f_\nu x_\nu y_\nu = 32\ 197\cdot9;$

C.F. $(x) = 88\ 711\cdot58;$ C.F. $(y) = 9818\cdot56;$ C.F. $(x, y) = 29\ 513\cdot04;$

$$\text{S.S.} (x) \equiv X_2 = 17\ 199\cdot42; \qquad \text{S.S.} (y) \equiv Y_2 = 548\cdot25;$$

$$\text{S.P.} (x, y) \equiv Z_2 = 2684\cdot86.$$

Therefore the least-squares estimates of α_2 and β_2 are

$$\alpha_2^* = 6\cdot3049; \qquad \beta_2^* = Z_2/X_2 = 0\cdot156\ 102;$$

and the estimated regression equation is $y_{2e}^* = 3\cdot3465 + 0\cdot156\ 102x.$
Error S.S. $= Y_2 - X_2\beta_2^{*2} = 129\cdot14$ with 245 d.f.

Comparison of regressions

Hence pooled error S.S. $= 166\cdot15$ with 432 d.f.

$$\text{and} \quad \text{s.e.} (\beta_1^* - \beta_2^*) = \left[\frac{166\cdot15}{432} \times \left(\frac{1}{X_1} + \frac{1}{X_2} \right) \right]^{\frac{1}{2}} = 10^{-2} \times 0\cdot714\ 816.$$

Therefore, to test the null hypothesis $H(\beta_1 - \beta_2 = 0)$, we have

$$t = 0\cdot046\ 702/(10^{-2} \times 0\cdot714\ 816) = 6\cdot53 \text{ with 432 d.f.}$$

The d.f. for t are effectively infinite, and so the null hypothesis is rejected at the one per cent level of significance.

The 95 per cent confidence interval for $\beta_1 - \beta_2$ is

$$\beta_1^* - \beta_2^* \pm 1\cdot96 \times \text{s.e.} (\beta_1^* - \beta_2^*) \quad \text{i.e.} \quad 0\cdot060\ 712, \ 0\cdot032\ 692.$$

This is the interval in terms of 32x intensity units; hence in terms of the original intensity units, the appropriate difference is $32(\beta_1 - \beta_2)$, which has the required limits $1\cdot9428, \ 1\cdot0461.$

60   Leicester, 1967.

$$\text{Let } E(y) = \alpha + \beta(x - \bar{x}).$$

It is convenient to make the transformations

$$u = 1000(x - 2\cdot500); \qquad v = 1000(y - 1\cdot900)$$

and then assume that $E(v) = \gamma + \delta(u - \bar{u})$. For the transformed u, v observations we have the following calculations.

$$\Sigma u_i = 3575; \quad \Sigma u_i^2 = 547\ 925; \quad \text{Mean } \bar{u} = 143\cdot00;$$

$$\text{C.F.} (u) = 511\ 225\cdot00; \qquad \text{S.S.} (u) \equiv U = 36\ 700\cdot00;$$

$$\Sigma v_i = 2620; \quad \Sigma v_i^2 = 292\ 650; \quad \text{Mean } \bar{v} = 104\cdot80;$$

$$\text{C.F.} (v) = 274\ 576\cdot00; \qquad \text{S.S.} (v) \equiv V = 18\ 074\cdot00;$$

$\Sigma u_i v_i = 398\ 225;$ C.F. $(u, v) = 374\ 660\cdot00;$ S.P. $(u, v) \equiv W = 23\ 565\cdot00.$

Therefore the least-squares estimates of γ and δ are

$$\gamma^* = 104\cdot80; \qquad \delta^* = W/U = 0\cdot642\ 098.$$
$$\text{Error S.S.} = V - U\delta^{*2} = 2942\cdot96 \text{ with 23 d.f.,}$$

and the best estimate of $\mathrm{var}(v)$ is $s_v^2 = 127\cdot9548$.

Transforming back to original variables, the estimate of α is

$$\alpha^* = \bar{y} = 10^{-3}\ \bar{v} + 1\cdot900 = 2\cdot0048$$

and that of β is $\beta^* = \delta^* = 0\cdot642\ 098$.

The best estimate of σ^2 is $10^{-6}\ s_v^2 = 10^{-3} \times 0\cdot127\ 955$.

Finally, $\bar{x} = 10^{-3}\bar{u} + 2\cdot500 = 2\cdot6430$.

Hence the estimated regression equation of y on x is

$$y_e^* = 0\cdot3077 + 0\cdot642\ 098x.$$

Therefore, corresponding to $x = 2\cdot60$, the estimated y value is

$$y_{e0}^* = 0\cdot3077 + 1\cdot6695 = 1\cdot9772$$

and $$\text{s.e.}(y_{e0}^*) = \left[10^{-3} \times 0\cdot127\ 955 \times \left\{\frac{1}{25} + \frac{(2\cdot60 - 2\cdot6430)^2}{10^{-6}\ U}\right\}\right]^{\frac{1}{2}}$$

$$= 10^{-2} \times 0\cdot3401.$$

61   Leicester, 1968.

Let x denote hardness and y abrasion loss. Then suppose that

$$E(y) = \alpha + \beta(x - \bar{x}).$$

$\Sigma\,x_i = 2108$; $\Sigma\,x_i^2 = 152\ 422$;   Mean $\bar{x} = 70\cdot2667$;

C.F.$(x) = 148\ 122\cdot13$;   S.S.$(x) \equiv X = 4299\cdot87$;

$\Sigma\,y_i = 5263$; $\Sigma\,y_i^2 = 1\ 148\ 317$;   Mean $\bar{y} = 175\cdot4333$;

C.F.$(y) = 923\ 305\cdot63$;   S.S.$(y) \equiv Y = 225\ 011\cdot37$;

$\Sigma\,x_iy_i = 346\ 867$;   C.F.$(x, y) = 369\ 813\cdot47$;

S.P.$(x, y) \equiv Z = -22\ 946\cdot47$.

Therefore the least-squares estimates of α and β are

$$\alpha^* = 175\cdot4333; \qquad \beta^* = Z/X = -5\cdot336\ 550;$$

and the estimated regression equation is $y_e^* = 550\cdot4151 - 5\cdot336\ 550x$.

$$\text{Error S.S.} = Y - X\beta^{*2} = 102\ 556\cdot38 \text{ with 28 d.f.}$$

The estimated abrasion value corresponding to $x = 90$ is

$$y_{e0}^* = 550\cdot4151 - 480\cdot2895 = 70\cdot1256$$

and $$\text{s.e.}(y_{e0}^*) = \left[\frac{102\ 556\cdot38}{28} \times \left\{\frac{1}{30} + \frac{(90 - \bar{x})^2}{X}\right\}\right]^{\frac{1}{2}} = 21\cdot3024.$$

The 5 per cent table value of Student's distribution with 28 d.f. is $2 \cdot 048$. Hence, if y_{e0} is the true abrasion value corresponding to $x = 90$, then the 95 per cent confidence interval for y_{e0} is

$$y_{e0}^* \pm 2 \cdot 048 \times \text{s.e.} (y_{e0}^*) \quad \text{i.e.} \quad 113 \cdot 75, \ 26 \cdot 50.$$

These limits are too wide to be of much practical use.

62   Leicester, 1970.

Let x denote time in minutes and y the common logarithm of the percentage water content at time x. Also, let the samples be denoted by the suffix i ($i = 1, 2, 3, 4$), the numbering being in the order from left to right in which the data are recorded. Then for the ith sample, we have

$$E(y) = \alpha_i + \beta_i (x - \bar{x}_i); \qquad \text{var}(y) = \sigma^2.$$

The necessary basic calculations are given in the table below, the symbols α_i^* and β_i^* being the least-squares estimates of α_i and β_i respectively, and $Y_i - X_i \beta_i^{*2}$ the error S.S. from the ith sample.

Expression	City Imperial 1 $i = 1$	City Imperial 2 $i = 2$	Manitoba 1 $i = 3$	Manitoba 2 $i = 4$
$\sum_{\nu} x_{i\nu}$	1114·75	1371·50	1221·50	1196·00
$\sum_{\nu} y_{i\nu}$	10·11	15·39	11·20	10·98
Mean $\bar{x}_i$	101·3410	85·7188	101·7917	99·6667
Mean $\bar{y}_i$	0·9191	0·9619	0·9333	0·9150
$\sum_{\nu} x_{i\nu}^2$	163 200·6875	183 724·7500	176 632·1250	171 351·5000
$\sum_{\nu} y_{i\nu}^2$	9·4407	15·1071	10·6560	10·2284
$\sum_{\nu} x_{i\nu} y_{i\nu}$	939·1075	1178·4850	1038·4925	998·3300
C.F. (x_i)	112 969·7784	117 563·2656	124 338·5208	119 201·3333
C.F. (y_i)	9·292 009	14·803 256	10·453 333	10·046 700
C.F. (x_i, y_i)	1024·5566	1319·2116	1140·0667	1094·3400
S.S. $(x_i) \equiv X_i$	50 230·9091	66 161·4844	52 293·6042	52 150·1667
S.S. $(y_i) \equiv Y_i$	0·148 691	0·303 844	0·202 667	0·181 700
S.P. $(x_i, y_i) \equiv Z_i$	$-85 \cdot 4491$	$-140 \cdot 7266$	$-101 \cdot 5742$	$-96 \cdot 0100$
α_i^*	0·9191	0·9619	0·9333	0·9150
β_i^*	$-0 \cdot 001 \ 701 \ 13$	$-0 \cdot 002 \ 127 \ 02$	$-0 \cdot 001 \ 942 \ 38$	$-0 \cdot 001 \ 841 \ 03$
$Y_i - X_i \beta_i^{*2}$	0·003 331	0·004 516	0·005 372	0·004 943
d.f.	9	14	10	10

$$\sum_{i=1}^{4} X_i \equiv X = 220 \ 836 \cdot 1644; \qquad \sum_{i=1}^{4} Z_i \equiv Z = -423 \cdot 7599.$$

If β is the true common value of the β_i when these are assumed equal, then the least-squares estimate of β is $\beta^* = Z/X = -0\cdot001\ 918\ 89$. Hence the S.S. due to regression is

$$\sum_{i=1}^{4} X_i \beta_i^{*2} - X\beta^{*2} = 0\cdot005\ 590 \text{ with 3 d.f.}$$

and the pooled error S.S. $= 0\cdot018\ 162$ with 43 d.f.

Therefore the F ratio for testing the equality of the regression co-efficients is

$$F = (43 \times 0\cdot005\ 590)/(3 \times 0\cdot018\ 162) = 4\cdot41 \text{ with 3, 43 d.f.}$$

The observed F is just significant at the one per cent level. Strictly, the null hypothesis is not tenable, but if we accept it, then

$$\text{s.e.}\,(\beta^*) = (0\cdot018\ 162/43X)^{\frac{1}{2}} = 10^{-4} \times 0\cdot437\ 334.$$

The 5 per cent table value of Student's distribution with 43 d.f. is $2\cdot016$. Hence the 95 confidence interval for β is

$$\beta^* \pm 2\cdot016 \times \text{s.e.}\,(\beta^*) \quad \text{i.e.} \quad -10^{-2} \times 0\cdot183\ 072,\ -10^{-2} \times 0\cdot200\ 706.$$

The grand means are

$$\bar{x} = 4903\cdot75/51 = 96\cdot1520; \quad \bar{y} = 47\cdot68/51 = 0\cdot9349.$$

Therefore the estimated pooled regression is

$$y_e^* = 1\cdot1194 - 10^{-2} \times 0\cdot191\ 889x.$$

63 Let y denote the units supplied and $x =$ heat consumed$/1000$.

(i) Let $E(y) = \alpha_1 + \beta_1(x - \bar{x}_1)$.

$\Sigma\,x_\nu = 3598\cdot307; \quad \Sigma\,x_\nu^2 = 646\ 856\cdot3898; \quad$ Mean $\bar{x}_1 = 163\cdot5594;$

$\text{C.F.}\,(x) = 588\ 536\cdot9666; \qquad \text{S.S.}\,(x) \equiv X_1 = 58\ 319\cdot4232;$

$\Sigma\,y_\nu = 240\cdot136; \quad \Sigma\,y_\nu^2 = 2925\cdot4792; \quad$ Mean $\bar{y}_1 = 10\cdot9153;$

$\text{C.F.}\,(y) = 2621\cdot1499; \qquad \text{S.S.}\,(y) \equiv Y_1 = 304\cdot3293;$

$\Sigma\,x_\nu y_\nu = 43\ 478\cdot3021; \qquad \text{C.F.}\,(x,\ y) = 39\ 276\cdot5023;$

$$\text{S.P.}\,(x,\ y) \equiv Z_1 = 4201\cdot7998.$$

Therefore the least-squares estimates of α_1 and β_1 are

$$\alpha_1^* = 10\cdot9153; \qquad \beta_1^* = Z_1/X_1 = 0\cdot072\ 048\ 0;$$

and the estimated regression equation is $y_{1e}^* = -0\cdot8688 + 0\cdot072\ 048\ 0x$.

$$\text{Error S.S.} = Y_1 - X_1\beta_1^{*2} = 1\cdot5984 \text{ with 20 d.f.}$$

(ii) Let $E(y) = \alpha_2 + \beta_2(x - \bar{x}_2)$.

$\Sigma\,x_\nu = 4331\cdot338; \quad \Sigma\,x_\nu^2 = 858\ 156\cdot4345; \quad$ Mean $\bar{x}_2 = 180\cdot4724;$

$\text{C.F.}\,(x) = 781\ 687\cdot0363; \qquad \text{S.S.}\,(x) \equiv X_2 = 76\ 469\cdot3982;$

$\Sigma\,y_\nu = 221\cdot442; \quad \Sigma\,y_\nu^2 = 2268\cdot3247; \quad$ Mean $\bar{y}_2 = 9\cdot2268;$

$$\text{C.F.}(y) = 2043 \cdot 1900; \qquad \text{S.S.}(y) \equiv Y_2 = 225 \cdot 1347;$$
$$\Sigma\, x_\nu y_\nu = 44\,096 \cdot 4956; \qquad \text{C.F.}(x, y) = 39\,964 \cdot 1729;$$
$$\text{S.P.}(x, y) \equiv Z_2 = 4132 \cdot 3227.$$

Therefore the least-squares estimates of α_2 and β_2 are

$$a_2^* = 9 \cdot 2268; \qquad \beta_2^* = Z_2/X_2 = 0 \cdot 054\,038\,9;$$

and the estimated regression equation is $y_{2e}^* = -0 \cdot 5257 + 0 \cdot 054\,038\,9x$.

Error S.S. $= Y_2 - X_2\beta_2^{*2} = 1 \cdot 8288$ with 22 d.f.

Comparison of regressions

Pooled error S.S. $= 3 \cdot 4272$ with 42 d.f.

Therefore s.e. $(\beta_1^* - \beta_2^*) = \left[\dfrac{3 \cdot 4272}{42} \times \left(\dfrac{1}{X_1} + \dfrac{1}{X_2} \right) \right]^{\frac{1}{2}} = 10^{-2} \times 0 \cdot 157\,04.$

Hence, to test the null hypothesis $H(\beta_1 - \beta_2 = 0)$, we have

$$t = 0 \cdot 018\,009\,1/0 \cdot 001\,570\,4 = 11 \cdot 47 \text{ with } 42 \text{ d.f.}$$

The observed t is evidently significant and so the hypothesis can be rejected.

64 (i) The observed frequency distribution is as follows:

Interval	Frequency	Interval	Frequency
265 –	9	310 –	17
270 –	13	315 –	21
275 –	9	320 –	22
280 –	18	325 –	9
285 –	8	330 –	2
290 –	31	335 –	0
295 –	29	340 –	4
300 –	39	345 –	1
305 –	13		
Total			245

For this distribution, we have

Mean $\bar{x} = 300 \cdot 9286;$ \qquad Variance $s^2 = 294 \cdot 959\,025.$

Therefore \qquad s.e. $(\bar{x}) = s/\sqrt{245} = 1 \cdot 0972.$

Hence the required 99 per cent confidence interval for the true mean is

$$\bar{x} \pm 2 \cdot 576 \times \text{s.e.}\,(\bar{x}) \quad \text{i.e.} \quad 303 \cdot 76,\ 298 \cdot 10.$$

(ii) Let x denote the measurements made by the short-period test and y the corresponding lifetime in hours. Then we assume that

$$E(y) = \alpha + \beta(x - \bar{x}); \qquad \text{var}(y) = \sigma^2.$$

$\Sigma\, x_i = 6477; \quad \Sigma\, x_i^2 = 2\,114\,591; \quad \text{Mean } \bar{x} = 323 \cdot 85;$

$$C.F.(x) = 2\ 097\ 576 \cdot 45; \qquad S.S.(x) \equiv X = 17\ 014 \cdot 55;$$

$$\Sigma\ y_i = 24\ 220; \quad \Sigma\ y_i^2 = 31\ 176\ 450; \quad \text{Mean } \bar{y} = 1211 \cdot 00;$$

$$C.F.(y) = 29\ 330\ 420 \cdot 00; \qquad S.S.(y) \equiv Y = 1\ 846\ 030 \cdot 00;$$

$$\Sigma\ x_i\ y_i = 7\ 706\ 700; \qquad C.F.(x, y) = 7\ 843\ 647 \cdot 00;$$

$$S.P.(x, y) \equiv Z = -\ 136\ 947 \cdot 00.$$

Therefore the least-squares estimates of α and β are

$$\alpha^* = 1211 \cdot 00; \qquad \beta^* = Z/X = -8 \cdot 048\ 817;$$

and the estimated regression equation is

$$y_e^* = 3817 \cdot 61 - 8 \cdot 048\ 817x.$$

From the first sample the average for the short-period test is $300 \cdot 9286$. Therefore, corresponding to this value, the estimated lifetime is

$$y_{e0}^* = 3817 \cdot 61 - 8 \cdot 048\ 817 \times 300 \cdot 9286 = 1395 \cdot 49.$$

Also,   error S.S. $= Y - X\beta^{*2} = 743\ 768 \cdot 67$ with 18 d.f.

Hence

$$\text{s.e.}(y_{e0}^*) = \left[\frac{743\ 768 \cdot 67}{18} \times \left\{ \frac{1}{20} + \frac{(300 \cdot 9286 - 323 \cdot 85)^2}{X} \right\} \right]^{\frac{1}{2}} = 57 \cdot 8096.$$

65 (i) *Yield analysis*

Raw S.S. $= 9\ 304\ 673;$ Grand total $= 29\ 097;$ C.F. $= 8\ 466\ 354 \cdot 09;$

$$\text{Total S.S.} = 838\ 318 \cdot 91 \text{ with } 99 \text{ d.f.}$$

$$\text{Between-years S.S.} = \frac{(15\ 055)^2 + (14\ 042)^2}{50} - C.F.$$

$$= 10\ 261 \cdot 69 \text{ with } 1 \text{ d.f.}$$

The analysis of variance is as follows:

Source of variation	d.f.	S.S.	M.S.	F.R.
Between years	1	10 261·69	10 261·69	1·21
Error	98	828 057·22	8 449·56	
Total	99	838 318·91		

The 5 per cent table value of the F distribution with 1, 98 d.f. is $> 3 \cdot 94$, and so the observed F ratio is not significant. There is no evidence of a difference in the yield for the two years. The average yields are $301 \cdot 10$ and $280 \cdot 84$ for 1930 and 1931 respectively.

Protein analysis

Raw S.S. $= 21\ 352 \cdot 68;$ Grand total $= 1459 \cdot 4;$ C.F. $= 21\ 298 \cdot 4836;$

$$\text{Total S.S.} = 54 \cdot 1964 \text{ with } 99 \text{ d.f.}$$

$$\text{Between-years S.S.} = \frac{(740 \cdot 6)^2 + (718 \cdot 8)^2}{50} - C.F.$$

$$= 4 \cdot 7524 \text{ with } 1 \text{ d.f.}$$

The analysis of variance is as follows:

Source of variation	d.f.	S.S.	M.S.	F.R.
Between years	1	4·7524	4·752 4	9·42
Error	98	49·4440	0·504 53	
Total	99	54·1964		

The observed F ratio is evidently significant, and so there is evidence for believing that the protein content of the wheat differed in the two years. The averages for the two years are 14·812 and 14·376.

(ii) Let x denote protein content and y the yield of wheat.

1930 analysis

$$\text{Let } E(y) = \alpha_1 + \beta_1(x - \bar{x}_1); \qquad \text{var}(y) = \sigma^2.$$

$$\Sigma x_\nu = 740\cdot6; \quad \Sigma x_\nu^2 = 11\ 003\cdot56; \quad \text{Mean } \bar{x}_1 = 14\cdot812;$$

$$\text{C.F.}(x) = 10\ 969\cdot7672; \qquad \text{S.S.}(x) \equiv X_1 = 33\cdot7928;$$

$$\Sigma y_\nu = 15\ 055; \quad \Sigma y_\nu^2 = 4\ 849\ 373; \quad \text{Mean } \bar{y}_1 = 301\cdot10;$$

$$\text{C.F.}(y) = 4\ 533\ 060\cdot50; \qquad \text{S.S.}(y) \equiv Y_1 = 316\ 312\cdot50;$$

$$\Sigma x_\nu y_\nu = 220\ 755\cdot8; \qquad \text{C.F.}(x, y) = 222\ 994\cdot66;$$

$$\text{S.P.}(x, y) \equiv Z_1 = -2238\cdot86.$$

Therefore the least-squares estimates of α_1 and β_1 are

$$\alpha_1^* = 301\cdot10; \qquad \beta_1^* = Z_1/X_1 = -66\cdot2526;$$

and the estimated regression equation is $y_{1e}^* = 1282\cdot43 - 66\cdot2526x$.

$$\text{Error S.S.} = Y_1 - X_1\beta_1^{*2} = 167\ 982\cdot15 \text{ with 48 d.f.}$$

1931 analysis

$$\text{Let } E(y) = \alpha_2 + \beta_2(x - \bar{x}_2); \qquad \text{var}(y) = \sigma^2.$$

$$\Sigma x_\nu = 718\cdot8; \quad \Sigma x_\nu^2 = 10\ 349\cdot12; \quad \text{Mean } \bar{x}_2 = 14\cdot376;$$

$$\text{C.F.}(x) = 10\ 333\cdot4688; \qquad \text{S.S.}(x) \equiv X_2 = 15\cdot6512;$$

$$\Sigma y_\nu = 14\ 042; \quad \Sigma y_\nu^2 = 4\ 455\ 300; \quad \text{Mean } \bar{y}_2 = 280\cdot84;$$

$$\text{C.F.}(y) = 3\ 943\ 555\cdot28; \qquad \text{S.S.}(y) \equiv Y_2 = 511\ 744\cdot72;$$

$$\Sigma x_\nu y_\nu = 200\ 673\cdot5; \qquad \text{C.F.}(x, y) = 201\ 867\cdot79;$$

$$\text{S.P.}(x, y) \equiv Z_2 = -1194\cdot29.$$

Therefore the least-squares estimates of α_2 and β_2 are

$$\alpha_2^* = 280\cdot84; \qquad \beta_2^* = Z_2/X_2 = -76\cdot3066;$$

and the estimated regression equation is $y_{2e}^* = 1377\cdot82 - 76\cdot3066x$.

$$\text{Error S.S.} = Y_2 - X_2\beta_2^{*2} = 420\ 612\cdot52 \text{ with 48 d.f.}$$

Comparison of regressions

Pooled error S.S. $= 588\ 594 \cdot 67$ with 96 d.f. Hence

$$\text{s.e.}\,(\beta_1^* - \beta_2^*) = \left[\frac{588\ 594 \cdot 67}{96} \times \left(\frac{1}{X_1} + \frac{1}{X_2}\right)\right]^{\frac{1}{2}} = 23 \cdot 941\ 05.$$

The 5 per cent table value of Student's distribution with 96 d.f. is $1 \cdot 995$. Therefore the 95 per cent confidence interval for $\beta_1 - \beta_2$ is

$$\beta_1^* - \beta_2^* \pm 1 \cdot 995 \times \text{s.e.}\,(\beta_1^* - \beta_2^*) \quad \text{i.e.} \quad 57 \cdot 8164,\ -37 \cdot 7084.$$

Since this interval includes the zero point, therefore the null hypothesis $H(\beta_1 - \beta_2 = 0)$ is acceptable.

$$\text{Again,}\quad \text{s.e.}\,(a_1^* - a_2^*) = \left[\frac{588\ 594 \cdot 67}{96} \times \left(\frac{1}{50} + \frac{1}{50}\right)\right]^{\frac{1}{2}} = 35 \cdot 0177.$$

Therefore the 95 per cent confidence interval for $a_1 - a_2$ is

$$a_1^* - a_2^* \pm 1 \cdot 995 \times \text{s.e.}\,(a_1^* - a_2^*) \quad \text{i.e.} \quad 90 \cdot 12,\ -49 \cdot 60.$$

Since this interval includes the zero point, therefore the null hypothesis $H(a_1 - a_2 = 0)$ is acceptable.

There is thus evidence for believing that there is no difference in the linear relationship between yield and protein content in the two years.

66 Let x denote total protein content and y the glutenin content of wheat. We then assume that for the ith study ($i = 1, 2, 3$),

$$E(y) = a_i + \beta_i (x - \bar{x}_i) \quad \text{and} \quad \text{var}(y) = \sigma^2.$$

Study no. 1 analysis

$$\Sigma\, x_\nu = 214 \cdot 25; \quad \Sigma\, x_\nu^2 = 2785 \cdot 8129; \quad \text{Mean}\ \bar{x}_1 = 12 \cdot 6029;$$
$$\text{C.F.}\,(x) = 2700 \cdot 1801; \quad \text{S.S.}\,(x) \equiv X_1 = 85 \cdot 6328;$$
$$\Sigma\, y_\nu = 87 \cdot 97; \quad \Sigma\, y_\nu^2 = 468 \cdot 6577; \quad \text{Mean}\ \bar{y}_1 = 5 \cdot 1747;$$
$$\text{C.F.}\,(y) = 455 \cdot 2189; \quad \text{S.S.}\,(y) \equiv Y_1 = 13 \cdot 4388;$$
$$\Sigma\, x_\nu\, y_\nu = 1140 \cdot 3907; \quad \text{C.F.}\,(x, y) = 1108 \cdot 6807;$$
$$\text{S.P.}\,(x, y) \equiv Z_1 = 31 \cdot 7100.$$

Therefore the least-squares estimates of a_1 and β_1 are

$$a_1^* = 5 \cdot 1747; \qquad \beta_1^* = Z_1/X_1 = 0 \cdot 370\ 302;$$

and the estimated regression equation is $y_{1e}^* = 0 \cdot 5078 + 0 \cdot 370\ 302x$.

$$\text{Error S.S.} = Y_1 - X_1 \beta_1^{*2} = 1 \cdot 6965 \text{ with 15 d.f.}$$

Study no. 2 analysis

$$\Sigma\, x_\nu = 211 \cdot 51; \quad \Sigma\, x_\nu^2 = 2666 \cdot 7467; \quad \text{Mean}\ \bar{x}_2 = 12 \cdot 4418;$$
$$\text{C.F.}\,(x) = 2631 \cdot 5577; \quad \text{S.S.}\,(x) \equiv X_2 = 35 \cdot 1890;$$

$\Sigma\, y_\nu = 80 \cdot 32;\quad \Sigma\, y_\nu^2 = 386 \cdot 8222;\quad$ Mean $\bar{y}_2 = 4 \cdot 7247;$

$\text{C.F.}(y) = 379 \cdot 4884;\qquad \text{S.S.}(y) \equiv Y_2 = 7 \cdot 3338;$

$\Sigma\, x_\nu\, y_\nu = 1014 \cdot 8622;\qquad \text{C.F.}(x, y) = 999 \cdot 3225;$

$\text{S.P.}(x, y) \equiv Z_2 = 15 \cdot 5397.$

Therefore the least-squares estimates of α_2 and β_2 are

$$\alpha_2^* = 4 \cdot 7247;\qquad \beta_2^* = Z_2/X_2 = 0 \cdot 441\,607;$$

and the estimated regression equation is $y_{2e}^* = -0 \cdot 7697 + 0 \cdot 441\,607x.$

Error S.S. $= Y_2 - X_2\beta_2^{*2} = 0 \cdot 4714$ with 15 d.f.

Study no. 3 analysis

$\Sigma\, x_\nu = 238 \cdot 95;\quad \Sigma\, x_\nu^2 = 3213 \cdot 7331;\quad$ Mean $\bar{x}_3 = 13 \cdot 2750;$

$\text{C.F.}(x) = 3172 \cdot 0612;\qquad \text{S.S.}(x) \equiv X_3 = 41 \cdot 6719;$

$\Sigma\, y_\nu = 87 \cdot 00;\quad \Sigma\, y_\nu^2 = 427 \cdot 5640;\quad$ Mean $\bar{y}_3 = 4 \cdot 8333;$

$\text{C.F.}(y) = 420 \cdot 5000;\qquad \text{S.S.}(y) \equiv Y_3 = 7 \cdot 0640;$

$\Sigma\, x_\nu\, y_\nu = 1168 \cdot 5302;\qquad \text{C.F.}(x, y) = 1154 \cdot 9250;$

$\text{S.P.}(x, y) \equiv Z_3 = 13 \cdot 6052.$

Therefore the least-squares estimates of α_3 and β_3 are

$$\alpha_3^* = 4 \cdot 8333;\qquad \beta_3^* = Z_3/X_3 = 0 \cdot 326\,484;$$

and the estimated regression equation is $y_{3e}^* = 0 \cdot 4992 + 0 \cdot 326\,484x.$

Error S.S. $= Y_3 - X_3\beta_3^{*2} = 2 \cdot 6221$ with 16 d.f.

Comparison of regressions

Pooled error S.S. $= 4 \cdot 7900$ with 46 d.f.

$$X \equiv \sum_{i=1}^{3} X_i = 162 \cdot 4937;\quad Y \equiv \sum_{i=1}^{3} Y_i = 27 \cdot 8366;\quad Z \equiv \sum_{i=1}^{3} Z_i = 60 \cdot 8549.$$

If β is the common value of the β_i, then the best estimate of β is

$$\beta^* = Z/X = 0 \cdot 374\,506 \quad \text{and} \quad X\beta^{*2} = 22 \cdot 7905.$$

Therefore the S.S. for testing the null hypothesis $H(\beta_1 = \beta_2 = \beta_3)$ is

$$\sum_{i=1}^{3} X_i\beta_i^{*2} - X\beta^{*2} = 0 \cdot 2561 \text{ with 2 d.f.}$$

The analysis of variance is as follows:

Source of variation	d.f.	S.S.	M.S.	F.R.
$H(\beta_1 = \beta_2 = \beta_3)$	2	0·2561	0·128 05	1·23
Error	46	4·7900	0·104 13	
Total	48	5·0461		

The 5 per cent table value of F with 1, 46 d.f. is 4·05, and so the null hypothesis is acceptable.

Grand mean $\bar{x} = 664 \cdot 71/52 = 12 \cdot 7829;$
Grand mean $\bar{y} = 255 \cdot 29/52 = 4 \cdot 9094.$

Therefore the estimated pooled regression equation is

$$y_e^* = 4\cdot9094 + 0\cdot374\ 506(x - 12\cdot7829) = 0\cdot1221 + 0\cdot374\ 506x.$$

Also,   s.e. $(\beta^*) = (4\cdot7900/46X)^{\frac{1}{2}} = 10^{-1} \times 0\cdot253\ 146.$

The 5 per cent table value of Student's distribution with 46 d.f. is $2\cdot012$. Hence the required confidence interval for β is

$$\beta^* \pm 2\cdot012 \times \text{s.e.}\,(\beta^*) \quad \text{i.e.} \quad 0\cdot425\ 439,\ 0\cdot323\ 573.$$

67 I   Let x denote age of culture and y its bacterial number.

Anaerobic cultures

$$\text{Let } E(y) = a_1 + \beta_1(x - \bar{x}_1); \qquad \text{var}(y) = \sigma^2.$$

$$\Sigma\ x_\nu = 70; \quad \Sigma\ x_\nu^2 = 906; \quad \text{Mean } \bar{x}_1 = 10\cdot0;$$

$$\text{C.F.}\,(x) = 700; \qquad \text{S.S.}\,(x) \equiv X_1 = 206;$$

$$\Sigma\ y_\nu = 8744; \quad \Sigma\ y_\nu^2 = 13\ 517\ 080; \quad \text{Mean } \bar{y}_1 = 1249\cdot14;$$

$$\text{C.F.}\,(y) = 10\ 922\ 505\cdot14; \qquad \text{S.S.}\,(y) \equiv Y_1 = 2\ 594\ 574\cdot86;$$

$$\Sigma\ x_\nu\ y_\nu = 107\ 104; \qquad \text{C.F.}\,(x, y) = 87\ 440;$$

$$\text{S.P.}\,(x, y) \equiv Z_1 = 19\ 664\cdot00.$$

Therefore the least-squares estimates of a_1 and β_1 are

$$a_1^* = 1249\cdot14; \qquad \beta_1^* = Z_1/X_1 = 95\cdot4563;$$

and the estimated regression equation is $y_{1e}^* = 294\cdot58 + 95\cdot4563x$.

$$\text{Error S.S.} = Y_1 - X_1\beta_1^{*2} = 717\ 522\cdot39 \text{ with 5 d.f.}$$

Control cultures

$$\text{Let } E(y) = a_2 + \beta_2(x - \bar{x}_2); \qquad \text{var}(y) = \sigma^2.$$

$$\Sigma\ x_\nu = 92; \quad \Sigma\ x_\nu^2 = 1018; \quad \text{Mean } \bar{x}_2 = 6\cdot5714;$$

$$\text{C.F.}\,(x) = 604\cdot5714; \qquad \text{S.S.}\,(x) \equiv X_2 = 413\cdot4286;$$

$$\Sigma\ y_\nu = 26\ 766; \quad \Sigma\ y_\nu^2 = 74\ 667\ 466; \quad \text{Mean } \bar{y}_2 = 1911\cdot86;$$

$$\text{C.F.}\,(y) = 51\ 172\ 768\cdot29; \qquad \text{S.S.}\,(y) \equiv Y_2 = 23\ 494\ 697\cdot71;$$

$$\Sigma\ x_\nu\ y_\nu = 263\ 607; \qquad \text{C.F.}\,(x, y) = 175\ 890\cdot86;$$

$$\text{S.P.}\,(x, y) \equiv Z_2 = 87\ 716\cdot14.$$

Therefore the least-squares estimates of a_2 and β_2 are

$$a_2^* = 1911\cdot86; \qquad \beta_2^* = Z_2/X_2 = 212\cdot1676;$$

and the estimated regression equation is $y_{2e}^* = 517\cdot62 + 212\cdot1676x$.

$$\text{Error S.S.} = Y_2 - X_2\beta_2^{*2} = 4\ 884\ 171\cdot87 \text{ with 12 d.f.}$$

Aerated cultures

$$\text{Let } E(y) = a_3 + \beta_3(x - \bar{x}_3); \qquad \text{var}(y) = \sigma^2.$$

$$\Sigma\ x_\nu = 64; \quad \Sigma\ x_\nu^2 = 626; \quad \text{Mean } \bar{x}_3 = 5\cdot3333;$$

C.F. $(x) = 341 \cdot 3333;$ S.S. $(x) \equiv X_3 = 284 \cdot 6667;$

$\Sigma\, y_\nu = 32\,083;$ $\Sigma\, y_\nu^2 = 112\,226\,965;$ Mean $\bar{y}_3 = 2673 \cdot 58;$

C.F. $(y) = 85\,776\,574 \cdot 08;$ S.S. $(y) \equiv Y_3 = 26\,450\,390 \cdot 92;$

$\Sigma\, x_\nu y_\nu = 238\,806;$ C.F. $(x, y) = 171\,109 \cdot 33;$

S.P. $(x, y) \equiv Z_3 = 67\,696 \cdot 67.$

Therefore the least-squares estimates of α_3 and β_3 are

$$\alpha_3^* = 2673 \cdot 58; \qquad \beta_3^* = Z_3/X_3 = 237 \cdot 8103;$$

and the estimated regression equation is $y_{3e}^* = 1405 \cdot 27 + 237 \cdot 8103x.$

Error S.S. $= Y_3 - X_3 \beta_3^{*2} = 10\,351\,424 \cdot 72$ with 10 d.f.

Comparison of regressions

$X \equiv \sum_{i=1}^{3} X_i = 904 \cdot 0953;$ $Y \equiv \sum_{i=1}^{3} Y_i = 52\,539\,663 \cdot 49;$ $Z \equiv \sum_{i=1}^{3} Z_i = 175\,076 \cdot 81.$

If β is the common value of the β_i, then the best estimate of β is

$$\beta^* = Z/X = 193 \cdot 6486; \qquad X\beta^{*2} = 33\,903\,375 \cdot 12;$$

S.S. due to regressions $= \sum_{i=1}^{3} X_i \beta_i^{*2} - X\beta^{*2} = 2\,683\,169 \cdot 39$ with 2 d.f.

The analysis of variance is as follows:

Source of variation	d.f.	S.S.	M.S.	F.R.
$H(\beta_1 = \beta_2 = \beta_3)$	2	2 683 169·39	1 341 584·695	2·27
Error	27	15 953 118·98	590 856·259	
Total	29	18 636 288·37		

The 5 per cent table value of F with 2, 27 d.f. is $3 \cdot 35$, and so the observed F ratio is not significant. The null hypothesis is acceptable.

Standard errors and confidence intervals

s.e. $(\beta_1^*) = (590\,856 \cdot 259/X_1)^{\frac{1}{2}} = 53 \cdot 555\,899;$

s.e. $(\beta_2^*) = (590\,856 \cdot 259/X_2)^{\frac{1}{2}} = 37 \cdot 804\,254;$

s.e. $(\beta_3^*) = (590\,856 \cdot 259/X_3)^{\frac{1}{2}} = 45 \cdot 558\,833.$

The 5 per cent table value of Student's distribution with 27 d.f. is $2 \cdot 052$. Hence the 95 per cent confidence intervals for β_1, β_2, and β_3 are

$205 \cdot 3530, -14 \cdot 4404;$ $289 \cdot 7419, 134 \cdot 5933;$ $331 \cdot 2970, 144 \cdot 3236.$

II   Let x denote bacterial number and y "protein" nitrogen.

Anaerobic cultures

Let $E(y) = \alpha_1 + \beta_1(x - \bar{x}_1);$ var $(y) = \sigma^2.$

$\Sigma\, x_\nu = 193 \cdot 20;$ $\Sigma\, x_\nu^2 = 6655 \cdot 3350;$ Mean $\bar{x}_1 = 27 \cdot 6000;$

C.F. $(x) = 5332 \cdot 3200;$ S.S. $(x) \equiv X_1 = 1323 \cdot 0150;$

$$\Sigma \, y_\nu = 272; \quad \Sigma \, y_\nu^2 = 13\,856; \quad \text{Mean } \bar{y}_1 = 38 \cdot 8571;$$

$$\text{C.F.} \, (y) = 10\,569 \cdot 1429; \quad \text{S.S.} \, (y) \equiv Y_1 = 3286 \cdot 8571;$$

$$\Sigma \, x_\nu \, y_\nu = 9458 \cdot 75; \quad \text{C.F.} \, (x, y) = 7507 \cdot 2000;$$

$$\text{S.P.} \, (x, y) \equiv Z_1 = 1951 \cdot 5500.$$

Therefore the least-squares estimates of α_1 and β_1 are

$$\alpha_1^* = 38 \cdot 8571; \qquad \beta_1^* = Z_1/X_1 = 1 \cdot 475\,078;$$

and the estimated regression equation is $y_{1e}^* = -1 \cdot 8551 + 1 \cdot 475\,078x$.

$$\text{Error S.S.} = Y_1 - X_1\beta_1^{*2} = 408 \cdot 1683 \text{ with 5 d.f.}$$

Control cultures

$$E(y) = \alpha_2 + \beta_2(x - \bar{x}_2); \qquad \text{var}(y) = \sigma^2.$$

$$\Sigma \, x_\nu = 245 \cdot 12; \quad \Sigma \, x_\nu^2 = 8153 \cdot 4184; \quad \text{Mean } \bar{x}_2 = 30 \cdot 6400;$$

$$\text{C.F.} \, (x) = 7510 \cdot 4768; \quad \text{S.S.} \, (x) \equiv X_2 = 642 \cdot 9416;$$

$$\Sigma \, y_\nu = 271; \quad \Sigma \, y_\nu^2 = 10\,093; \quad \text{Mean } \bar{y}_2 = 33 \cdot 8750;$$

$$\text{C.F.} \, (y) = 9180 \cdot 1250; \quad \text{S.S.} \, (y) \equiv Y_2 = 912 \cdot 8750;$$

$$\Sigma \, x_\nu \, y_\nu = 8481 \cdot 45; \quad \text{C.F.} \, (x, y) = 8303 \cdot 4400;$$

$$\text{S.P.} \, (x, y) \equiv Z_2 = 178 \cdot 0100.$$

Therefore the least-squares estimates of α_2 and β_2 are

$$\alpha_2^* = 33 \cdot 8750; \qquad \beta_2^* = Z_2/X_2 = 0 \cdot 276\,868;$$

and the estimated regression equation is $y_{2e}^* = 25 \cdot 3918 + 0 \cdot 276\,868x$.

$$\text{Error S.S.} = Y_2 - X_2\beta_2^{*2} = 863 \cdot 5897 \text{ with 6 d.f.}$$

Aerated cultures

$$\text{Let } E(y) = \alpha_3 + \beta_3(x - \bar{x}_3); \qquad \text{var}(y) = \sigma^2.$$

$$\Sigma \, x_\nu = 192 \cdot 65; \quad \Sigma \, x_\nu^2 = 7538 \cdot 7225; \quad \text{Mean } \bar{x}_3 = 38 \cdot 5300;$$

$$\text{C.F.} \, (x) = 7422 \cdot 8045; \quad \text{S.S.} \, (x) \equiv X_3 = 115 \cdot 9180;$$

$$\Sigma \, y_\nu = 237; \quad \Sigma \, y_\nu^2 = 12\,419; \quad \text{Mean } \bar{y}_3 = 47 \cdot 4000;$$

$$\text{C.F.} \, (y) = 11\,233 \cdot 8000; \quad \text{S.S.} \, (y) \equiv Y_3 = 1185 \cdot 2000;$$

$$\Sigma \, x_\nu \, y_\nu = 8896 \cdot 90; \quad \text{C.F.} \, (x, y) = 9131 \cdot 6100;$$

$$\text{S.P.} \, (x, y) \equiv Z_3 = -234 \cdot 7100.$$

Therefore the least-squares estimates of α_3 and β_3 are

$$\alpha_3^* = 47 \cdot 4000; \qquad \beta_3^* = Z_3/X_3 = -2 \cdot 024\,793;$$

and the estimated regression equation is $y_{3e}^* = 125 \cdot 4153 - 2 \cdot 024\,793x$.

$$\text{Error S.S.} = Y_3 - X_3\beta_3^{*2} = 709 \cdot 9609 \text{ with 3 d.f.}$$

Comparison of regressions

$$X \equiv \sum_{i=1}^{3} X_i = 2081 \cdot 8746; \quad Y \equiv \sum_{i=1}^{3} Y_i = 5384 \cdot 9321; \quad Z \equiv \sum_{i=1}^{3} Z_i = 1894 \cdot 8500.$$

If β is the common value of the β_i, then the best estimate of β is

$$\beta^* = Z/X = 0\cdot910\ 165; \qquad X\beta^{*2} = 1724\cdot6256;$$

S.S. due to regressions $= \sum_{i=1}^{3} X_i\beta_i^{*2} - X\beta^{*2} = 1678\cdot5876$ with 2 d.f.

The analysis of variance is as follows:

Source of variation	d.f.	S.S.	M.S.	F.R.
$H(\beta_1 = \beta_2 = \beta_3)$	2	1678·5876	839·293 80	5·93
Error	14	1981·7189	141·551 35	
Total	16	3660·3065		

The one and 5 per cent table values of F with 2, 14 d.f. are 6·51 and 3·74 respectively. The observed F ratio is not significant at the one per cent level.

Standard errors and confidence intervals

$$\text{s.e.}(\beta_1^*) = (141\cdot551\ 35/X_1)^{\frac{1}{2}} = 0\cdot327\ 095;$$
$$\text{s.e.}(\beta_2^*) = (141\cdot551\ 35/X_2)^{\frac{1}{2}} = 0\cdot469\ 214;$$
$$\text{s.e.}(\beta_3^*) = (141\cdot551\ 35/X_3)^{\frac{1}{2}} = 1\cdot105\ 049.$$

The 5 per cent table value of Student's distribution with 14 d.f. is 2·145. Hence the 95 per cent confidence intervals for β_1, β_2 and β_3 are

2·176 697, 0·773 459;   1·283 332, −0·729 596;   0·345 537, −4·395 123.

68   The data in this exercise are a typical example of an "asymptotic regression" (see, for example, *Statistical Methods* by G.W. Snedecor and W.G. Cochran, 6th edition, Iowa State University Press, 1967; p. 447). The dual transformation suggested in the exercise linearises the asymptotic regression remarkably.

The transformed x and y values are given in the following table. The y values are most readily obtained from a table of C.I. Bliss which is reproduced in the above-mentioned text of Snedecor and Cochran (pp. 569–71).

	Group (i)		Group (ii)		Group (iii)	
x	Marquis (y)	Mindum (y)	Marquis (y)	Reward (y)	Mindum (y)	Kubanka (y)
0·60	28·8	26·7	28·9	28·3	25·4	26·8
0·90	36·0	33·5	36·6	35·8	33·5	35·7
1·08	40·4	37·5	40·4	39·3	37·6	39·8
1·20	42·6	39·8	43·0	41·8	40·2	42·4
1·30	44·3	41·8	44·7	43·7	41·8	44·5
1·38	46·1	43·6	46·2	45·6	44·0	46·7
1·45	–	–	48·1	47·1	46·4	48·2
1·51	48·5	46·1	48·9	48·0	47·0	49·5
1·56	49·6	47·5	50·1	49·5	48·0	50·2
1·60	50·7	48·5	51·1	50·5	49·3	51·6
1·64	51·5	49·4	51·9	51·4	50·1	52·2
1·68	53·1	50·5	53·1	52·7	51·5	54·0
1·78	54·9	52·5	55·1	55·0	54·7	56·9
1·86	56·2	53·7	56·4	56·4	55·8	57·9
1·98	59·4	57·5	58·5	59·5	60·2	62·0
2·08	60·8	59·1	59·6	61·0	62·5	64·9
2·16	60·9	60·8	59·6	62·1	66·7	68·9
2·23	60·3	61·2	58·8	62·3	67·3	69·1

Group (i) analysis

$$\Sigma \, x_\nu = 26{\cdot}54; \quad \Sigma \, x_\nu^2 = 44{\cdot}6698; \quad \text{Mean } \bar{x}_1 = 1{\cdot}5612;$$
$$\text{C.F.}(x) = 41{\cdot}4336; \quad \text{S.S.}(x) \equiv X_1 = 3{\cdot}2362.$$

Marquis data

$$\text{Let } E(y) = \alpha_1 + \beta_1(x - \bar{x}_1); \quad \text{var}(y) = \sigma_1^2.$$
$$\Sigma \, y_\nu = 844{\cdot}1; \quad \Sigma \, y_\nu^2 = 43\,257{\cdot}17; \quad \text{Mean } \bar{y}_1 = 49{\cdot}6529;$$
$$\text{C.F.}(y) = 41\,912{\cdot}0476; \quad \text{S.S.}(y) \equiv Y_1 = 1345{\cdot}1224;$$
$$\Sigma \, x_\nu \, y_\nu = 1383{\cdot}382; \quad \text{C.F.}(x, y) = 1317{\cdot}7891;$$
$$\text{S.P.}(x, y) \equiv Z_1 = 65{\cdot}5929.$$

Therefore the least-squares estimates of α_1 and β_1 are

$$\alpha_1^* = 49{\cdot}6529; \quad \beta_1^* = Z_1/X_1 = 20{\cdot}2685;$$

and the estimated regression equation is $y_{1e}^* = 18{\cdot}0097 + 20{\cdot}2685x$.

Error S.S. $= Y_1 - X_1 \beta_1^{*2} = 15{\cdot}6523$ with 15 d.f.

Mindum data

$$\text{Let } E(y) = \alpha_2 + \beta_2(x - \bar{x}_1); \qquad \text{var}(y) = \sigma_1^2.$$

$$\Sigma\, y_\nu = 809\cdot7; \quad \Sigma\, y_\nu^2 = 40\ 079\cdot03; \quad \text{Mean } \bar{y}_2 = 47\cdot6294;$$

$$\text{C.F.}\,(y) = 38\ 565\cdot5347; \qquad \text{S.S.}\,(y) \equiv Y_2 = 1513\cdot4953;$$

$$\Sigma\, x_\nu\, y_\nu = 1334\cdot019; \qquad \text{C.F.}\,(x, y) = 1264\cdot0846;$$

$$\text{S.P.}\,(x, y) \equiv Z_2 = 69\cdot9344.$$

Therefore the least-squares estimates of α_2 and β_2 are

$$\alpha_2^* = 47\cdot6294; \qquad \beta_2^* = Z_2/X_1 = 21\cdot6100;$$

and the estimated regression equation is $y_{2e}^* = 13\cdot8919 + 21\cdot6100x$.

$$\text{Error S.S.} = Y_2 - X_1\beta_2^{*2} = 2\cdot2155 \text{ with 15 d.f.}$$

Comparison of regressions and confidence interval

The pooled estimate of σ_1^2 is

$$\sigma_1^{*2} = 17\cdot8678/30 = 0\cdot595\ 593,$$

and $\qquad \text{s.e.}\,(\beta_2^* - \beta_1^*) = (2\sigma_1^{*2}/X_1)^{\frac{1}{2}} = 0\cdot606\ 698.$

Hence, to test the null hypothesis $H(\beta_1 - \beta_2 = 0)$, we have

$$t = 1\cdot3415/0\cdot606\ 698 = 2\cdot211 \text{ with 30 d.f.}$$

The 5 per cent table value of Student's distribution with 30 d.f. is $2\cdot042$, and so the observed result is significant. The null hypothesis can be rejected.

The 95 per cent confidence interval for $\beta_2 - \beta_1$ is

$$\beta_2^* - \beta_1^* \pm 2\cdot042 \times \text{s.e.}\,(\beta_2^* - \beta_1^*) \quad \text{i.e.} \quad 2\cdot5804,\ 0\cdot1026.$$

Group (ii) analysis

$$\Sigma\, x_\nu = 27\cdot99; \quad \Sigma\, x_\nu^2 = 46\cdot7723; \quad \text{Mean } \bar{x}_2 = 1\cdot5550;$$

$$\text{C.F.}\,(x) = 43\cdot524\ 45; \qquad \text{S.S.}\,(x) \equiv X_2 = 3\cdot247\ 85.$$

Marquis data

$$\text{Let } E(y) = \alpha_3 + \beta_3(x - \bar{x}_2); \qquad \text{var}(y) = \sigma_2^2.$$

$$\Sigma\, y_\nu = 891\cdot0; \quad \Sigma\, y_\nu^2 = 45\ 328\cdot70; \quad \text{Mean } \bar{y}_3 = 49\cdot5000;$$

$$\text{C.F.}\,(y) = 44\ 104\cdot50; \qquad \text{S.S.}\,(y) \equiv Y_3 = 1224\cdot20;$$

$$\Sigma\, x_\nu\, y_\nu = 1447\cdot842; \qquad \text{C.F.}\,(x, y) = 1385\cdot505;$$

$$\text{S.P.}\,(x, y) \equiv Z_3 = 62\cdot337.$$

Therefore the least-squares estimates of α_3 and β_3 are

$$\alpha_3^* = 49\cdot5000; \qquad \beta_3^* = Z_3/X_2 = 19\cdot1933;$$

and the estimated regression equation is $y_{3e}^* = 19\cdot6544 + 19\cdot1933x$.

$$\text{Error S.S.} = Y_3 - X_2\beta_3^{*2} = 27\cdot7479 \text{ with 16 d.f.}$$

Reward data

$$\text{Let } E(y) = \alpha_4 + \beta_4(x - \bar{x}_2); \qquad \text{var}(y) = \sigma_2^2.$$

$$\Sigma\, y_\nu = 890 \cdot 0; \quad \Sigma\, y_\nu^2 = 45\,510 \cdot 38; \quad \text{Mean } \bar{y}_4 = 49 \cdot 4444;$$

$$\text{C.F.}(y) = 44\,005 \cdot 5556; \qquad \text{S.S.}(y) \equiv Y_4 = 1504 \cdot 8244;$$

$$\Sigma\, x_\nu\, y_\nu = 1453 \cdot 728; \qquad \text{C.F.}(x, y) = 1383 \cdot 950;$$

$$\text{S.P.}(x, y) \equiv Z_4 = 69 \cdot 778.$$

Therefore the least-squares estimates of α_4 and β_4 are

$$\alpha_4^* = 49 \cdot 4444; \qquad \beta_4^* = Z_4/X_2 = 21 \cdot 4844;$$

and the estimated regression equation is $y_{4e}^* = 16 \cdot 0362 + 21 \cdot 4844x$.

$$\text{Error S.S.} = Y_4 - X_2\beta_4^{*2} = 5 \cdot 6837 \text{ with 16 d.f.}$$

Comparison of regressions and confidence interval

The pooled estimate of σ_2^2 is

$$\sigma_2^{*2} = 33 \cdot 4316/32 = 1 \cdot 044\,737\,5,$$

and $\qquad \text{s.e.}(\beta_4^* - \beta_3^*) = (2\sigma_2^{*2}/X_2)^{\frac{1}{2}} = 0 \cdot 802\,085.$

Hence, to test the null hypothesis $H(\beta_4 - \beta_3 = 0)$, we have

$$t = 2 \cdot 2911/0 \cdot 802\,085 = 2 \cdot 856 \text{ with 32 d.f.}$$

The 5 per cent table value of Student's distribution with 32 d.f. is $2 \cdot 037$, and so the observed result is significant. The null hypothesis can be rejected.

The 95 per cent confidence interval for $\beta_4 - \beta_3$ is

$$\beta_4^* - \beta_3^* \pm 2 \cdot 037 \times \text{s.e.}(\beta_4^* - \beta_3^*) \quad \text{i.e.} \quad 3 \cdot 9249,\ 0 \cdot 6573.$$

Group (iii) analysis

The x values are the same as in Group (ii).

Mindum data

$$\text{Let } E(y) = \alpha_5 + \beta_5(x - \bar{x}_2); \qquad \text{var}(y) = \sigma_3^2.$$

$$\Sigma\, y_\nu = 882 \cdot 0; \quad \Sigma\, y_\nu^2 = 45\,353 \cdot 36; \quad \text{Mean } \bar{y}_5 = 49 \cdot 0000;$$

$$\text{C.F.}(y) = 43\,218 \cdot 00; \qquad \text{S.S.}(y) \equiv Y_5 = 2135 \cdot 36;$$

$$\Sigma\, x_\nu\, y_\nu = 1454 \cdot 493; \qquad \text{C.F.}(x, y) = 1371 \cdot 510;$$

$$\text{S.P.}(x, y) \equiv Z_5 = 82 \cdot 983.$$

Therefore the least-squares estimates of α_5 and β_5 are

$$\alpha_5^* = 49 \cdot 0000; \qquad \beta_5^* = Z_5/X_2 = 25 \cdot 5501;$$

and the estimated regression equation is $y_{5e}^* = 9 \cdot 2696 + 25 \cdot 5501x$.

$$\text{Error S.S.} = Y_5 - X_2\beta_5^{*2} = 15 \cdot 1388 \text{ with 16 d.f.}$$

Kubanka data

$$\text{Let } E(y) = \alpha_6 + \beta_6(x - \bar{x}_2); \qquad \text{var}(y) = \sigma_3^2.$$

$$\Sigma\, y_\nu = 921 \cdot 3; \quad \Sigma\, y_\nu^2 = 49\ 300 \cdot 65; \quad \text{Mean } \bar{y}_6 = 51 \cdot 1833;$$

$$\text{C.F.}(y) = 47\ 155 \cdot 2050; \qquad \text{S.S.}(y) \equiv Y_6 = 2145 \cdot 4450;$$

$$\Sigma\, x_\nu\, y_\nu = 1515 \cdot 850; \qquad \text{C.F.}(x, y) = 1432 \cdot 6215;$$

$$\text{S.P.}(x, y) \equiv Z_6 = 83 \cdot 2285.$$

Therefore the least-squares estimates of α_6 and β_6 are

$$\alpha_6^* = 51 \cdot 1833; \qquad \beta_6^* = Z_6/X_2 = 25 \cdot 6257;$$

and the estimated regression equation is $y_{6e}^* = 11 \cdot 3353 + 25 \cdot 6257x$.

$$\text{Error S.S.} = Y_6 - X_2\beta_6^{*2} = 12 \cdot 6582 \text{ with } 16 \text{ d.f.}$$

Comparison of regressions and confidence interval

The pooled estimate of σ_3^2 is

$$\sigma_3^{*2} = 27 \cdot 7970/32 = 0 \cdot 868\ 656,$$

and $\text{s.e.}(\beta_6^* - \beta_5^*) = (2\sigma_3^{*2}/X_2)^{\frac{1}{2}} = 0 \cdot 731\ 376.$

Hence, to test the null hypothesis $H(\beta_6 - \beta_5 = 0)$, we have

$$t = 0 \cdot 0756/0 \cdot 731\ 376 = 0 \cdot 103 \text{ with } 32 \text{ d.f.}$$

The 5 per cent table value of Student's distribution with 32 d.f. is 2·037, and so the observed result is not significant. The null hypothesis can be accepted.

The 95 per cent confidence interval for $\beta_6 - \beta_5$ is

$$\beta_6^* - \beta_5^* \pm 2 \cdot 037 \times \text{s.e.}(\beta_6^* - \beta_5^*) \quad \text{i.e.} \quad 1 \cdot 5645, -1 \cdot 4142.$$

69   I $\text{Let } E(y) = \alpha + \beta(x - \bar{x}); \qquad \text{var}(y) = \sigma_0^2.$

$$\Sigma\, x_i = 416 \cdot 4; \quad \Sigma\, x_i^2 = 5979 \cdot 42; \quad \text{Mean } \bar{x} = 13 \cdot 8800;$$

$$\text{C.F.}(x) = 5779 \cdot 6320; \qquad \text{S.S.}(x) \equiv X = 199 \cdot 7880;$$

$$\Sigma\, y_i = 9451; \quad \Sigma\, y_i^2 = 3\ 049\ 557; \quad \text{Mean } \bar{y} = 315 \cdot 0333;$$

$$\text{C.F.}(y) = 2\ 977\ 380 \cdot 03; \qquad \text{S.S.}(y) \equiv Y = 72\ 176 \cdot 97;$$

$$\Sigma\, x_i\, y_i = 134\ 699 \cdot 7; \qquad \text{C.F.}(x, y) \equiv 131\ 179 \cdot 88;$$

$$\text{S.P.}(x, y) \equiv Z = 3519 \cdot 82.$$

Therefore the least-squares estimates of α and β are

$$\alpha^* = 315 \cdot 0333; \qquad \beta^* = Z/X = 17 \cdot 6178;$$

and the estimated regression equation is $y_e^* = 70 \cdot 4982 + 17 \cdot 6178x$.

$$\text{Error S.S.} = Y - X\beta^{*2} = 10\ 165 \cdot 3920 \text{ with } 28 \text{ d.f.}$$

Therefore $\text{s.e.}(\beta^*) = (10\ 165 \cdot 3920/28X)^{\frac{1}{2}} = 1 \cdot 348\ 026.$

The 5 per cent table value of Student's distribution with 28 d.f. is 2·048. Hence the 95 per cent confidence interval for β is

$$\beta^* \pm 2\cdot048 \times \text{s.e.}(\beta^*) \quad \text{i.e.} \quad 20\cdot3799, \ 14\cdot8557.$$

II   The dough totals are as follows:

178·5	189·9	184·4	174·6	186·7	173·5	186·5	169·3	184·2
186·7	146·3	163·8	170·0	166·6	164·5	165·0	149·5	191·1
181·1	188·0	182·0	173·2	168·9	186·3	190·3		

Grand total = 4400·9;   Raw S.S. = 78 094·47;   C.F. = 77 471·6832; Total S.S. = 622·7868 with 249 d.f.;   Grand mean = 17·6036. Between-doughs S.S. = 77 836·0030 − C.F. = 364·3198 with 24 d.f.

The analysis of variance is as follows:

Source of variation	d.f.	S.S.	M.S.	F.R.
Between doughs	24	364·3198	15·1800	13·21
Error	225	258·4670	1·1487	
Total	249	622·7868		

The one per cent table value of F with 24, 225 d.f. is < 1·88 and so the observed F ratio is obviously highly significant. There are almost certainly real differences in the extensibility of the doughs.

If μ is the true mean of the x population, then the best estimate of μ is $\mu^* = 17\cdot6036$. Also,

$$\text{s.e.}(\mu^*) = (1\cdot1487/250)^{\frac{1}{2}} = 0\cdot067\,785.$$

Therefore the 95 per cent confidence interval for μ is

$$\mu^* \pm 1\cdot96 \times \text{s.e.}(\mu^*) \quad \text{i.e.} \quad 17\cdot7365, \ 17\cdot4707.$$

If σ^2 denotes the population error variance, then $258\cdot4670/\sigma^2$ is χ^2 with 225 d.f. Furthermore, if χ_1^2 and χ_2^2 are the lower and upper 2·5 per cent points of the χ^2 distribution with 225 d.f., then

$$\sqrt{2\chi_1^2} - \sqrt{449} = -1\cdot96; \qquad \sqrt{2\chi_2^2} - \sqrt{449} = 1\cdot96.$$

Hence $\chi_1^2 = 184\cdot8888$ and $\chi_2^2 = 267\cdot9520$. Therefore the 95 per cent confidence interval for σ^2 is

$$184\cdot888 \leqslant 258\cdot4670/\sigma^2 \leqslant 267\cdot9520,$$

whence the limits for σ^2 are 1·3980, 0·9646.

70   Leicester, 1969.

(i)

Group	1	2	3	4
Total	219	355	407	714
$\bar{y}_i$	43·8	71·0	81·4	142·8
x_i	− 1·5	− 0·5	0·0	2·0

Grand total = 1695;   Raw S.S. = 181 445;   C.F. = 143 651·25;

Grand mean $\bar{y}$ = 84·75;   Total S.S. = 37 793·75 with 19 d.f.

Between-groups S.S. = 169 886·20 − C.F. = 26 234·95 with 3 d.f.

The analysis of variance is as follows:

Source of variation	d.f.	S.S.	M.S.	F.R.
Between groups	3	26 234·95	8744·9833	12·11
Error	16	11 558·80	722·4250	
Total	19	37 793·75		

The one per cent table value of F with 3, 16 d.f. is 5·29. Therefore the observed F ratio is highly significant, and the null hypothesis of no difference between groups is rejected.

The best estimate of σ^2 is s^2 = 722·4250.

(ii) The least-squares estimates of α and β are

$$\alpha^* = \bar{y} = 84\cdot75; \qquad \beta^* = \Sigma\, x_i \bar{y}_i / \Sigma\, x_i^2 = 184\cdot4/6\cdot5 = 28\cdot3692.$$

The 5 per cent table value of Student's distribution with 16 d.f. is 2·120. Also,

$$\text{s.e.}(\beta^*) = \{722\cdot4250/(5 \times 6\cdot5)\}^{\frac{1}{2}} = 4\cdot714\ 707.$$

Hence the 95 per cent confidence interval for β is

$$\beta^* \pm 2\cdot120 \times \text{s.e.}(\beta^*) \quad \text{i.e.} \quad 38\cdot3644,\ 18\cdot3740.$$

The reconstructed estimates of the treatment means are

$$42\cdot1962; \quad 70\cdot5654; \quad 84\cdot7500; \quad 141\cdot4884.$$

S.S. due to regression = $5 \times 6\cdot5\beta^{*2}$ = 26 156·37 with 1 d.f.

Therefore S.S. due to deviations from linear regression

$$= 26\ 234\cdot95 - 26\ 156\cdot37 = 78\cdot58 \text{ with 2 d.f.}$$

Clearly, the M.S. due to deviations from linear regression is less than the error M.S. Hence the hypothesis of the specified linear regression of means is adequate to describe the variation between the group means.

71 Leicester, 1964.

Let x denote carotene in wheat and y that in flour.

$$\Sigma\, x_i = 31\cdot03; \quad \Sigma\, x_i^2 = 49\cdot2981; \quad \text{Mean } \bar{x} = 1\cdot5515;$$

$$\text{C.F.}(x) = 48\cdot143\ 045; \qquad \text{S.S.}(x) \equiv X = 1\cdot155\ 055;$$

$$\Sigma\, y_i = 43\cdot62; \quad \Sigma\, y_i^2 = 96\cdot8644; \quad \text{Mean } \bar{y} = 2\cdot1810;$$

$$\text{C.F.}(y) = 95\cdot135\ 220; \qquad \text{S.S.}(y) \equiv Y = 1\cdot729\ 180;$$

$$\Sigma\, x_i\, y_i = 68\cdot4335; \qquad \text{C.F.}(x, y) = 67\cdot676\ 430;$$

$$\text{S.P.}(x, y) \equiv Z = 0\cdot757\ 070.$$

(i) If $u = x - y$ and $v = x + y$, then

$$\text{S.S.}(u) \equiv U = X + Y - 2Z = 1 \cdot 370\ 095;$$
$$\text{S.S.}(v) \equiv V = X + Y + 2Z = 4 \cdot 398\ 375;$$
$$\text{S.P.}(u, v) \equiv W = X - Y = -0 \cdot 574\ 125.$$

Hence, if r_{uv} denotes the sample correlation coefficient between u and v, then

$$r_{uv}^2 = W^2/UV = 0 \cdot 054\ 698.$$

The test for the equality of the variances of x and y is equivalent to testing that $\text{corr}(u, v) = 0$. Hence the required test is that

$$F = 18 r_{uv}^2 / (1 - r_{uv}^2) = 1 \cdot 042$$

has the F distribution with 1, 18 d.f. The 5 per cent table value of F with 1, 18 d.f. is 4·41, and so the observed F ratio is not significant. We may therefore accept the hypothesis that the experimental error variances of the carotene measurements made on wheat and flour respectively are the same.

(ii) The best estimate of the true difference between the flour and wheat measurements is $\bar{u} = \bar{y} - \bar{x} = 0 \cdot 6295$; and

$$\text{s.e.}(u) = (1 \cdot 370\ 095/19)^{\frac{1}{2}} = 10^{-1} \times 0 \cdot 600\ 459.$$

(iii) The estimate of the linear regression coefficient of y on x is

$$\beta_y^* = Z/X = 0 \cdot 655\ 441;$$

and that of the linear regression coefficient of x on y is

$$\beta_x^* = Z/Y = 0 \cdot 437\ 820.$$

Hence the estimate of $\text{corr}(x, y)$ is $(\beta_y^* \times \beta_x^*)^{\frac{1}{2}} = 0 \cdot 5357$.

72   Let $x = 100 \times$ measurement by Method I;

$y = 100 \times$ measurement by Method II;

and put $u = x - y$; $v = x + y$. Then for the u and v observations we have the following calculations:

$$\Sigma\ u_i = 303; \quad \Sigma\ u_i^2 = 4149; \quad \text{Mean } \bar{u} = 6 \cdot 3125;$$
$$\text{C.F.}(u) = 1912 \cdot 6875; \quad \text{S.S.}(u) \equiv U = 2236 \cdot 3125;$$
$$\Sigma\ v_i = 13\ 281; \quad \Sigma\ v_i^2 = 3\ 993\ 441; \quad \text{Mean } \bar{v} = 276 \cdot 6875;$$
$$\text{C.F.}(v) = 3\ 674\ 686 \cdot 6875; \quad \text{S.S.}(v) \equiv V = 318\ 754 \cdot 3125;$$
$$\Sigma\ u_i v_i = 77\ 729; \quad \text{C.F.}(u, v) = 83\ 836 \cdot 3125;$$
$$\text{S.P.}(u, v) \equiv W = -6107 \cdot 3125.$$

(i) If δ denotes the true difference between the measurements made by Methods I and II, then the best estimate of δ is $\bar{u} = 6 \cdot 3125$; and

$$\text{s.e.}(\bar{u}) = \{U/(47 \times 48)\}^{\frac{1}{2}} = 0 \cdot 995\ 627.$$

Hence to test the null hypothesis $H(\delta = 0)$, we have

$$t = 6 \cdot 3125/0 \cdot 995\ 627 = 6 \cdot 340 \text{ with } 47 \text{ d.f.}$$

The one per cent table value of Student's distribution with 47 d.f. is $< 2 \cdot 423$. Therefore the observed result is highly significant and the null hypothesis can be rejected confidently.

(ii) The sample $\operatorname{corr}(u, v) = W/(UV)^{\frac{1}{2}} = -0 \cdot 228\ 747$.
The test for the equality of $\operatorname{var}(x)$ and $\operatorname{var}(y)$ is equivalent to that for testing $\operatorname{corr}(u, v) = 0$. Therefore the required test ratio is

$$F = 46 \times (0 \cdot 228\ 747)^2/0 \cdot 947\ 675 = 2 \cdot 54 \text{ with } 1, 46 \text{ d.f.}$$

The 5 per cent table value of F with 1, 46 d.f. is $4 \cdot 05$, and so the null hypothesis can be accepted.

(iii) Let x_1, x_2, y denote measurements made by Methods I, II and the A.O.A.C. Method respectively. Assume that

$$E(y) = \alpha_1 + \beta_1(x_1 - \bar{x}_1); \qquad \operatorname{var}(y) = \sigma_1^2;$$

$$E(y) = \alpha_2 + \beta_2(x_2 - \bar{x}_2); \qquad \operatorname{var}(y) = \sigma_2^2;$$

$$\sum_\nu x_{1\nu} = 67 \cdot 92; \quad \sum_\nu x_{1\nu}^2 = 103 \cdot 8262; \quad \text{Mean } \bar{x}_1 = 1 \cdot 4150;$$

$$\text{C.F.}(x_1) = 96 \cdot 1068; \qquad \text{S.S.}(x_1) \equiv X_1 = 7 \cdot 7194;$$

$$\sum_\nu x_{2\nu} = 64 \cdot 89; \quad \sum_\nu x_{2\nu}^2 = 96 \cdot 0533; \quad \text{Mean } \bar{x}_2 = 1 \cdot 3519;$$

$$\text{C.F.}(x_2) = 87 \cdot 7232; \qquad \text{S.S.}(x_2) \equiv X_2 = 8 \cdot 3301;$$

$$\sum_\nu y_\nu = 62 \cdot 87; \quad \sum_\nu y_\nu^2 = 91 \cdot 2613; \quad \text{Mean } \bar{y} = 1 \cdot 3098;$$

$$\text{C.F.}(y) = 82 \cdot 3466; \qquad \text{S.S.}(y) \equiv Y = 8 \cdot 9147;$$

$$\sum_\nu x_{1\nu} y_\nu = 97 \cdot 0556; \qquad \text{C.F.}(x_1, y) = 88 \cdot 9610;$$

$$\text{S.P.}(x_1, y) \equiv Z_1 = 8 \cdot 0946;$$

$$\sum_\nu x_{2\nu} y_\nu = 93 \cdot 4846; \qquad \text{C.F.}(x_2, y) = 84 \cdot 9924;$$

$$\text{S.P.}(x_2, y) \equiv Z_2 = 8 \cdot 4922.$$

Hence the least-squares estimates of α_1, α_2, β_1, β_2 are

$$\alpha_1^* = \alpha_2^* = 1 \cdot 3098; \quad \beta_1^* = Z_1/X_1 = 1 \cdot 048\ 605; \quad \beta_2^* = Z_2/X_2 = 1 \cdot 019\ 460;$$

and the estimated regression lines are

$$y_{1e}^* = -0 \cdot 1740 + 1 \cdot 048\ 605 x_1; \quad y_{2e}^* = -0 \cdot 0684 + 1 \cdot 019\ 460 x_2;$$

The error S.S. in the x_1, y analysis is

$$Y - X_1\beta_1^{*2} = 0 \cdot 4267 \text{ with } 46 \text{ d.f.}$$

Therefore $\text{s.e.}(\beta_1^*) = (0 \cdot 4267/46 X_1)^{\frac{1}{2}} = 10^{-1} \times 0 \cdot 346\ 650.$

Similarly, the error S.S. in the x_2, y analysis is

$$Y - X_2\beta_2^{*2} = 0 \cdot 2572 \text{ with } 46 \text{ d.f.}$$

Therefore $\text{s.e.}(\beta_2^*) = (0 \cdot 2572/46 X_2)^{\frac{1}{2}} = 10^{-1} \times 0 \cdot 259\ 079.$

The 5 per cent table value of Student's distribution with 46 d.f. is 2·012. Therefore the 95 per cent confidence intervals for β_1 and β_2 are

$$\beta_1^* \pm 2·012 \times \text{s.e.}(\beta_1^*) \quad \text{i.e.} \quad 1·118\ 351,\ 0·978\ 859;$$
$$\beta_2^* \pm 2·012 \times \text{s.e.}(\beta_2^*) \quad \text{i.e.} \quad 1·071\ 587,\ 0·967\ 333.$$

73   Let x and y denote the moisture measurements made by the carbide and oven methods respectively.

$$\text{Put} \quad u = 100(y - x) \quad \text{and} \quad v = 100(y + x).$$

Then for the u and v observations we have the following calculations:

$$\Sigma u_i = 31; \quad \Sigma u_i^2 = 13\ 915; \quad \text{Mean } \bar{u} = 0·7381;$$
$$\text{C.F.}(u) = 22·8810; \quad \text{S.S.}(u) \equiv U = 13\ 892·1190;$$
$$\Sigma v_i = 91\ 575; \quad \Sigma v_i^2 = 207\ 706\ 935; \quad \text{Mean } \bar{v} = 2180·3571;$$
$$\text{C.F.}(v) = 199\ 666\ 205·36; \quad \text{S.S.}(v) \equiv V = 8\ 040\ 729·64;$$
$$\Sigma u_i v_i = 77\ 845; \quad \text{C.F.}(u, v) = 67\ 591·0714;$$
$$\text{S.P.}(u, v) \equiv W = 10\ 253·9286.$$

(i) If δ denotes the true difference between the y and x measurements, then the best estimate of δ is $\bar{u} = 0·7381$; and
$$\text{s.e.}(\bar{u}) = \{U/(41 \times 42)\}^{\frac{1}{2}} = 2·8403.$$
Therefore, to test the null hypothesis $H(\delta = 0)$, we have

$$t = 0·7381/2·8403 = 0·260 \text{ with } 41 \text{ d.f.}$$

The observed value of t is obviously not significant and so the null hypothesis can be accepted.

The estimate of corr(u, v) is $r_{uv} = W/(UV)^{\frac{1}{2}} = 0·030\ 680\ 2$.
The test for the equality of var(x) and var(y) is equivalent to testing that corr$(u, v) = 0$. Hence the test ratio is

$$F = 40r_{uv}^2/(1 - r_{uv}^2) = 0·038 \text{ with } 1, 40 \text{ d.f.}$$

The observed F ratio is clearly not significant and so the null hypothesis is acceptable.

(ii) Assume that
$$E(y) = \alpha_1 + \beta_1(x - \bar{x}); \quad \text{var}(y) = \sigma_1^2.$$
$$\Sigma x_i = 457·72; \quad \Sigma x_i^2 = 5189·1290; \quad \text{Mean } \bar{x} = 10·8981;$$
$$\text{C.F.}(x) = 4988·2762; \quad \text{S.S.}(x) \equiv X = 200·8528;$$
$$\Sigma y_i = 458·03; \quad \Sigma y_i^2 = 5196·9135; \quad \text{Mean } \bar{y} = 10·9055;$$
$$\text{C.F.}(y) = 4995·0353; \quad \text{S.S.}(y) \equiv Y = 201·8782;$$
$$\Sigma x_i y_i = 5192·3255; \quad \text{C.F.}(x, y) = 4991·6546;$$
$$\text{S.P.}(x, y) \equiv Z = 200·6709.$$

Therefore the least-squares estimates of α_1 and β_1 are
$$\alpha_1^* = 10·9055; \quad \beta_1^* = Z/X = 0·999\ 094;$$

and the estimated regression equation is $y_{1e}^* = 0·0173 + 0·999\ 094x$.

(iii) Let w denote manometric pressure readings. Assume that

$$E(y) = \alpha_2 + \beta_2(w - \bar{w}); \qquad \text{var}(y) = \sigma_2^2.$$

$\Sigma w_i = 2935 \cdot 0; \quad \Sigma w_i^2 = 232\ 926 \cdot 50; \quad \text{Mean } \bar{w} = 69 \cdot 8810;$

$\text{C.F.}(w) = 205\ 100 \cdot 60; \qquad \text{S.S.}(w) \equiv S = 27\ 825 \cdot 90;$

$\Sigma w_i y_i = 34\ 364 \cdot 345; \qquad \text{C.F.}(w, y) = 32\ 007 \cdot 573;$

$$\text{S.P.}(w, y) \equiv T = 2356 \cdot 772.$$

Therefore the least-squares estimates of α_2 and β_2 are

$$\alpha_2^* = 10 \cdot 9055; \qquad \beta_2^* = T/S = 0 \cdot 084\ 697\ 1;$$

and the estimated regression equation is $y_{2e}^* = 4 \cdot 9868 + 0 \cdot 084\ 697\ 1w.$

Error S.S. $= Y - S\beta_2^{*2} = 2 \cdot 2664$ with 40 d.f.

For $w = 98 \cdot 6$, the estimated y value is

$$y_{eo}^* = 4 \cdot 9868 + 0 \cdot 084\ 697\ 1 \times 98 \cdot 6 = 13 \cdot 3379;$$

and   s.e. $(y_{eo}^*) = \left[\dfrac{2 \cdot 2664}{40} \times \left\{ \dfrac{1}{42} + \dfrac{(98 \cdot 6 - \bar{w})^2}{S} \right\} \right]^{\frac{1}{2}} = 10^{-1} \times 0 \cdot 550\ 317.$

The one per cent table value of Student's distribution with 40 d.f. is $2 \cdot 704$. Therefore the 99 per cent confidence interval for y_{eo} is

$$y_{eo}^* \pm 2 \cdot 704 \times \text{s.e.}(y_{eo}^*) \quad \text{i.e.} \quad 13 \cdot 4867,\ 13 \cdot 1891.$$

74 Leicester, 1969.
(i) Let x and y denote the "onset to bath" and "eruption to bath" variables respectively. Then the two frequency distributions give

$$\text{Mean } \bar{x} = 15 \cdot 1150; \qquad \text{Variance } s_1^2 = 27 \cdot 6746;$$
$$\text{Mean } \bar{y} = 13 \cdot 0123; \qquad \text{Variance } s_2^2 = 29 \cdot 4562;$$

(ii) The best estimates of θ_1 and θ_2 are $\bar{x}$ and $\bar{y}$ respectively. Also,

s.e. $(\bar{x}) = s_1/\sqrt{826} = 0 \cdot 183\ 042; \qquad$ s.e. $(\bar{y}) = s_2/\sqrt{855} = 0 \cdot 185\ 612.$

The required 95 per cent confidence interval for θ_1 is

$$\bar{x} \pm 1 \cdot 96 \times \text{s.e.}(\bar{x}) \quad \text{i.e.} \quad 15 \cdot 47,\ 14 \cdot 76;$$

and that for θ_2 is

$$\bar{y} \pm 1 \cdot 96 \times \text{s.e.}(\bar{y}) \quad \text{i.e.} \quad 13 \cdot 38,\ 12 \cdot 65.$$

(iii) The best estimate of $\theta_1 - \theta_2$ is $\bar{x} - \bar{y}$, but the two means are not independent, since the frequency distributions of x and y are based largely on observations made on the same patients. Therefore

$$\text{var}(\bar{x} - \bar{y}) = \text{var}(\bar{x}) + \text{var}(\bar{y}) - 2\,\text{cov}(\bar{x}, \bar{y})$$

$$= \dfrac{\sigma_1^2}{n_1} + \dfrac{\sigma_2^2}{n_2} - 2\rho \dfrac{\sigma_1 \sigma_2}{\sqrt{n_1 n_2}},$$

where in the exercise $n_1 = 826$, $n_2 = 855$ and $\rho = \text{corr}(\bar{x}, \bar{y})$.

If r is some estimate of ρ, then

$$\text{s.e.}(\bar{x} - \bar{y}) = \left[\frac{s_1^2}{n_1} + \frac{s_2^2}{n_2} - \frac{2r\,s_1 s_2}{\sqrt{n_1 n_2}}\right]^{\frac{1}{2}},$$

and for large samples

$$\frac{\bar{x} - \bar{y} - (\theta_1 - \theta_2)}{\text{s.e.}(\bar{x} - \bar{y})} \quad \text{is} \quad \sim N(0, 1).$$

Therefore a 95 per cent large-sample confidence interval for $\theta_1 - \theta_2$ is

$$\bar{x} - \bar{y} \pm 1{\cdot}96 \times \text{s.e.}(\bar{x} - \bar{y}).$$

The data do not give sufficient information for estimating ρ, but we know that, in general, $-1 \leqslant r \leqslant 1$. Then for known s_1, s_2, n_1, n_2

$$\left[\frac{s_1^2}{n_1} + \frac{s_2^2}{n_2} - \frac{2s_1 s_2}{\sqrt{n_1 n_2}}\right]^{\frac{1}{2}} \leqslant \text{s.e.}(\bar{x} - \bar{y}) \leqslant \left[\frac{s_1^2}{n_1} + \frac{s_2^2}{n_2} + \frac{2s_1 s_2}{\sqrt{n_1 n_2}}\right]^{\frac{1}{2}}$$

or

$$\left|\frac{s_1}{\sqrt{n_1}} - \frac{s_2}{\sqrt{n_2}}\right| \leqslant \text{s.e.}(\bar{x} - \bar{y}) \leqslant \frac{s_1}{\sqrt{n_1}} + \frac{s_2}{\sqrt{n_2}}.$$

Hence the smallest and largest values for the confidence interval for $\theta_1 - \theta_2$ are

$$\bar{x} - \bar{y} \pm 1{\cdot}96 \times \left|\frac{s_1}{\sqrt{n_1}} - \frac{s_2}{\sqrt{n_2}}\right| \quad \text{i.e.} \quad 2{\cdot}1027 \pm 0{\cdot}0050$$

and

$$\bar{x} - \bar{y} \pm 1{\cdot}96 \times \left(\frac{s_1}{\sqrt{n_1}} + \frac{s_2}{\sqrt{n_2}}\right) \quad \text{i.e.} \quad 2{\cdot}1027 \pm 0{\cdot}7226.$$

These give

$$2{\cdot}0977 \leqslant \theta_1 - \theta_2 \leqslant 2{\cdot}1077; \qquad 1{\cdot}3801 \leqslant \theta_1 - \theta_2 \leqslant 2{\cdot}8253.$$

The wider interval gives the greater safety margin for ρ unknown.

75   Let $n = 18$ denote the number of years. Also, for $k \geqslant 0$, set

$$X_k = \sum_{\nu=1}^{n} (x_\nu - \bar{x})^k; \qquad Y_k = \sum_{\nu=1}^{n} y_\nu (x_\nu - \bar{x})^k.$$

Clearly $X_1 = X_3 = 0$ and so the least-squares equations for the estimates α^*, β^*, γ^* are found to be

$$Y_0 = n\alpha^* + \gamma^* X_2; \quad Y_1 = \beta^* X_2; \quad Y_2 = \alpha^* X_2 + \gamma^* X_4.$$

The solution of these equations gives

$$\alpha^* = \frac{Y_0 - \gamma^* X_2}{n}; \quad \beta^* = \frac{Y_1}{X_2}; \quad \gamma^* = \frac{nY_2 - Y_0 X_2}{nX_4 - X_2^2};$$

and the estimated regression equation is

$$y_{e\nu}^* = a^* + \beta^*(x_\nu - \bar{x}) + \gamma^*(x_\nu - \bar{x})^2$$
$$= (a^* - \beta^*\bar{x} + \gamma^*\bar{x}^2) + (\beta^* - 2\gamma^*\bar{x})x_\nu + \gamma^{*2}x_\nu^2.$$

For numerical calculation,

$$X_2 = \Sigma\ x_\nu^2 - \frac{1}{n}(\Sigma\ x_\nu)^2;$$

$$X_4 = \Sigma\ x_\nu^4 - \frac{4}{n}(\Sigma\ x_\nu)(\Sigma\ x_\nu^3) + \frac{6}{n^2}(\Sigma\ x_\nu)^2(\Sigma\ x_\nu^2) - \frac{3}{n^3}(\Sigma\ x_\nu)^4;$$

$$Y_0 = \Sigma\ y_\nu; \qquad Y_1 = \Sigma\ x_\nu\ y_\nu - \frac{1}{n}(\Sigma\ x_\nu)(\Sigma\ y_\nu);$$

$$Y_2 = \Sigma\ x_\nu^2\ y_\nu - \frac{2}{n}(\Sigma\ x_\nu)(\Sigma\ x_\nu\ y_\nu) + \frac{1}{n^2}(\Sigma\ x_\nu)^2(\Sigma\ y_\nu).$$

Transfer the time origin to 1948 so that the x values are 0, 1, 2, ... , 17. Hence

$$\Sigma\ x_\nu = 153; \quad \Sigma\ x_\nu^2 = 1785; \quad \Sigma\ x_\nu^3 = 23\ 409; \quad \Sigma\ x_\nu^4 = 327\ 369;$$
$$\Sigma\ y_\nu = 3258; \quad \Sigma\ x_\nu\ y_\nu = 33\ 100; \quad \Sigma\ x_\nu^2\ y_\nu = 418\ 580.$$

Therefore $\quad \bar{x} = 8 \cdot 5; \qquad X_2 = 484 \cdot 5; \qquad X_4 = 23\ 377 \cdot 125;$

$$Y_0 = 3258; \qquad Y_1 = 5407; \qquad Y_2 = 91\ 270 \cdot 5;$$

whence $\quad a^* = 171 \cdot 687\ 500; \qquad \beta^* = 11 \cdot 159\ 959; \qquad \gamma^* = 0 \cdot 345\ 975.$

The estimated regression equation is

$$y_{e\nu}^* = 101 \cdot 824\ 542 + 5 \cdot 278\ 384x_\nu + 0 \cdot 345\ 975x_\nu^2.$$

The observed and estimated y values are given in the table below.

x_ν	y_ν	$y_{e\nu}^*$	x_ν	y_ν	$y_{e\nu}^*$
0	100	101·82	9	180	177·35
1	106	107·45	10	185	189·21
2	112	113·77	11	196	201·75
3	124	120·77	12	214	214·99
4	130	128·47	13	231	228·91
5	136	136·87	14	243	243·53
6	146	145·95	15	251	258·84
7	159	155·73	16	272	274·85
8	172	166·19	17	301	291·54

76   Leicester, 1960.

(i) For $n = 74$, put

$$X = \sum_{\nu=1}^{n} (x_\nu - \bar{x})^2; \quad Y = \sum_{\nu=1}^{n} (y_\nu - \bar{y})^2; \quad Z = \sum_{\nu=1}^{n} (z_\nu - \bar{z})^2;$$

$$U = \sum_{\nu=1}^{n} (x_\nu - \bar{x})(y_\nu - \bar{y}); \qquad V = \sum_{\nu=1}^{n} (y_\nu - \bar{y})(z_\nu - \bar{z});$$

$$W = \sum_{\nu=1}^{n} (z_\nu - \bar{z})(x_\nu - \bar{x}).$$

Then the equations for the least-squares estimates α^*, β^*, γ^* are found to be as follows:

$$\alpha^* = \bar{z}; \quad W - X\beta^* - U\gamma^* = 0; \quad V - U\beta^* - Y\gamma^* = 0.$$

The solution of these equations gives

$$\alpha^* = \bar{z}; \quad \beta^* = \frac{UV - WY}{U^2 - XY}; \quad \gamma^* = \frac{UW - VX}{U^2 - XY}.$$

Direct computation gives

$$\Sigma\, x_\nu = 2493; \quad \Sigma\, y_\nu = 2890; \quad \Sigma\, z_\nu = 3933;$$
$$\Sigma\, x_\nu^2 = 97\ 275; \quad \Sigma\, y_\nu^2 = 126\ 702; \quad \Sigma\, z_\nu^2 = 232\ 801;$$
$$\Sigma\, x_\nu y_\nu = 92\ 701; \quad \Sigma\, y_\nu z_\nu = 145\ 157; \quad \Sigma\, z_\nu x_\nu = 139\ 273;$$

C.F.$(x) = 83\ 987{\cdot}15;$ C.F.$(y) = 112\ 866{\cdot}22;$ C.F.$(z) = 209\ 033{\cdot}64;$

C.F.$(x, y) = 97\ 361{\cdot}76;$ C.F.$(y, z) = 153\ 599{\cdot}59;$ C.F.$(z, x) = 132\ 499{\cdot}58;$

$$X = 13\ 287{\cdot}85; \quad Y = 13\ 835{\cdot}78; \quad Z = 23\ 767{\cdot}36;$$
$$U = -4660{\cdot}76; \quad V = -8442{\cdot}59; \quad W = 6773{\cdot}42.$$

Hence $\alpha^* = \dfrac{3933}{74} = 53{\cdot}1486; \quad \beta^* = \dfrac{54\ 366\ 663{\cdot}20}{162\ 125\ 085{\cdot}50} = 0{\cdot}335\ 338;$

$$\gamma^* = -\frac{80\ 614\ 584{\cdot}53}{162\ 125\ 085{\cdot}50} = -0{\cdot}497\ 237.$$

(ii) If $\gamma = 0$, then the least-squares estimate of β is

$$\beta^{**} = W/X = 0{\cdot}509\ 745.$$

S.S. due to regression $= X\beta^{**2} = 3452{\cdot}71$ with 1 d.f.

Error S.S. $= Z - X\beta^{**2} = 20\ 314{\cdot}65$ with 72 d.f.

Therefore, to test the null hypothesis $H(\beta = 0)$, we have

$$F = (72 \times 3452{\cdot}71)/20\ 314{\cdot}65 = 12{\cdot}24 \text{ with } 1, 72 \text{ d.f.}$$

The observed result is obviously highly significant and so the null hypothesis is rejected.

77 $\Sigma\, x_i = 5442{\cdot}2; \quad \Sigma\, x_i^2 = 122\ 155{\cdot}04; \quad$ Mean $\bar{x} = 22{\cdot}3959;$

C.F.$(x) = 121\ 882{\cdot}8840; \qquad$ S.S.$(x) \equiv X = 272{\cdot}1560;$

$\Sigma\, y_i = 4019{\cdot}6; \quad \Sigma\, y_i^2 = 66\ 588{\cdot}92; \quad$ Mean $\bar{y} = 16{\cdot}5416;$

C.F.$(y) = 66\ 490{\cdot}4698; \qquad$ S.S.$(y) \equiv Y = 98{\cdot}4502;$

$\Sigma\, x_i y_i = 90\ 113{\cdot}83; \qquad$ C.F.$(x, y) = 90\ 022{\cdot}4984;$

$$\text{S.P.}(x, y) \equiv Z = 91{\cdot}3316.$$

Let $u = x - y$ and $v = x + y$. Then for the u and v observations we have the following calculations:

$$\text{S.S.}(u) \equiv U = X + Y - 2Z = 187{\cdot}9430;$$
$$\text{S.S.}(v) \equiv V = X + Y + 2Z = 553{\cdot}2694;$$
$$\text{S.P.}(u, v) \equiv W = X - Y = 173{\cdot}7058.$$

(i) If θ is the true mean difference between the length and breadth of the eggs, then the best estimate of θ is $\theta^* = \bar{x} - \bar{y} = 5\cdot8543$; and

$$\text{s.e.}(\theta^*) = \{U/(242 \times 243)\}^{\frac{1}{2}} = 10^{-1} \times 0\cdot565\ 330.$$

Hence the 95 per cent confidence interval for θ is

$$\theta^* \pm 1\cdot96 \times \text{s.e.}(\theta^*) \quad \text{i.e.} \quad 5\cdot9651,\ 5\cdot7435.$$

(ii) The estimate of $\text{corr}(u, v)$ is r_{uv}, where

$$r_{uv}^2 = W^2/(UV) = 0\cdot290\ 179.$$

The test for the equality of $\text{var}(x)$ and $\text{var}(y)$ is equivalent to testing that $\text{corr}(u, v) = 0$. Hence the required test ratio is

$$F = 241 r_{uv}^2/(1 - r_{uv}^2) = 98\cdot52 \text{ with } 1,\ 241 \text{ d.f.}$$

The observed F ratio is highly significant and so the null hypothesis is rejected confidently.

(iii) The estimate of $\text{corr}(x, y)$ is r_{xy} where

$$r_{xy}^2 = Z^2/(XY) = 0\cdot311\ 320.$$

To test the null hypothesis $H[\text{corr}(x, y) = 0]$, we have

$$F = 241 r_{xy}^2/(1 - r_{xy}^2) = 108\cdot94 \text{ with } 1,\ 241 \text{ d.f.}$$

The observed F ratio is highly significant and so the null hypothesis can be rejected confidently.

The estimates of the linear regression coefficients β_y and β_x are

$$\beta_y^* = Z/X = 0\cdot335\ 585; \qquad \beta_x^* = Z/Y = 0\cdot927\ 693;$$

and the estimated linear regressions are

$$y_e^* = 9\cdot0259 + 0\cdot335\ 585x; \quad x_e^* = 7\cdot0504 + 0\cdot927\ 693y.$$

78 (i) Let x denote area and y transparency. Make the transformations

$$u = (y - 0\cdot4955)/0\cdot03; \qquad v = (x - 43)/3.$$

Next, we assume that

$$E(u) = \alpha + \beta(v - \bar{v});$$

and that f_{ij} denotes the observed frequency in the cell (u_i, v_j). The marginal distribution of u gives

$$\sum_i f_{i.}\, u_i = 413; \quad \sum_i f_{i.}\, u_i^2 = 2127; \quad \text{Mean } \bar{u} = 1\cdot0325;$$
$$\text{C.F.}(u) = 426\cdot4225; \qquad \text{S.S.}(u) \equiv U = 1700\cdot5775.$$

Similarly, the marginal distribution of v gives

$$\sum_j f_{.j}\, v_j = -199; \quad \sum_j f_{.j}\, v_j^2 = 1325; \quad \text{Mean } \bar{v} = -0\cdot4975;$$
$$\text{C.F.}(v) = 99\cdot0025; \qquad \text{S.S.}(v) \equiv V = 1225\cdot9975.$$

Finally, the bivariate distribution gives

$$\sum_i \sum_j f_{ij} u_i v_j = 460; \qquad C.F.(u, v) = -205 \cdot 4675;$$
$$S.P.(u, v) \equiv W = 665 \cdot 4675.$$

Hence the estimate of $\text{corr}(u, v)$ is r_{uv}, where

$$r_{uv}^2 = W^2/(UV) = 0 \cdot 212\,406.$$

Therefore $\qquad\qquad r_{uv} = 0 \cdot 460\,875.$

(ii) The least-squares estimate of β is

$$\beta^* = W/V = 0 \cdot 542\,797;$$

and the linear regression of u on v is

$$u_e^* = 1 \cdot 3025 + 0 \cdot 542\,797v.$$
$$\text{Error S.S.} = U - V\beta^{*2} = 1339 \cdot 3636 \text{ with } 398 \text{ d.f.}$$

Therefore
$$\text{s.e.}(\beta^*) = (1339 \cdot 3636/398V)^{\frac{1}{2}} = 10^{-1} \times 0 \cdot 523\,918.$$

Hence the 95 per cent confidence interval for β is

$$\beta^* \pm 1 \cdot 96 \times \text{s.e.}(\beta^*) \quad \text{i.e.} \quad 0 \cdot 645\,485,\ 0 \cdot 440\,109.$$

In terms of the original x and y units, the estimated regression equation is

$$\frac{y_e^* - 0 \cdot 4955}{0 \cdot 03} = 1 \cdot 3025 + 0 \cdot 542\,797 \times \left(\frac{x - 43}{3} \right),$$

which reduces to

$$y_e^* = 0 \cdot 3012 + 10^{-2} \times 0 \cdot 542\,797x.$$

Also, the 95 per cent confidence interval of the regression coefficient of y on x is $10^{-2} \times 0 \cdot 645\,485,\ 10^{-2} \times 0 \cdot 440\,109$.

(iii) For graphical representation, it is convenient to work in terms of the u, v variables. Then for any v_j, the observed array mean is

$$\bar{u}_j = \sum_i f_{ij} u_i / f_{.j}.$$

The v_j and $\bar{u}_j$ values are given in the table below.

v_j	$\bar{u}_j$	v_j	$\bar{u}_j$
−5	−2·0000	1	1·7547
−4	−0·6250	2	2·1935
−3	−0·3889	3	2·5000
−2	0·1538	4	4·5000
−1	0·8125	5	4·0000
0	1·4333	6	5·0000

For drawing the estimated regression line, we have

$$u_{ej}^* = -1 \cdot 4115 \text{ for } v_j = -5; \qquad u_{ej}^* = 4 \cdot 5593 \text{ for } v_j = 6.$$

79 (i)   Let x denote wage and y rent. Make the transformations $u = (x - 26)/2$; $v = (y - 4 \cdot 5)/0 \cdot 5$. Next assume that

$$E(u) = \alpha_1 + \beta_1(v - \bar{v}); \qquad E(v) = \alpha_2 + \beta_2(u - \bar{u});$$

and that f_{ij} denotes the observed frequency in the cell (u_i, v_j).

The marginal distribution of u gives

$$\sum_i f_{i.} u_i = -1243; \quad \sum_i f_{i.} u_i^2 = 10\ 413; \quad \text{Mean } \bar{u} = -0 \cdot 627\ 144;$$
$$\text{C.F.}(u) = 779 \cdot 5404; \qquad \text{S.S.}(u) \equiv U = 9633 \cdot 4596.$$

Similarly, the marginal distribution of v gives

$$\sum_j f_{.j} v_j = -1562; \quad \sum_j f_{.j} v_j^2 = 6106; \quad \text{Mean } \bar{v} = -0 \cdot 788\ 093;$$
$$\text{C.F.}(v) = 1231 \cdot 0010; \qquad \text{S.S.}(v) \equiv V = 4874 \cdot 9990.$$

Finally, the bivariate distribution gives

$$\sum_i \sum_j f_{ij} u_i v_j = 3521; \qquad \text{C.F.}(u, v) = 979 \cdot 5994;$$
$$\text{S.P.}(u, v) \equiv W = 2541 \cdot 4006.$$

Hence the estimate of $\text{corr}(u, v)$ is r_{uv}, where

$$r_{uv}^2 = W^2/(UV) = 0 \cdot 137\ 527.$$

Therefore
$$r_{uv} = 0 \cdot 370\ 846.$$

To test the null hypothesis $H[\text{corr}(u, v) = 0]$, the test ratio is
$$F = 1980 r_{uv}^2/(1 - r_{uv}^2) = 315 \cdot 7 \text{ with } 1, 1980 \text{ d.f.}$$
The observed F ratio is highly significant and so the null hypothesis is rejected.

(ii) The estimates of the parameters of the regression equations are

$$\alpha_1^* = -0 \cdot 6271; \qquad \beta_1^* = W/V = 0 \cdot 521\ 313;$$
$$\alpha_2^* = -0 \cdot 7881; \qquad \beta_2^* = W/U = 0 \cdot 263\ 810.$$

Hence the two estimated regression equations are

$$u_e^* = -0 \cdot 2163 + 0 \cdot 521\ 313v; \tag{1}$$
$$v_e^* = -0 \cdot 6227 + 0 \cdot 263\ 810u. \tag{2}$$

In terms of the original x and y units, the estimated regression equations become

$$x_e^* = 16 \cdot 1838 + 2 \cdot 085\ 252y; \qquad y_e^* = 2 \cdot 4738 + 0 \cdot 065\ 952\ 5x.$$

For graphical representation, it is convenient to work in terms of the u, v variables. Then, for any v_j, the observed array mean is

$$\bar{u}_j = \sum_i f_{ij} u_i / f_{.j}.$$

The v_j and $\bar{u}_j$ values are given in the table below.

v_j	$\bar{u}_j$	v_j	$\bar{u}_j$
-5	$-2\cdot3333$	1	$0\cdot4082$
-4	$-1\cdot7826$	2	$0\cdot8559$
-3	$-1\cdot6927$	3	$2\cdot1600$
-2	$-1\cdot2214$	4	$2\cdot5000$
-1	$-0\cdot8844$	5	$5\cdot2500$
0	$-0\cdot2697$	6	$0\cdot8000$

For drawing the estimated regression line (1), we have

$$u^*_{ej} = -2\cdot8229 \text{ for } v_j = -5; \qquad u^*_{ej} = 2\cdot9116 \text{ for } v_j = 6.$$

Similarly, for any u_i, the observed array mean is

$$\bar{v}_i = \sum_j f_{ij}\, v_j / f_{i.}.$$

The u_i and $\bar{v}_i$ values are given in the table below.

u_i	$\bar{v}_i$	u_i	$\bar{v}_i$
-9	$-3\cdot0000$	1	$-0\cdot4000$
-7	$-2\cdot6667$	2	$0\cdot1442$
-6	$-1\cdot6000$	3	$0\cdot2727$
-5	$-2\cdot0000$	4	$0\cdot6000$
-4	$-1\cdot8118$	5	$0\cdot4717$
-3	$-1\cdot2680$	6	$0\cdot5000$
-2	$-1\cdot1387$	7	$2\cdot2667$
-1	$-0\cdot9302$	10	$-1\cdot0000$
0	$-0\cdot7646$	12	$2\cdot0000$

For drawing the estimated regression line (2), we have

$$v^*_{ei} = -2\cdot9970 \text{ for } u_i = -9; \qquad v^*_{ei} = 2\cdot5430 \text{ for } u_i = 12.$$

80   *Analysis of the data of Table 1*

Let x denote age and y temperature. Make the transformations

$$u = (x - 52)/3; \qquad v = (y - 98\cdot3)/0\cdot2.$$

Next assume that

$$E(u) = \alpha_1 + \beta_1(v - \bar{v}); \qquad E(v) = \alpha_2 + \beta_2(u - \bar{u});$$

and that f_{ij} denotes the observed frequency in the cell (u_i, v_j).
The marginal distribution of u gives

$$\sum_i f_{i.}\, u_i = -2710; \quad \sum_i f_{i.}\, u_i^2 = 30\,010; \quad \text{Mean } \bar{u} = -2\cdot9234;$$
$$\text{C.F.}(u) = 7922\cdot4380; \qquad \text{S.S.}(u) \equiv U = 22\,087\cdot5620.$$

Similarly, the marginal distribution of v gives

$$\sum_j f_{.j}\, v_j = 108; \quad \sum_j f_{.j}\, v_j^2 = 5568; \quad \text{Mean } \bar{v} = 0\cdot1165;$$
$$\text{C.F.}(v) = 12\cdot5825; \qquad \text{S.S.}(v) \equiv V = 5555\cdot4175.$$

Finally, the bivariate distribution gives

$$\sum_i \sum_j f_{ij} u_i v_j = -1907; \qquad C.F.(u, v) = -315 \cdot 7282;$$
$$S.P.(u, v) \equiv W = -1591 \cdot 2718.$$

Hence the estimate of $\mathrm{corr}(u, v)$ is

$$r_{uv} = W/(UV)^{\frac{1}{2}} = -0 \cdot 143\ 652.$$

The estimates of the parameters of the regression equations are

$$\alpha_1^* = -2 \cdot 9234; \qquad \beta_1^* = W/V = -0 \cdot 286\ 436;$$
$$\alpha_2^* = 0 \cdot 1165; \qquad \beta_2^* = W/U = -0 \cdot 072\ 043\ 8.$$

Hence the two estimated regression equations are

$$u_e^* = -2 \cdot 8900 - 0 \cdot 286\ 436v; \tag{1}$$
$$v_e^* = -0 \cdot 0941 - 0 \cdot 072\ 043\ 8u. \tag{2}$$

In terms of the original x and y units, the estimated regression equations become

$$x_e^* = 465 \cdot 6799 - 4 \cdot 296\ 540y; \quad y_e^* = 98 \cdot 5310 - 10^{-2} \times 0 \cdot 480\ 292x.$$

For graphical representation, it is convenient to work in terms of the u, v variables. Then, for any v_j, the observed array mean is

$$\bar{u}_j = \sum_i f_{ij} u_i / f_{.j}.$$

The v_j and $\bar{u}_j$ values are given in the table below.

v_j	$\bar{u}_j$	v_j	$\bar{u}_j$
-9	$5 \cdot 0000$	0	$-2 \cdot 3007$
-8	$0 \cdot 0000$	1	$-4 \cdot 2000$
-7	$-3 \cdot 5000$	2	$-3 \cdot 5530$
-6	$-0 \cdot 3333$	3	$-3 \cdot 2055$
-5	$-4 \cdot 1364$	4	$-5 \cdot 2439$
-4	$-2 \cdot 5588$	5	$-3 \cdot 7778$
-3	$-1 \cdot 6667$	6	$-2 \cdot 8571$
-2	$-1 \cdot 9286$	7	$-7 \cdot 6667$
-1	$-2 \cdot 0080$	8	$0 \cdot 0000$

For drawing the estimated regression line (1), we have

$$u_{ej}^* = -0 \cdot 3121 \text{ for } v_j = -9; \qquad u_{ej}^* = -4 \cdot 8951 \text{ for } v_j = 7.$$

Similarly, for any u_i, the observed array mean is

$$\bar{v}_i = \sum_j f_{ij} v_j / f_{i.}.$$

The u_i and $\bar{v}_i$ values are given in the table below.

u_i	$\bar{v}_i$	u_i	$\bar{v}_i$
– 10	1·4118	1	– 0·5294
– 9	0·3651	2	0·2308
– 8	– 0·1059	3	– 0·2195
– 7	0·4953	4	– 0·1569
– 6	0·5169	5	– 1·5667
– 5	0·3704	6	0·3600
– 4	0·5949	7	– 0·6190
– 3	0·1639	8	– 1·0000
– 2	0·2368	9	– 1·8000
– 1	– 0·2121	10	– 0·5000
0	– 0·6333	11	– 1·0000

For drawing the estimated regression line (2), we have

$$v^*_{ei} = 0·6263 \text{ for } u_i = -10; \qquad v^*_{ei} = -0·8145 \text{ for } u_i = 10.$$

Analysis of the data of Table 2

Let x denote reaction time and y head length. Make the transformations

$$u = (x - 19·995)/2; \qquad v = (y - 7·805)/0·1.$$

Next assume that

$$E(u) = \alpha_1 + \beta_1(v - \bar{v}); \qquad E(v) = \alpha_2 + \beta_2(u - \bar{u});$$

and that f_{ij} denotes the observed frequency in the cell (u_i, v_j).
The marginal distribution of u gives

$$\sum_i f_{i.} u_i = -3016; \quad \sum_i f_{i.} u_i^2 = 14\,030; \quad \text{Mean } \bar{u} = -0·6431;$$
$$\text{C.F.}(u) = 1939·5002; \qquad \text{S.S.}(u) \equiv U = 12\,090·4998.$$

Similarly, the marginal distribution of v gives

$$\sum_j f_{.j} v_j = -4953; \quad \sum_j f_{.j} v_j^2 = 34\,119; \quad \text{Mean } \bar{v} = -1·0561;$$
$$\text{C.F.}(v) = 5230·7482; \qquad \text{S.S.}(v) \equiv V = 28\,888·2518.$$

Finally, the bivariate distribution gives

$$\sum_i \sum_j f_{ij} u_i v_j = 2570; \qquad \text{C.F.}(u, v) = 3185·1275;$$
$$\text{S.P.}(u, v) \equiv W = -615·1275.$$

Hence the estimate of $\operatorname{corr}(u, v)$ is

$$r_{uv} = W/(UV)^{\frac{1}{2}} = -10^{-1} \times 0·329\,141.$$

The estimates of the parameters of the regression equations are

$$\alpha_1^* = -0·6431; \qquad \beta_1^* = W/V = -10^{-1} \times 0·212\,933;$$
$$\alpha_2^* = -1·0561; \qquad \beta_2^* = W/U = -10^{-1} \times 0·508\,769.$$

Hence the two estimated regression equations are

$$u_e^* = -0.6656 - 10^{-1} \times 0.212\ 933v; \tag{1}$$
$$v_e^* = -1.0888 - 10^{-1} \times 0.508\ 769u. \tag{2}$$

In terms of the original x and y units, the estimated regression equations become

$$x_e^* = 21.988 - 0.425\ 866y; \quad y_e^* = 7.747 - 10^{-2} \times 0.254\ 384x.$$

For graphical representation, it is convenient to work in terms of the u, v variables. Then, for any v_j, the observed array mean is

$$\bar{u}_j = \sum_i f_{ij}\, u_i / f_{.j}.$$

The v_j and $\bar{u}_j$ values are given in the table below.

v_j	$\bar{u}_j$	v_j	$\bar{u}_j$
−10	0.0000	0	−0.6174
−9	−0.5714	1	−0.7600
−8	−0.8421	2	−0.7562
−7	−0.5814	3	−0.5642
−6	−0.3269	4	−0.7629
−5	−0.5714	5	−0.7714
−4	−0.4894	6	−1.1667
−3	−0.7218	7	−0.5000
−2	−0.5456	8	0.0000
−1	−0.7141		

For drawing the estimated regression line (1), we have

$$u_{ej}^* = -0.4527 \text{ for } v_j = -10; \qquad u_{ej}^* = -0.8359 \text{ for } v_j = 8.$$

Similarly, for any u_i, the observed array mean is

$$\bar{v}_i = \sum_j f_{ij}\, v_j / f_{i.}.$$

The u_i and $\bar{v}_i$ values are given in the table below.

u_i	$\bar{v}_i$	u_i	$\bar{v}_i$
−7	−3.0000	2	−1.0963
−6	−3.3333	3	−1.6170
−5	−1.3158	4	−1.3200
−4	−0.3494	5	−0.3333
−3	−1.1490	6	−1.5000
−2	−0.9121	7	−4.0000
−1	−0.9966	8	0.0000
0	−1.1231	10	−5.0000
1	−1.1699	12	−12.0000

For drawing the estimated regression line (2), we have

$$v_{ei}^* = -0.7327 \text{ for } u_i = -7; \qquad v_{ei}^* = -1.6993 \text{ for } u_i = 12.$$

Analysis of the data of Table 3

Let x denote maximum temperature and y minimum temperature. Make the transformations $u = (x - 69)/2$; $v = (y - 51)/2$. Next assume that

$$E(u) = \alpha_1 + \beta_1(v - \bar{v}); \qquad E(v) = \alpha_2 + \beta_2(u - \bar{u});$$

and that f_{ij} denotes the observed frequency in the cell (u_i, v_j). The marginal distribution of u gives

$$\sum_i f_{i.} u_i = -531; \quad \sum_i f_{i.} u_i^2 = 13\ 383; \quad \text{Mean } \bar{u} = -0.3498;$$
$$\text{C.F.}(u) = 185.7451; \qquad \text{S.S.}(u) \equiv U = 13\ 197.2549.$$

Similarly, the marginal distribution of v gives

$$\sum_j f_{.j} v_j = 411; \quad \sum_j f_{.j} v_j^2 = 8271; \quad \text{Mean } \bar{v} = 0.2708;$$
$$\text{C.F.}(v) = 111.2787; \qquad \text{S.S.}(v) \equiv V = 8159.7213.$$

Finally, the bivariate distribution gives

$$\sum_i \sum_j f_{ij} u_i v_j = 3013; \qquad \text{C.F.}(u, v) = -143.7688;$$
$$\text{S.P.}(u, v) \equiv W = 3156.7688.$$

Hence the estimate of $\text{corr}(u, v)$ is

$$r_{uv} = W/(UV)^{\frac{1}{2}} = 0.304\ 203.$$

The estimates of the parameters of the regression equations are

$$\alpha_1^* = -0.3498; \qquad \beta_1^* = W/V = 0.386\ 872;$$
$$\alpha_2^* = 0.2708; \qquad \beta_2^* = W/U = 0.239\ 199.$$

Hence the two estimated regression equations are

$$u_e^* = -0.4546 + 0.386\ 872v; \tag{1}$$
$$v_e^* = 0.3545 + 0.239\ 199u. \tag{2}$$

In terms of the original x and y units, the estimated regression equations become

$$x_e^* = 57.5099 + 0.239\ 199y; \quad y_e^* = 23.3966 + 0.386\ 872x.$$

For graphical representation, it is convenient to work in terms of the u, v variables. Then, for any v_j, the observed array mean is

$$\bar{u}_j = \sum_i f_{ij} u_i / f_{.j}.$$

The v_j and $\bar{u}_j$ values are given in the table below.

v_j	$\bar{u}_j$	v_j	$\bar{u}_j$
-7	-1.0000	0	-1.0805
-6	-2.0000	1	-0.4047
-5	-1.3333	2	0.2949
-4	-0.9310	3	0.9801
-3	-1.1100	4	1.9383
-2	-1.1172	5	3.1538
-1	-1.2596	6	3.2000

For drawing the estimated regression line (1), we have

$$u_{ej}^* = -3\cdot1627 \text{ for } v_j = -7; \qquad u_{ej}^* = 1\cdot8666 \text{ for } v_j = 6.$$

Similarly, for any u_i, the observed array mean is

$$\bar{v}_i = \sum_j f_{ij}\, v_j / f_{i.}.$$

The u_i and $\bar{v}_i$ values are given in the table below.

u_i	$\bar{v}_i$	u_i	$\bar{v}_i$
-9	$-3\cdot0000$	3	$0\cdot7160$
-7	$-0\cdot8333$	4	$1\cdot2407$
-6	$-1\cdot5238$	5	$0\cdot8696$
-5	$-1\cdot2941$	6	$1\cdot7143$
-4	$-0\cdot7451$	7	$1\cdot6154$
-3	$-0\cdot3864$	8	$1\cdot7500$
-2	$-0\cdot1472$	9	$2\cdot6000$
-1	$0\cdot2775$	10	$2\cdot3333$
0	$0\cdot4404$	11	$4\cdot0000$
1	$0\cdot8248$	12	$3\cdot0000$
2	$1\cdot1920$		

For drawing the estimated regression line (2), we have

$$v_{ei}^* = -1\cdot7983 \text{ for } u_i = -9; \qquad v_{ei}^* = 2\cdot9857 \text{ for } u_i = 11.$$

Analysis of the data of Table 4

Let x denote number of pistils and y number of stamens. Make the transformations $u = x - 15$; $v = y - 27$. Next assume that

$$E(u) = \alpha_1 + \beta_1(v - \bar{v}); \qquad E(v) = \alpha_2 + \beta_2(u - \bar{u});$$

and that f_{ij} denotes the observed frequency in the cell (u_i, v_j).
The marginal distribution of u gives

$$\sum_i f_{i.}\, u_i = 656; \quad \sum_i f_{i.}\, u_i^2 = 5670; \quad \text{Mean } \bar{u} = 2\cdot4478;$$
$$\text{C.F.}(u) = 1605\cdot7313; \qquad \text{S.S.}(u) \equiv U = 4064\cdot2687.$$

Similarly, the marginal distribution of v gives

$$\sum_j f_{.j}\, v_j = -72; \quad \sum_j f_{.j}\, v_j^2 = 3810; \quad \text{Mean } \bar{v} = -0\cdot2687;$$
$$\text{C.F.}(v) = 19\cdot3433; \qquad \text{S.S.}(v) \equiv V = 3790\cdot6567.$$

Finally, the bivariate distribution gives

$$\sum_i \sum_j f_{ij}\, u_i\, v_j = 1812; \qquad \text{C.F.}(u, v) = -176\cdot2388;$$
$$\text{S.P.}(u, v) \equiv W = 1988\cdot2388.$$

Hence the estimate of corr (u, v) is

$$r_{uv} = W/(UV)^{\frac{1}{2}} = 0\cdot506\ 547.$$

The estimates of the parameters of the regression equations are

$$\alpha_1^* = 2\cdot4478; \qquad \beta_1^* = W/V = 0\cdot524\ 510;$$
$$\alpha_2^* = -0\cdot2687; \qquad \beta_2^* = W/U = 0\cdot489\ 200.$$

Hence the two estimated regression equations are

$$u_e^* = 2\cdot5887 + 0\cdot524\ 510v; \qquad (1)$$
$$v_e^* = -1\cdot4662 + 0\cdot489\ 200u. \qquad (2)$$

In terms of the original x and y units, the estimated regression equations become

$$x_e^* = 3\cdot4269 + 0\cdot524\ 510y; \quad y_e^* = 18\cdot1958 + 0\cdot489\ 200x.$$

For graphical representation, it is convenient to work in terms of the u, v variables. Then, for any v_j, the observed array mean is

$$\bar{u}_j = \sum_i f_{ij}\, u_i / f_{\cdot j}.$$

The v_j and $\bar{u}_j$ values are given in the table below.

v_j	$\bar{u}_j$	v_j	$\bar{u}_j$
-9	$1\cdot0000$	1	$1\cdot5714$
-8	$-1\cdot8333$	2	$2\cdot9130$
-7	$-1\cdot7500$	3	$4\cdot1500$
-6	$-0\cdot8889$	4	$4\cdot8000$
-5	$0\cdot3125$	5	$6\cdot6923$
-4	$0\cdot5833$	6	$5\cdot0000$
-3	$0\cdot8636$	7	$13\cdot0000$
-2	$0\cdot9615$	8	$10\cdot0000$
-1	$3\cdot0000$	10	$-13\cdot0000$
0	$3\cdot0000$	11	$10\cdot0000$

For drawing the estimated regression line (1), we have

$$u_{ej}^* = -2\cdot1319 \text{ for } v_j = -9; \qquad u_{ej}^* = 5\cdot7358 \text{ for } v_j = 6.$$

Similarly, for any u_i, the observed array mean is $\bar{v}_i = \sum_j f_{ij}\, v_j / f_i$. The u_i and $\bar{v}_i$ values are given in the table below.

u_i	$\bar{v}_i$	u_i	$\bar{v}_i$
-13	$10\cdot0000$	4	$0\cdot4286$
-8	$2\cdot0000$	5	$1\cdot3684$
-6	$2\cdot0000$	6	$2\cdot0000$
-5	$-5\cdot0000$	7	$2\cdot2000$
-4	$-4\cdot6667$	8	$3\cdot6000$
-3	$-3\cdot9231$	9	$0\cdot7500$
-2	$-3\cdot9167$	10	$6\cdot2500$
-1	$-3\cdot5909$	11	$4\cdot3333$
0	$-2\cdot0286$	12	$3\cdot5000$
1	$-0\cdot1935$	13	$7\cdot0000$
2	$0\cdot0800$	16	$2\cdot0000$
3	$-0\cdot1481$		

For drawing the estimated regression line (2), we have

$$v_{ei}^* = -4 \cdot 0414 \text{ for } u_i = -6; \qquad v_{ei}^* = 4 \cdot 4042 \text{ for } u_i = 12.$$

Analysis of the data of Table 5

Let x denote number of pistils and y number of stamens. Make the transformations $u = x - 13$; $v = y - 18$. Next assume that

$$E(u) = \alpha_1 + \beta_1(v - \bar{v}); \qquad E(v) = \alpha_2 + \beta_2(u - \bar{u});$$

and that f_{ij} denotes the observed frequency in the cell (u_i, v_j).
The marginal distribution of u gives

$$\sum_i f_{i \cdot} u_i = -318; \quad \sum_i f_{i \cdot} u_i^2 = 4552; \quad \text{Mean } \bar{u} = -0 \cdot 8525;$$
$$\text{C.F.}(u) = 271 \cdot 1099; \qquad \text{S.S.}(u) \equiv U = 4280 \cdot 8901.$$

Similarly, the marginal distribution of v gives

$$\sum_j f_{\cdot j} v_j = -51; \quad \sum_j f_{\cdot j} v_j^2 = 4065; \quad \text{Mean } \bar{v} = -0 \cdot 1367;$$
$$\text{C.F.}(v) = 6 \cdot 9732; \qquad \text{S.S.}(v) \equiv V = 4058 \cdot 0268.$$

Finally, the bivariate distribution gives

$$\sum_i \sum_j f_{ij} u_i v_j = 3165; \qquad \text{C.F.}(u, v) = 43 \cdot 4799;$$
$$\text{S.P.}(u, v) \equiv W = 3121 \cdot 5201.$$

Hence the estimate of $\text{corr}(u, v)$ is

$$r_{uv} = W/(UV)^{\frac{1}{2}} = 0 \cdot 748 \ 930.$$

The estimates of the parameters of the regression equations are

$$\alpha_1^* = -0 \cdot 8525; \qquad \beta_1^* = W/V = 0 \cdot 769 \ 221;$$
$$\alpha_2^* = -0 \cdot 1367; \qquad \beta_2^* = W/U = 0 \cdot 729 \ 175.$$

Hence the two estimated regression equations are

$$u_e^* = -0 \cdot 7473 + 0 \cdot 769 \ 221v; \qquad (1)$$
$$v_e^* = \quad 0 \cdot 4849 + 0 \cdot 729 \ 175u. \qquad (2)$$

In terms of the original x and y units, the estimated regression equations become

$$x_e^* = -1 \cdot 5933 + 0 \cdot 769 \ 221y; \quad y_e^* = 9 \cdot 0056 + 0 \cdot 729 \ 175x.$$

For graphical representation, it is convenient to work in terms of the u, v variables. Then, for any v_j, the observed array mean is

$$\bar{u}_j = \sum_i f_{ij} u_i / f_{\cdot j}.$$

The v_j and $\bar{u}_j$ values are given in the table below.

v_j	$\bar{u}_j$	v_j	$\bar{u}_j$
−10	−2·0000	2	0·6429
−8	−5·0000	3	2·0000
−7	−4·6667	4	3·8235
−6	−5·6000	5	4·1429
−5	−5·1818	6	2·0000
−4	−3·8696	7	5·1667
−3	−3·8605	8	5·6667
−2	−2·4571	9	3·5000
−1	−0·8710	10	6·0000
0	−0·9091	11	5·0000
1	−0·1163		

For drawing the estimated regression line (1), we have

$$u_{ej}^* = -6\cdot9011 \text{ for } v_j = -8; \qquad u_{ej}^* = 7\cdot7141 \text{ for } v_j = 11.$$

Similarly, for any u_i, the observed array mean is

$$\bar{v}_i = \sum_j f_{ij} v_j / f_{i\cdot}.$$

The u_i and $\bar{v}_i$ values are given in the table below.

u_i	$\bar{v}_i$	u_i	$\bar{v}_i$
−8	−7·0000	2	1·2857
−7	−4·8333	3	3·0435
−6	−4·8750	4	2·2500
−5	−3·1143	5	5·5455
−4	−2·4286	6	5·2222
−3	−0·8947	7	6·5000
−2	−0·9500	8	5·0000
−1	−0·2857	9	4·0000
0	0·3333	10	8·0000
1	1·2778	11	7·0000

For drawing the estimated regression line (2), we have

$$v_{ei}^* = -5\cdot3485 \text{ for } u_i = -8; \qquad v_{ei}^* = 8\cdot5058 \text{ for } u_i = 11.$$

Analysis of the data of Table 6

Let x denote longitudinal girth and y length. Make the transformations

$$u = (x - 11\cdot25)/0\cdot1; \qquad v = (y - 4\cdot125)/0\cdot05.$$

Next assume that

$$E(u) = \alpha_1 + \beta_1(v - \bar{v}); \qquad E(v) = \alpha_2 + \beta_2(u - \bar{u});$$

and that f_{ij} denotes the observed frequency in the cell (u_i, v_j).
The marginal distribution of u gives

$$\sum_i f_{i\cdot} u_i = 1226; \quad \sum_i f_{i\cdot} u_i^2 = 12\,224; \quad \text{Mean } \bar{u} = 1\cdot2838;$$
$$\text{C.F.}(u) = 1573\cdot9016; \qquad \text{S.S.}(u) \equiv U = 10\,650\cdot0984.$$

Similarly, the marginal distribution of v gives

$$\sum_j f_{.j}\, v_j = 1241; \quad \sum_j f_{.j}\, v_j^2 = 12\,543; \quad \text{Mean } \bar{v} = 1\cdot2995;$$
$$\text{C.F.}(v) = 1612\cdot6503; \quad \text{S.S.}(v) \equiv V = 10\,930\cdot3497.$$

Finally, the bivariate distribution gives

$$\sum_i \sum_j f_{ij}\, u_i\, v_j = 11\,119; \quad \text{C.F.}(u,\,v) = 1593\cdot1581;$$
$$\text{S.P.}(u,\,v) \equiv W = 9525\cdot8419.$$

Hence the estimate of $\mathrm{corr}\,(u,\,v)$ is $r_{uv} = W/(UV)^{\frac{1}{2}} = 0\cdot882\ 896.$
The estimates of the parameters of the regression equations are

$$\alpha_1^* = 1\cdot2838; \qquad \beta_1^* = W/V = 0\cdot871\ 504;$$
$$\alpha_2^* = 1\cdot2995; \qquad \beta_2^* = W/U = 0\cdot894\ 437.$$

Hence the two estimated regression equations are

$$u_e^* = 0\cdot1513 + 0\cdot871\ 504v; \tag{1}$$
$$v_e^* = 0\cdot1512 + 0\cdot894\ 437u. \tag{2}$$

In terms of the original x and y units, the estimated regression equations become

$$x_e^* = 4\cdot0752 + 1\cdot743\ 008y; \quad y_e^* = -0\cdot8986 + 0\cdot447\ 218x.$$

For graphical representation, it is convenient to work in terms of the u, v variables. Then, for any v_j, the observed array mean is

$$\bar{u}_j = \sum_i f_{ij}\, u_i/f_{.j}\ .$$

The v_j and $\bar{u}_j$ values are given in the table below.

v_j	$\bar{u}_j$	v_j	$\bar{u}_j$
−8	−5·5000	2	1·8878
−7	−5·8333	3	2·6374
−6	−5·3571	4	3·9057
−5	−3·6875	5	4·4630
−4	−3·5000	6	5·3750
−3	−2·2909	7	5·8800
−2	−1·7246	8	7·4000
−1	−0·8235	9	7·5833
0	0·1953	10	8·5000
1	1·0495	11	9·6000

For drawing the estimated regression line (1), we have

$$u_{ej}^* = -6\cdot8207 \text{ for } v_j = -8; \qquad u_{ej}^* = 9\cdot7378 \text{ for } v_j = 11.$$

Similarly, for any u_i, the observed array mean is

$$\bar{v}_i = \sum_j f_{ij}\, v_j/f_{i.}\ .$$

The u_i and $\bar{v}_i$ values are given in the table below.

u_i	$\bar{v}_i$	u_i	$\bar{v}_i$
-9	$-6\cdot0000$	2	$1\cdot8317$
-8	$-6\cdot0000$	3	$2\cdot8980$
-7	$-5\cdot3333$	4	$3\cdot7500$
-6	$-5\cdot0000$	5	$4\cdot2879$
-5	$-4\cdot3000$	6	$5\cdot5208$
-4	$-3\cdot3429$	7	$6\cdot4138$
-3	$-2\cdot8947$	8	$7\cdot6250$
-2	$-1\cdot7848$	9	$10\cdot0000$
-1	$-0\cdot7222$	10	$9\cdot3333$
0	$0\cdot0756$	11	$11\cdot0000$
1	$1\cdot2368$		

For drawing the estimated regression line (2), we have

$$v^*_{ei} = -6\cdot0893 \text{ for } u_i = -7; \qquad v^*_{ei} = 9\cdot9577 \text{ for } u_i = 11.$$

81   Leicester, 1962.

In the tossing of an unbiased die, define a success (S) as the showing of a 5 or 6 on the uppermost face, and let a failure (F) be the complementary event. Then $P(S) = \frac{1}{3}$ and $P(F) = \frac{2}{3}$. With one throw of 12 unbiased dice, the probability of obtaining r successes is

$$B(r) = \binom{12}{r}(\tfrac{1}{3})^r(\tfrac{2}{3})^{12-r}, \text{ for } 0 \leqslant r \leqslant 12.$$

For $r = 0$,

$$B(0) = (\tfrac{2}{3})^{12} = 10^{-2} \times 0\cdot770\ 735;$$

and for $r \geqslant 1$ we have the recurrence relation

$$B(r) = \left(\frac{13-r}{2r}\right)B(r-1).$$

Hence the following model distribution:

No. of successes (r)	$B(r)$	No. of successes (r)	$B(r)$
0	$10^{-2} \times 0\cdot770\ 735$	7	$10^{-1} \times 0\cdot476\ 893$
1	$10^{-1} \times 0\cdot462\ 441$	8	$10^{-1} \times 0\cdot149\ 029$
2	$0\cdot127\ 171$	9	$10^{-2} \times 0\cdot331\ 176$
3	$0\cdot211\ 952$	10	$10^{-3} \times 0\cdot496\ 764$
4	$0\cdot238\ 446$	11	$10^{-4} \times 0\cdot451\ 604$
5	$0\cdot190\ 757$	12	$10^{-5} \times 0\cdot188\ 168$
6	$0\cdot111\ 275$		
Total			$1\cdot000\ 000$

Pool the frequencies for $r \geqslant 10$ to compare the observed and expected distributions. The goodness-of-fit $\chi^2 = 35 \cdot 4871$ with 10 d.f. The $0 \cdot 1$ per cent table value of χ^2 with 10 d.f. is $29 \cdot 588$ and so the observed result is highly significant. The hypothesis of unbiased dice is definitely rejected.

82   Leicester, 1964.

The observed frequency distribution of the digits is as follows:

Digit	Frequency	Digit	Frequency
0	72	5	62
1	79	6	69
2	76	7	52
3	72	8	73
4	72	9	80
Total			707

On the hypothesis of the randomness of the distribution of the digits, the expected frequency of each digit is $70 \cdot 7$. Hence the goodness-of-fit $\chi^2 = 8 \cdot 799$ with 9 d.f. The 5 per cent table value of χ^2 with 9 d.f. is $18 \cdot 307$, and so the observed value of χ^2 is not significant. The hypothesis of the randomness of the distribution of the digits can therefore be accepted.

(i) Let p_5 and p_7 be the true proportions of 5's and 7's in the population. The observed frequencies are $n_5 = 62$ and $n_7 = 52$. Then, since these have a multinomial distribution, we obtain

$$\text{var}(n_5 - n_7) = N[p_5(1 - p_5) + p_7(1 - p_7) + 2p_5 p_7], \quad \text{where } N = 707.$$

To test the null hypothesis $H(p_5 = p_7)$, the pooled estimate of the common value is $p^* = 114/1414$; and if the null hypothesis is true, then

$$\text{estimate var}(n_5 - n_7) = 2 \times 707[p^*(1 - p^*) + p^{*2}] = 114.$$

Therefore a large-sample test for the null hypothesis is that

$$z = (62 - 52)/\sqrt{114} = 0 \cdot 937 \text{ is } \sim N(0, 1).$$

The observed z is obviously not significant and so the null hypothesis is acceptable.

(ii) Again, $n_5 + n_7 = 114$ and $E(n_5 + n_7) = 2Np$, where $p = 0 \cdot 1$. Hence if the null hypothesis $H(p = 0 \cdot 1)$ is true, then

$$\text{var}(n_5 + n_7) = 2Np(1 - 2p) = 113 \cdot 12.$$

Therefore a large-sample test for the null hypothesis is that

$$w = (114 - 141 \cdot 4)/\sqrt{113 \cdot 12} = -2 \cdot 576 \text{ is } \sim N(0, 1).$$

The observed value of w is significant at the 5 per cent level and so the null hypothesis is not tenable.

The χ^2 test is one of overall agreement between the observed and expected frequencies of the 10 digits. This masks the disparity between the observed and expected frequencies of 5's and 7's.

83 (i)   Sample S.S. = 1937·6686;   sample mean = 11·9112. Therefore a test for the Poisson hypothesis is that

$$\chi^2 = 1937 \cdot 6686/11 \cdot 9112 = 162 \cdot 676 \text{ with 168 d.f.}$$

For 168 d.f. the 5 per cent point of the χ^2 distribution is χ_1^2, where

$$\sqrt{2\chi_1^2} - \sqrt{335} \sim 1 \cdot 96 \quad \text{i.e.} \quad \chi_1^2 = 133 \cdot 50.$$

The observed distribution of corpuscles is therefore significantly different from that expected on the Poisson hypothesis.

A second method for testing the Poisson hypothesis is by goodness of fit. If μ is the mean of the Poisson population, then its estimate is $\mu* = 11 \cdot 9112$. Let w be the random variable denoting the number of corpuscles in a cell of the haemocytometer. Then the probability that there are w corpuscles in a cell is

$$Q(w) = e^{-\mu^*}(\mu^*)^w/w!, \quad \text{for } w \geqslant 0.$$

In particular, $Q(0) = e^{-11 \cdot 9112} = 10^{-5} \times 0 \cdot 671\ 478;$

and for $w \geqslant 1$ $Q(w) = \left(\dfrac{11 \cdot 9112}{w}\right) Q(w - 1).$

The model probability distribution is as follows:

w	$Q(w)$	w	$Q(w)$
0	$10^{-5} \times 0 \cdot 671\ 478$	12	$0 \cdot 114\ 329$
1	$10^{-4} \times 0 \cdot 799\ 811$	13	$0 \cdot 104\ 754$
2	$10^{-3} \times 0 \cdot 476\ 335$	14	$10^{-1} \times 0 \cdot 891\ 247$
3	$10^{-2} \times 0 \cdot 189\ 124$	15	$10^{-1} \times 0 \cdot 707\ 721$
4	$10^{-2} \times 0 \cdot 563\ 173$	16	$10^{-1} \times 0 \cdot 526\ 863$
5	$10^{-1} \times 0 \cdot 134\ 161$	17	$10^{-1} \times 0 \cdot 369\ 151$
6	$10^{-1} \times 0 \cdot 266\ 336$	18	$10^{-1} \times 0 \cdot 244\ 280$
7	$10^{-1} \times 0 \cdot 453\ 197$	19	$10^{-1} \times 0 \cdot 153\ 140$
8	$10^{-1} \times 0 \cdot 674\ 765$	20	$10^{-2} \times 0 \cdot 912\ 041$
9	$10^{-1} \times 0 \cdot 893\ 029$	21	$10^{-2} \times 0 \cdot 517\ 310$
10	$0 \cdot 106\ 370$	$\geqslant 22$	$0 \cdot 005\ 598$
11	$0 \cdot 115\ 181$		
Total			$1 \cdot 000\ 000$

Pool the frequencies for $w \leqslant 6$ and $w \geqslant 18$. Then the goodness-of-fit $\chi^2 = 4 \cdot 1299$ with 11 d.f. The 5 per cent table value of χ^2 with 11 d.f. is 19·675, and so the observed result is not significant. The observed frequency distribution could have arisen from a Poisson population with mean 11·9112.

The two tests lead to opposite conclusions. The first test is to be preferred because it is more powerful than the second.

(ii) We have   s.e. $(\mu^*) = \sqrt{\mu^*}/13 = 0 \cdot 265\ 485.$ Therefore the 95 per cent confidence interval for μ is

$$\mu^* \pm 1 \cdot 96 \times \text{s.e.} (\mu^*) \quad \text{i.e.} \quad 12 \cdot 4316, \ 11 \cdot 3908.$$

(iii) We now have

$$\text{s.e.}(\mu^*) = \{1937 \cdot 6686/(168 \times 169)\}^{\frac{1}{2}} = 0 \cdot 261\ 241.$$

Therefore the 95 per cent confidence interval for μ is

$$\mu^* \pm 1 \cdot 96 \times \text{s.e.}(\mu^*) \quad \text{i.e.} \quad 12 \cdot 4232,\ 11 \cdot 3992.$$

84 Let x denote the mean log sentence-length. Then for the observed frequency distribution we have

sample mean $\bar{x} = 1 \cdot 1506$; sample standard deviation $s_x = 0 \cdot 272\ 88$.

If u_i denotes the upper limit of the ith class-interval, then the cumulative distribution of the fitted normal curve is as follows:

u_i	$z_i = \dfrac{u_i - \bar{x}}{\cdot s_x}$	Area from $-\infty$ to z_i	u_i	$z_i = \dfrac{u_i - \bar{x}}{s_x}$	Area from $-\infty$ to z_i
$-\infty$	$-\infty$	0·000 00	1·2	0·181	0·571 81
0·3	$-3 \cdot 117$	0·000 91	1·3	0·547	0·707 81
0·4	$-2 \cdot 751$	0·002 97	1·4	0·914	0·819 64
0·5	$-2 \cdot 384$	0·008 56	1·5	1·280	0·899 73
0·6	$-2 \cdot 018$	0·021 80	1·6	1·647	0·950 22
0·7	$-1 \cdot 651$	0·049 37	1·7	2·013	0·977 94
0·8	$-1 \cdot 285$	0·099 40	1·8	2·380	0·991 34
0·9	$-0 \cdot 918$	0·179 31	1·9	2·746	0·996 98
1·0	$-0 \cdot 552$	0·290 48	2·0	3·113	0·999 07
1·1	$-0 \cdot 185$	0·426 62	∞	∞	1·000 00

Pool the frequencies for $u \leqslant 0 \cdot 7$ and $u \geqslant 1 \cdot 7$. Then the goodness-of-fit $\chi^2 = 16 \cdot 1844$ with 8 d.f. The 5 per cent table value of χ^2 with 8 d.f. is $15 \cdot 507$, and so the observed result is significant. The normal distribution is not a good fit for the observed frequency distribution.

85 (i) Assuming that there is no association between the sex and religion of the patients, the expected frequencies estimated from the data are as follows:

Religion	Boys	Girls
CE	554·2484	293·7516
RC	81·6993	43·3007
NC	30·7190	16·2810
J	15·0327	7·9673
O	18·3007	9·6993
Total	700·0001	370·9999

Hence the contingency χ^2 for testing independence of the two traits is $3 \cdot 0111$ with 4 d.f. The 5 per cent table value of χ^2 with 4 d.f. is $9 \cdot 488$, and so the observed result is not significant. There is no association between religion and sex amongst the patient populations.

(ii) Suppose the true proportion of RC patients is p. Then the best estimate of p is $p^* = 125/1071 = 0.116\ 713$. If the null hypothesis $H(p = 0.07)$ is true, then

$$\text{s.d.}(p^*) = \{(0.07 \times 0.93)/1071\}^{\frac{1}{2}} = 10^{-2} \times 0.7796.$$

Hence to test the above null hypothesis we have

$$z = (0.116\ 713 - 0.07)/0.007\ 796 = 5.992 \text{ is } \sim N(0, 1).$$

The observed z is highly significant and we conclude that the true proportion of RC patients differs from the proportion of Roman Catholics in the country.

86   Leicester, 1961.

(i) Assuming that there is no association between the disease distribution of "gainfully employed" women and other women, the expected frequencies estimated from the data are as follows:

Occupation	Appendicitis	Hernia	Peptic ulcer	Other diseases
Women "gainfully employed"	40.000	27.600	18.000	18.400
Other women	60.000	41.400	27.000	27.600

Hence the contingency χ^2 for testing the independence of the two traits is 30.884 with 3 d.f. The 0.1 per cent table value of χ^2 with 3 d.f. is 16.268, and so the observed result is highly significant. There is an association between occupation and the disease distribution of the women.

(ii) Pool the frequencies for peptic ulcer and other diseases. Assuming that there is no association between the disease distribution of the male occupational groups, the expected frequencies estimated from the data are as follows:

Occupation	Appendicitis	Hernia	Peptic ulcer and other diseases
Iron moulders	29.436	37.846	14.718
Other metal workers	20.103	25.846	10.051
Transport and communication workers	17.949	23.077	8.974
Coal miners	16.513	21.231	8.256

Hence the contingency χ^2 for testing the independence of the two traits is 23.692 with 6 d.f. The 0.1 per cent table value of χ^2 with 6 d.f. is 22.457, and so the observed result is highly significant. There is an association between occupation and the disease distribution of the men.

(iii) Assuming that there is no association between sex and the disease distribution, the expected frequencies estimated from the data are as follows:

Sex	Appendicitis	Hernia	Peptic ulcer	Other diseases
Female	96·842	93·158	29·474	40·526
Male	87·158	83·842	26·526	36·474

Hence the contingency χ^2 for testing the independence of the two traits is 32·271 with 3 d.f. This is significant at the 0·1 per cent level. There is an association between sex and the disease distribution.

(iv) Let p_1 and p_2 be the true proportions of appendicitis patients in the female and male populations. Then for the best estimates, we have

$$p_1^* = 100/260 = 0.384\ 615; \qquad p_2^* = 84/234 = 0.358\ 974;$$

and $\quad$ s.e. $(p_1^* - p_2^*) = \left[\dfrac{p_1^*(1 - p_1^*)}{259} + \dfrac{p_2^*(1 - p_2^*)}{233}\right]^{\frac{1}{2}} = 10^{-1} \times 0.436\ 061.$

Hence the 95 per cent large-sample confidence interval for $p_1 - p_2$ is

$$p_1^* - p_2^* \pm 1.96 \times \text{s.e.}(p_1^* - p_2^*) \quad \text{i.e.} \quad 0.1111, -0.0598.$$

Therefore the null hypothesis $H(p_1 - p_2 = 0)$ can be accepted at the 5 per cent level of signficance.

87   Let the time variable be $x = $ Year $- 1839$. Assuming that the sex proportion of the workers remained the same during the years considered, the expected frequencies estimated from the data are as follows:

x	Males	Females	x	Males	Females
0	103 149·90	156 186·10	35	190 727·63	288 793·37
8	125 817·85	190 509·15	39	192 072·81	290 830·19
11	131 623·75	199 300·25	46	200 491·50	303 577·50
17	150 830·51	228 382·49	51	210 326·18	318 468·82
22	179 609·83	271 959·17	56	214 338·64	324 544·36
31	179 020·37	271 066·63	59	209 257·03	316 849·97
Total				2 087 266·00	3 160 468·00

Hence the contingency χ^2 for testing the equality of the sex-proportion over the years considered is 6544·70 with 12 d.f., an undoubtedly significant result. The sex proportion almost certainly changed over the years.

Let y_i denote the observed proportion of male workers at time x_i. The y_i are not of equal precision, but if this complication is ignored, a linear regression can be fitted by the simplest application of the method of least squares. The equation of the estimated regression is

$$y_e^* = 0.4311 - 10^{-3} \times 0.946\ 183x.$$

The table below gives the values of x, y and y_e^* for comparative purposes. The linear regression is an excellent fit.

x	y	y_e^*	x	y	y_e^*
0	0·4355	0·4311	35	0·3913	0·3980
8	0·4239	0·4235	39	0·3841	0·3942
11	0·4276	0·4207	46	0·3896	0·3876
17	0·4145	0·4150	51	0·3937	0·3828
22	0·4043	0·4103	56	0·3808	0·3781
31	0·3964	0·4018	59	0·3758	0·3753

Considered over the 60-year period, the percentage of male workers in the cotton industry declined steadily by about 0·1 for each year.

88   (i)   Let the age-groups be numbered from 1 to 7. If r_i is the number of boys and n_i the total number of children born to mothers in the ith age-group, then the test for no sex difference is obtained by using the fact that the function $\chi_i^2 = (2r_i - n_i)^2/n_i$ has the χ^2 distribution with 1 d.f. We thus have

$$\chi_1^2 = 1\cdot508; \quad \chi_2^2 = 136\cdot731; \quad \chi_3^2 = 147\cdot309; \quad \chi_4^2 = 216\cdot178;$$
$$\chi_5^2 = 131\cdot090; \quad \chi_6^2 = 117\cdot619; \quad \chi_7^2 = 17\cdot854.$$

The 5 per cent table value of χ^2 with 1 d.f. is 3·841. Therefore, except for χ_1^2, all the other observed χ_i^2's are highly significant. Thus, except for the age-group of mothers aged less than 20 years, there is definite evidence that the proportion of boys is not 0·5.

(ii)   Ignoring the age of the mothers, suppose p is the true proportion of boys in the entire population of births. Then, to test the null hypothesis $H(p = 0\cdot5)$, we have

$$\chi^2 = \frac{(2 \times 477\,533 - 928\,570)^2}{928\,570} = 756\cdot042 \text{ with 1 d.f.}$$

The observed χ^2 value is highly significant and so the null hypothesis is almost certainly rejected.

Alternatively, the maximum-likelihood estimate of p is

$$\hat{p} = 477\,533/928\,570 = 0\cdot514\,267;$$

and $$\text{s.e.}(\hat{p}) = \left[\frac{\hat{p}(1 - \hat{p})}{928\,569}\right]^{\frac{1}{2}} = 10^{-3} \times 0\cdot518\,664.$$

If the null hypothesis $H(p = 0\cdot5)$ is true, then

$$\text{s.d.}(\hat{p}) = (0\cdot25/928\,570)^{\frac{1}{2}} = 10^{-3} \times 0\cdot518\,875.$$

Hence, if the above null hypothesis is true, we have

$$z = \frac{0\cdot514\,267 - 0\cdot5}{10^{-3} \times 0\cdot518\,875} = 27\cdot50 \text{ is } \sim N(0,\,1).$$

The significance of the observed result is beyond doubt, and so the null hypothesis is again rejected.

(iii) The heterogeneity χ^2 is, in fact, the contingency χ^2

$$= \sum_{i=1}^{7} (r_i - n_i\hat{p})^2/n_i\hat{p}(1 - \hat{p})$$

$$= 2\cdot9495 + 1\cdot7887 + 3\cdot2155 + 0\cdot8618 + 0\cdot8364 + 2\cdot6724 + 0\cdot0048$$

$$= 12\cdot3291 \text{ with 6 d.f.}$$

The 5 per cent table value of χ^2 with 6 d.f. is $12\cdot592$, and so the observed result is just not significant. There is thus some suspicion that the sex of children may be associated with the age of the mothers.

(iv) On the assumption of homogeneity, if p is the true proportion of boys, then its maximum-likelihood estimate is

$$\hat{p} = 0\cdot514\ 267 \quad \text{and} \quad \text{s.e.}\,(\hat{p}) = 10^{-3} \times 0\cdot518\ 664.$$

Hence the 95 per cent large-sample confidence interval for p is

$$\hat{p} \pm 1\cdot96 \times \text{s.e.}\,(\hat{p}) \quad \text{i.e.} \quad 0\cdot515\ 284,\ 0\cdot513\ 250.$$

(v) Suppose p_1 and p_2 are the true proportions of boys born to mothers who are > 20 and < 20 years of age. Then the maximum-likelihood estimates of p_1 and p_2 are

$$\hat{p}_1 = 472\ 037/917\ 709 = 0\cdot514\ 365; \quad \hat{p}_2 = 5496/10\ 861 = 0\cdot506\ 031;$$

and $\quad$ s.e. $(\hat{p}_1 - \hat{p}_2) = \left[\dfrac{\hat{p}_1(1 - \hat{p}_1)}{917\ 708} + \dfrac{\hat{p}_2(1 - \hat{p}_2)}{10\ 860} \right]^{\frac{1}{2}} = 10^{-2} \times 0\cdot482\ 588.$

Hence the 95 per cent large-sample confidence interval for $p_1 - p_2$ is

$$\hat{p}_1 - \hat{p}_2 \pm 1\cdot96 \times \text{s.e.}\,(\hat{p}_1 - \hat{p}_2) \quad \text{i.e.} \quad 0\cdot017\ 793,\ -\ 0\cdot001\ 125.$$

The null hypothesis $H(p_1 - p_2 = 0)$ is acceptable at the 5 per cent level.

89 Leicester, 1966.

(i) If r_i is the number of male births and n_i the total number of births in the ith month, then to test the null hypothesis $H[E(r_i) = \frac{1}{2}n_i]$, we have

$$\chi_i^2 = (2\ r_i - n_i)^2/n_i, \text{ a } \chi^2 \text{ with 1 d.f.}$$

The individual χ^2's for the 12 months are as follows:

$$\chi_1^2 = 5\cdot829; \quad \chi_2^2 = 2\cdot939; \quad \chi_3^2 = 2\cdot892; \quad \chi_4^2 = 27\cdot073;$$
$$\chi_5^2 = 14\cdot821; \quad \chi_6^2 = 10\cdot230; \quad \chi_7^2 = 15\cdot511; \quad \chi_8^2 = 5\cdot465;$$
$$\chi_9^2 = 6\cdot781; \quad \chi_{10}^2 = 2\cdot121; \quad \chi_{11}^2 = 8\cdot215; \quad \chi_{12}^2 = 21\cdot326.$$

Since the 5 per cent table value of χ^2 with 1 d.f. is $3\cdot841$, all the χ_i^2's except χ_2^2, χ_3^2 and χ_{10}^2 are significant. There is thus considerable evidence that in the individual months the male and female births are not equally frequent.

(ii) Over the year as a whole, the number of male births is 45 682 out of a total of 88 273. Hence, if p denotes the true proportion of male births, then the χ^2 for testing the null hypothesis $H(p = 0\cdot5)$ is

$$\chi^2_{mt} = \frac{(2 \times 45\ 682 - 88\ 273)^2}{88\ 273} = 108\cdot236 \text{ with 1 d.f.}$$

This χ^2 is highly significant, and there is no doubt that $p \neq 0\cdot5$.

(iii) However, we now have the heterogeneity χ^2

$$= \sum_{i=1}^{12} \chi^2_i - \chi^2_m = 14\cdot967 \text{ with 11 d.f.}$$

The 5 per cent table value of χ^2 with 11 d.f. is $19\cdot675$, and so the heterogeneity χ^2 is not significant. This implies that, on the whole, the proportion of male births is consistent over the 12 months.

Hence the best estimate of p is

$$\hat{p} = 45\ 682/88\ 273 = 0\cdot517\ 508;$$

and
$$\text{s.e.}(\hat{p}) = \left[\frac{\hat{p}(1-\hat{p})}{88\ 272}\right]^{\frac{1}{2}} = 10^{-2} \times 0\cdot168\ 187.$$

Therefore the large-sample 95 per cent confidence interval for p is

$$\hat{p} \pm 1\cdot96 \times \text{s.e.}(\hat{p}) \quad \text{i.e.} \quad 0\cdot5208,\ 0\cdot5142.$$

90   Leicester, 1969.

(i)   Assuming that there is no difference in the success rates in the different subjects, suppose that p is the true proportion of successes. Then the maximum-likelihood estimate of p is

$$\hat{p} = 115\ 290/166\ 108 = 0\cdot694\ 067.$$

Number the subjects from 1 to 5. Denote by n_i and r_i the number of entries and the number of passes in the ith subject. Then the contingency χ^2 is

$$= \sum_{i=1}^{5} (r_i - n_i\hat{p})^2/n_i\hat{p}(1-\hat{p})$$

$$= 95\cdot5791 + 0\cdot3201 + 24\cdot9612 + 38\cdot1475 + 0\cdot7586$$

$$= 156\cdot7665 \text{ with 4 d.f.}$$

There is no doubt about the high significance of the observed χ^2, and so the success rates are almost certainly not the same in the different subjects.

(ii)   The χ^2 for testing the null hypothesis that there is no difference in the success rates in mathematics and the other subjects taken as a whole is

$$\frac{166\ 108\,(41\ 894 \times 33\ 901 - 16\ 917 \times 73\ 396)^2}{115\ 290 \times 50\ 818 \times 58\ 811 \times 107\ 297} = 143\text{·}33 \text{ with 1 d.f.}$$

This χ^2 is also highly significant, and so the success rates in mathematics and the other subjects taken as a whole are almost definitely not the same.

(iii) Suppose that p_m and p_o are the true proportions of successes in mathematics and the other subjects taken as a whole. Then the maximum-likelihood estimates of p_m and p_o are

$$\hat{p}_m = 41\ 894/58\ 811 = 0\text{·}712\ 350; \quad \hat{p}_o = 73\ 396/107\ 297 = 0\text{·}684\ 045.$$

Also, $\qquad$ s.e.$(\hat{p}_m) = \left[\dfrac{\hat{p}_m\,(1 - \hat{p}_m)}{58\ 810}\right]^{\frac{1}{2}} = 10^{-2} \times 0\text{·}186\ 662$

and $\quad$ s.e.$(\hat{p}_m - \hat{p}_o) = \left[\dfrac{\hat{p}_m(1 - \hat{p}_m)}{58\ 810} + \dfrac{\hat{p}_o(1 - \hat{p}_o)}{107\ 296}\right]^{\frac{1}{2}} = 10^{-2} \times 0\text{·}234\ 490.$

Hence the 95 per cent large-sample confidence interval for p_m is

$$\hat{p}_m \pm 1\text{·}96 \times \text{s.e.}(\hat{p}_m) \quad \text{i.e.} \quad 0\text{·}7160,\ 0\text{·}7087;$$

and that for $p_m - p_o$ is

$$\hat{p}_m - \hat{p}_o \pm 1\text{·}96 \times \text{s.e.}(\hat{p}_m - \hat{p}_o) \quad \text{i.e.} \quad 0\text{·}0329,\ 0\text{·}0237.$$

The above analysis is strictly not valid because achievements in different subjects taken by the same student are related.

91 (i) Assuming that there is no association between parental type and down colour of chicks, the estimated expected frequencies are as follows:

Type of parent	Coloured down	White down
A	227·3741	66·6259
B	168·5971	49·4029
C	34·0288	9·9712
Total	430·0000	126·0000

Hence the contingency χ^2 for testing the independence of the two traits is 3·9770 with 2 d.f. The 5 per cent table value of χ^2 with 2 d.f. is 5·991, and so the observed result is not significant. The three types of parents give rise to progenies with the same relative frequencies of coloured and white-down chicks.

(ii) If p is the true proportion of chicks with white down, then the maximum-likelihood estimate of p is

$$\hat{p} = 126/556 = 0\text{·}226\ 619;$$

and $\qquad$ s.e. $(\hat{p}) = \left[\dfrac{\hat{p}(1 - \hat{p})}{555} \right]^{\frac{1}{2}} = 10^{-1} \times 0\cdot177\ 705.$

Hence the 95 per cent large-sample confidence interval for p is

$$\hat{p} \pm 1\cdot96 \times \text{s.e.}\,(\hat{p}) \quad \text{i.e.} \quad 0\cdot2614,\ 0\cdot1918.$$

(iii) Since the point $p = 0\cdot25$ is contained in the above confidence interval, it follows that the null hypothesis $H(p = 0\cdot25)$ is acceptable at the 5 per cent level.

(iv) The appropriate χ^2 for testing the null hypothesis $H(p = 0\cdot25)$ is

$$(126 \times 4 - 556)^2/3 \times 556 = 1\cdot621 \text{ with 1 d.f.}$$

The 5 per cent table value of χ^2 with 1 d.f. is $3\cdot841$, and so the observed χ^2 is not significant. The null hypothesis is acceptable.

92 Leicester, 1960.

By "standard error" the author means "standard deviation" evaluated when the null hypothesis is true.

(i) (a) The probability of a member being found affected is $p = 0\cdot5$. There are a total of 30 members and of these $r = 19$ are known to be affected. Therefore

$$E(r) = 15; \quad \text{var}(r) = 30/4; \quad \text{and} \quad \text{s.d.}\,(r) = \tfrac{1}{2}\sqrt{30} = 2\cdot74.$$

Hence, if the null hypothesis $H(p = 0\cdot5)$ is true, we have

$$z = \frac{r - E(r)}{\text{s.d.}\,(r)} \sim N(0,\ 1).$$

Therefore $r - E(r) \geqslant 2$ s.d. (r) approximately indicates significance at the 5 per cent level. In other words, for significance,

$$r - E(r) \geqslant \sqrt{30} = 5\cdot48.$$

But the observed value of $r - E(r)$ is 4, and so the result actually obtained is not significant at the 5 per cent level. Therefore we may accept the null hypothesis $H(p = 0\cdot5)$.

(b) There are in all 13 males in the sample. If a is the probability of a male being affected, then the null hypothesis is $H(a = 0\cdot5)$. Suppose that s denotes the observed number of affected males. Then if $H(a = 0\cdot5)$ is true, we have

$$E(s) = 6\cdot5; \quad \text{var}(s) = 13/4; \quad \text{and} \quad \text{s.d.}\,(s) = \tfrac{1}{2}\sqrt{13} = 1\cdot80.$$

Hence, if $H(a = 0\cdot5)$ is true, we have

$$w = \frac{s - E(s)}{\text{s.d.}\,(s)} \sim N(0,\ 1).$$

Therefore for significance at the 5 per cent level, we must have

$$s - E(s) \geqslant 2 \text{ s.d.}\,(s) \text{ approximately.}$$

In other words, for significance $s - E(s) \geqslant 3\cdot 61$. But the observed value of $s - E(s)$ is $4\cdot 5 > 3\cdot 61$. Thus the null hypothesis $H(\alpha = 0\cdot 5)$ is rejected and we conclude that the proportion of affected males is significantly different from $0\cdot 5$.

The above method is not the best one for analysing the data. The interest of the analysis is to determine whether or not there is a significant association between sex and the incidence of the trait. This is equivalent to testing that there is no difference in the true proportions of affected males and females. Ignoring the continuity correction, the appropriate contingency χ^2 is

$$\frac{30\,(11 \times 9 - 2 \times 16)^2}{19 \times 11 \times 17 \times 13} = 4\cdot 474 \text{ with 1 d.f.}$$

The 5 per cent table value of χ^2 with 1 d.f. is $3\cdot 841$, and so the observed χ^2 is significant. But since the cell frequencies of the 2×2 table are rather small, we need to use the continuity correction. Accordingly, the corrected χ^2 is

$$\frac{30\,(68)^2}{19 \times 11 \times 17 \times 13} = 3\cdot 003 \text{ with 1 d.f.}$$

The corrected χ^2 is no longer significant, and so there is no evidence for any difference between males and females as regards the incidence of the genetic characteristic.

(ii) There are 13 males in all, of whom 11 are affected. On the null hypothesis that the probability of a male being affected is $0\cdot 5$, we reject this hypothesis if

$$P(|s - 6\cdot 5| \geqslant 4\cdot 5) < 0\cdot 025.$$

The required probability is that of the events $s = 0, 1, 2$ and $11, 12, 13$

$$= 2\left[\binom{13}{0}(\tfrac{1}{2})^{13} + \binom{13}{1}(\tfrac{1}{2})^{13} + \binom{13}{2}(\tfrac{1}{2})^{13}\right] = \frac{92}{4096} = 0\cdot 0225 < 0\cdot 025.$$

Hence the null hypothesis is rejected.

93 (i)   For each of the ten age-groups, a contingency χ^2 with 1 d.f. can be evaluated to test the equality of the death rates in the two cities. The χ^2 values are found to be as follows:

$0 -$ age-group:

$$\chi^2 = \frac{10^{12} \times 7\,152\,393}{10^{12} \times 560\,769\cdot 8} = 12\cdot 7546;$$

$5 -$ age-group:

$$\chi^2 = \frac{10^{12} \times 5\,758\,941}{10^{12} \times 306\,553\cdot 4} = 18\cdot 7861;$$

$15 -$ age-group:

$$\chi^2 = \frac{10^{12} \times 2\,561\,118}{10^{12} \times 158\,473\cdot 2} = 16\cdot 1612;$$

25 — age-group:

$$\chi^2 = \frac{10^{12} \times 6\ 418\ 319}{10^{12} \times 209\ 494 \cdot 6} = 30 \cdot 6372;$$

35 — age-group:

$$\chi^2 = \frac{10^{12} \times 2\ 806\ 152}{10^{12} \times 155\ 096 \cdot 1} = 18 \cdot 0930;$$

45 — age-group:

$$\chi^2 = \frac{10^{12} \times 1\ 383\ 253}{10^{12} \times 58\ 054 \cdot 70} = 23 \cdot 8267;$$

55 — age-group:

$$\chi^2 = \frac{10^{11} \times 3\ 010\ 316}{10^{11} \times 160\ 774 \cdot 9} = 18 \cdot 7238;$$

65 — age-group:

$$\chi^2 = \frac{10^{10} \times 4\ 497\ 402}{10^{10} \times 177\ 908 \cdot 7} = 25 \cdot 2793;$$

75 — age-group:

$$\chi^2 = \frac{10^{7} \times 1\ 163\ 856}{10^{7} \times 974\ 239 \cdot 0} = 1 \cdot 1946;$$

85 — age-group:

$$\chi^2 = \frac{45\ 101\ 056}{967\ 499\ 040} = 0 \cdot 0466.$$

The 5 per cent table value of χ^2 with 1 d.f. is $3 \cdot 841$. Therefore, except for the last two groups, the death rates are significantly different.

The total $\chi^2 = 165 \cdot 5031$ with 10 d.f.

The mean χ^2 calculated from the totals ignoring age is

$$\chi_m^2 = (10^{17} \times 1\ 114\ 332)/(10^{17} \times 7818 \cdot 469) = 142 \cdot 5256 \text{ with 1 d.f.}$$

Hence the heterogeneity $\chi^2 = 22 \cdot 9775$ with 9 d.f.

Since this χ^2 is significant, the test of the equality of the overall mortality in the two cities is obtained by using

$$F = (9 \times 142 \cdot 5256)/22 \cdot 9775 = 55 \cdot 8255 \text{ with 1, 9 d.f.}$$

The $0 \cdot 1$ per cent table value of F with 1, 9 d.f. is $22 \cdot 86$, and so the observed F is highly significant. The overall mortality in the two cities is almost certainly different.

(ii) If p_1 and p_2 are the true proportions of overall mortality in Birmingham and Liverpool, then for their maximum-likelihood estimates we have

$$\hat{p}_1 = 4879/275\ 212 = 0 \cdot 017\ 728\ 2; \quad \hat{p}_2 = 7880/358\ 547 = 0 \cdot 021\ 977\ 6.$$

Therefore

$$\text{s.e.}(\hat{p}_2 - \hat{p}_1) = \left[\frac{\hat{p}_1(1 - \hat{p}_1)}{275\ 211} + \frac{\hat{p}_2(1 - \hat{p}_2)}{358\ 546} \right]^{\frac{1}{2}} = 10^{-3} \times 0 \cdot 351\ 033.$$

Hence the 95 per cent large-sample confidence interval for $p_2 - p_1$ is

$$\hat{p}_2 - \hat{p}_1 \pm 1 \cdot 96 \times \text{s.e.}(\hat{p}_2 - \hat{p}_1) \quad \text{i.e.} \quad 0 \cdot 004\ 937\ 4, \ 0 \cdot 003\ 561\ 4.$$

Therefore the required confidence interval for $1000(p_2 - p_1)$ is $3\cdot5614$, $4\cdot9374$.

94 From each of the six 2×2 tables given, a contingency χ^2 with 1 d.f. can be calculated to test the equality of the death rates amongst vaccinated and unvaccinated persons who got smallpox. The χ^2 values are found to be as follows:

Gloucester:
$$\chi^2 = \frac{10^8 \times 1\ 642\ 533}{10^8 \times 6236\cdot251} = 263\cdot3847;$$

Leicester:
$$\chi^2 = \frac{10^3 \times 4\ 286\ 222}{10^3 \times 221\ 854\cdot8} = 19\cdot3199;$$

Sheffield:
$$\chi^2 = \frac{10^9 \times 4\ 960\ 139}{10^9 \times 4593\cdot120} = 1079\cdot9063;$$

Glasgow:
$$\chi^2 = \frac{10^5 \times 1\ 686\ 310}{10^5 \times 11\ 248\cdot56} = 149\cdot9134;$$

London hospitals:
$$\chi^2 = \frac{10^{11} \times 7\ 435\ 479}{10^{11} \times 3887\cdot606} = 1912\cdot6113;$$

London (1901):
$$\chi^2 = \frac{10^7 \times 1\ 684\ 768}{10^7 \times 8593\cdot925} = 196\cdot0417.$$

The 5 per cent table value of χ^2 with 1 d.f. is $3\cdot841$. There is a highly significant difference in mortality in each table.

The total $\chi^2 = 3621\cdot1773$ with 6 d.f.

The mean χ^2 calculated from the pooled 2×2 table
$$= \frac{10^{13} \times 1\ 312\ 651}{10^{13} \times 374\cdot5881} = 3504\cdot2517 \text{ with 1 d.f.}$$

Therefore the heterogeneity $\chi^2 = 116\cdot9256$ with 5 d.f.

The $0\cdot1$ per cent table value of χ^2 with 5 d.f. is $20\cdot517$, and so the heterogeneity χ^2 is highly significant. Therefore the test for the overall comparison of deaths amongst vaccinated and unvaccinated patients is given by
$$F = (5 \times 3504\cdot2517)/116\cdot9256 = 149\cdot85 \text{ with 1, 5 d.f.}$$

The $0\cdot1$ per cent table value of F with 1, 5 d.f. is $47\cdot04$, and so the observed F ratio is highly significant. Therefore the proportions of deaths amongst the two groups of patients are almost certainly different.

If p_1 and p_2 denote the true overall proportions of deaths amongst vaccinated and unvaccinated patients, then their maximum-likelihood estimates are
$$\hat{p}_1 = 1188/16\ 103 = 0\cdot073\ 775\ 1; \quad \hat{p}_2 = 1894/4346 = 0\cdot435\ 803\ 0.$$

Also,

$$\text{s.e.}(\hat{p}_2 - \hat{p}_1) = \left[\frac{\hat{p}_1(1 - \hat{p}_1)}{16\ 102} + \frac{\hat{p}_2(1 - \hat{p}_2)}{4345} \right]^{\frac{1}{2}} = 10^{-2} \times 0.779\ 869.$$

Hence the 95 per cent large-sample confidence interval for $p_2 - p_1$ is

$$\hat{p}_2 - \hat{p}_1 \pm 1.96 \times \text{s.e.}(\hat{p}_2 - \hat{p}_1) \quad \text{i.e.} \quad 0.377\ 313,\ 0.346\ 743.$$

Therefore the required confidence interval for $100(p_2 - p_1)$ is 37.7313, 34.6743.

95   The joint sample likelihood is

$$L = \binom{n_1}{x_{11}}(1 - \theta)^{x_{11}}\, \theta^{x_{12}} \times \binom{n_2}{x_{21}}(1 - \theta\ )^{x_{21}}\theta^{2x_{22}},$$

so that

$$\log L = \text{constant} + x_{11} \log (1 - \theta) + (x_{12} + 2x_{22}) \log \theta + x_{21} \log (1 - \theta^2).$$

Therefore

$$\frac{d \log L}{d\theta} = - \frac{x_{11}}{1 - \theta} + \frac{x_{12} + 2x_{22}}{\theta} - \frac{2\theta x_{21}}{1 - \theta^2},$$

whence the equation for $\hat{\theta}$ is obtained as

$$x_{12} + 2x_{22} - x_{11}\hat{\theta} - (n_1 + 2n_2)\hat{\theta}^2 = 0.$$

Use $\hat{\theta}$ to estimate the expected frequencies in the four classes. Then the heterogeneity χ^2

$$= \frac{[x_{11} - n_1(1 - \hat{\theta})]^2}{n_1(1 - \hat{\theta})} + \frac{(x_{12} - n_1\hat{\theta})^2}{n_1\hat{\theta}} + \frac{[x_{21} - n_2(1 - \hat{\theta}^2)]^2}{n_2(1 - \hat{\theta}^2)} + \frac{(x_{22} - n_2\hat{\theta}^2)^2}{n_2\hat{\theta}^2}$$

$$= \frac{(x_{12} - n_1\hat{\theta})^2}{n_1\hat{\theta}(1 - \hat{\theta})} + \frac{(x_{22} - n_2\hat{\theta}^2)^2}{n_2\hat{\theta}^2(1 - \hat{\theta}^2)}, \quad \text{on reduction.}$$

The χ^2 has 1 d.f. since θ is estimated from the two samples and the sizes of the two samples are fixed.

If the null hypothesis $H(\theta = 0.5)$ is true, then

$$E(x_{11}) = E(x_{12}) = \tfrac{1}{2}n_1; \quad \text{var}(x_{11}) = \text{var}(x_{12}) = \tfrac{1}{4}n_1; \quad \text{cov}(x_{11}, x_{12}) = -\tfrac{1}{4}n_1.$$

Similarly,

$$E(x_{21}) = \tfrac{3}{4}n_2; \quad E(x_{22}) = \tfrac{1}{4}n_2; \quad \text{var}(x_{21}) = \text{var}(x_{22}) = \tfrac{3}{16}n_2;$$

$$\text{cov}(x_{21}, x_{22}) = -\tfrac{3}{16}n_2.$$

Next, by substituting $\theta = 0.5$ in the expression for $d \log L/d\theta$, we obtain

$$- \frac{x_{11}}{1/2} + \frac{x_{12}}{1/2} + \frac{2x_{22}}{1/2} - \frac{x_{21}}{3/4}.$$

Clearly, if the null hypothesis $H(\theta = 0.5)$ is true, then the above linear function has zero expectation. Therefore, if the null hypothesis is true, the above function differs from its expectation because of sampling fluctuations. Therefore

$$2(x_{12} - x_{11}) + \tfrac{4}{3}(3x_{22} - x_{21}) \quad \text{or} \quad D = (x_{12} - x_{11}) + \tfrac{2}{3}(3x_{22} - x_{21})$$

is a suitable measure of the deviations from the null hypothesis of the observed frequencies in the two experiments. Also, in large samples, D is normally distributed due to the Central Limit Theorem. Finally,

$$\text{var}(D) = \text{var}(x_{12}) + \text{var}(x_{11}) - 2\,\text{cov}(x_{12}, x_{11}) +$$

$$+ \tfrac{4}{9}[9\,\text{var}(x_{22}) + \text{var}(x_{21}) - 6(x_{22}, x_{21})]$$

$$= n_1 + \tfrac{4}{3}n_2, \text{ on reduction.}$$

Hence the χ^2 for testing jointly the null hypothesis $H(\theta = 0\cdot 5)$ is

$$\chi_m^2 = D^2/\text{var}(D) \text{ with 1 d.f.}$$

Moffet's data

The equation for $\hat{\theta}$ is $160 - 81\hat{\theta} - 391\hat{\theta}^2 = 0$, whence $\hat{\theta} = 0\cdot 544\ 444$.

The heterogeneity χ^2

$$= \frac{10\cdot 7446}{45\cdot 8846} + \frac{6\cdot 4070}{21\cdot 4811} = 0\cdot 5325 \text{ with 1 d.f.}$$

The fit is evidently good.

Again, $\chi_m^2 = 2523/967 = 2\cdot 609$ with 1 d.f.

The 5 per cent table value of χ^2 with 1 d.f. is $3\cdot 841$. Therefore χ_m^2 is also not significant, and so the two sample data are in conformity with the null hypothesis $H(\theta = 0\cdot 5)$.

96 (i) The joint likelihood of the four samples is

$$L = \binom{145}{57}\left(\frac{27\lambda}{64}\right)^{57}\left(1 - \frac{27\lambda}{64}\right)^{88} \times \binom{168}{60}\left(\frac{27\lambda}{64}\right)^{60}\left(1 - \frac{27\lambda}{64}\right)^{108}$$

$$\times \binom{124}{44}\left(\frac{27\lambda}{64}\right)^{44}\left(1 - \frac{27\lambda}{64}\right)^{80} \times \binom{60}{5}\left(1 - \frac{37}{64 - 27\lambda}\right)^{5}\left(\frac{37}{64 - 27\lambda}\right)^{55}.$$

Therefore

$$\log L = \text{constant} + 161 \log \lambda + 216 \log(64 - 27\lambda) + 5 \log(1 - \lambda),$$

so that

$$\frac{d \log L}{d\lambda} = \frac{161}{\lambda} - \frac{216 \times 27}{64 - 27\lambda} - \frac{5}{1 - \lambda},$$

whence the equation for $\hat{\lambda}$ is found to be

$$10\ 304 - 20\ 803\hat{\lambda} + 10\ 314\hat{\lambda}^2 = 0.$$

The appropriate root of this equation gives $\hat{\lambda} = 0\cdot 874\ 287$.

(ii)   We have $\dfrac{27\hat{\lambda}}{64} = 0 \cdot 368\ 840$; $\dfrac{27(1 - \hat{\lambda})}{64 - 27\lambda} = 10^{-1} \times 0 \cdot 840\ 281$.

Hence the estimated expected frequencies in the experiments are as follows:

Experiment no.	Fired	Normal
I	53·4818	91·5182
II	61·9651	106·0349
III	45·7362	78·2638
I + II + III	161·1831	275·8169
IV	5·0417	54·9583

Hence the goodness-of-fit χ^2 for the first three experiments is $0 \cdot 5697$ with 2 d.f. This χ^2 is obviously not significant. There is good agreement between the observed and expected frequencies of the three experiments.

It is to be noted that, strictly, we should have used an estimated value of λ, calculated from the data of the three experiments only, for estimating the expected frequencies.

(iii) The second goodness-of-fit χ^2 is $0 \cdot 0006$ with 1 d.f., and is not significant. There is excellent agreement between the total frequencies of the first three experiments and those of the fourth experiment as compared with their expected values.

97   Since $E(a_1) = n(2 + \theta)/4$; $E(a_2 + a_3) = n(1 - \theta)/2$; $E(a_4) = n\theta/4$; we have

$$\mathrm{var}(a_1) = \frac{n}{16}(4 - \theta^2); \quad \mathrm{var}(a_2 + a_3) = \frac{n}{4}(1 - \theta^2); \quad \mathrm{var}(a_4) = \frac{n}{16}\theta(4 - \theta);$$

$$\mathrm{cov}(a_1, a_2 + a_3) = -\frac{n}{8}(2 + \theta)(1 - \theta); \quad \mathrm{cov}(a_1, a_4) = -\frac{n}{16}\theta(2 + \theta);$$

$$\mathrm{cov}(a_4, a_2 + a_3) = -\frac{n}{8}\theta(1 - \theta).$$

Hence

$$\begin{aligned}
\mathrm{var}(X) =\ & \theta^2(1 - \theta)^2\, \mathrm{var}(a_1) + \theta^2(2 + \theta)^2\, \mathrm{var}(a_2 + a_3) + \\
& + (2 + \theta)^2(1 - \theta)^2\, \mathrm{var}(a_4) - 2\theta^2(1 - \theta)(2 + \theta)\, \mathrm{cov}(a_1, a_2 + a_3) + \\
& + 2\theta(1 - \theta)^2(2 + \theta)\, \mathrm{cov}(a_1, a_4) - 2\theta(2 + \theta)^2(1 - \theta)\, \mathrm{cov}(a_4, a_2 + a_3) \\
=\ & \frac{n}{2}\theta(1 - \theta)(2 + \theta)(1 + 2\theta), \text{ on reduction.}
\end{aligned}$$

Hence the expression stated for $X^2/\mathrm{var}(X)$.

(i) To test that the observed frequencies of the character long v. round are in the ratio of $3:1$, the χ^2's for the two experiments are:

Experiment I: $\chi^2 = \dfrac{(323 - 3 \times 104)^2}{3 \times 427} = 0 \cdot 0945$ with 1 d.f.

Experiment II: $\chi^2 = \dfrac{(607 - 3 \times 196)^2}{3 \times 803} = 0 \cdot 1499$ with 1 d.f.

Similarly, to test that the observed frequencies of the character purple $v.$ red are in the ratio of $3:1$, the χ^2's for the two experiments are:

Experiment I: $\chi^2 = \dfrac{(315 - 3 \times 112)^2}{3 \times 427} = 0 \cdot 3443$ with 1 d.f.

Experiment II: $\chi^2 = \dfrac{(609 - 3 \times 194)^2}{3 \times 803} = 0 \cdot 3026$ with 1 d.f.

The 5 per cent table value of χ^2 with 1 d.f. is $3 \cdot 841$, and so all the four observed χ^2's are not significant. Therefore the single factor ratios in the two experiments are separately in agreement with the Mendelian expectation of $3:1$.

(ii) For given observed frequencies a_1, a_2, a_3, a_4 ($\Sigma\, a_i = n$) in any one experiment, the sample likelihood is

$$L = \frac{n!}{a_1!\, a_2!\, a_3!\, a_4!}\, [\tfrac{1}{4}(2 + \theta)]^{\alpha_1}[\tfrac{1}{4}(1 - \theta)]^{\alpha_2 + \alpha_3}[\tfrac{1}{4}\theta]^{\alpha_4},$$

so that

$$\log L = \text{constant} + a_1 \log(2 + \theta) + (a_2 + a_3)\log(1 - \theta) + a_4 \log \theta,$$

whence

$$\frac{d \log L}{d\theta} = \frac{a_1}{2 + \theta} - \frac{a_2 + a_3}{1 - \theta} + \frac{a_4}{\theta}$$

and

$$\frac{d^2 \log L}{d\theta^2} = -\frac{a_1}{(2 + \theta)^2} - \frac{a_2 + a_3}{(1 - \theta)^2} - \frac{a_4}{\theta^2}.$$

Equating $d \log L/d\theta$ to zero, the equation for $\hat{\theta}$ is found to be

$$2a_4 + (3a_1 + a_4 - 2n)\hat{\theta} - n\hat{\theta}^2 = 0.$$

Also, for large samples,

$$\text{var}(\hat{\theta}) = -1/E(d^2 \log L/d\theta^2) = \frac{2\theta(1 - \theta)(2 + \theta)}{n(1 + 2\theta)}.$$

If $\hat{\theta}_1$ denotes the maximum-likelihood estimate of θ obtained from Experiment I, then the equation for $\hat{\theta}_1$ is

$$170 + 119\hat{\theta}_1 - 427\hat{\theta}_1^2 = 0.$$

The relevant root of this equation gives $\hat{\theta}_1 = 0 \cdot 785\ 520$. Hence

$$\text{estimate var}(\hat{\theta}_1) = 10^{-3} \times 0 \cdot 854\ 956.$$

Similarly, if $\hat{\theta}_2$ denotes the maximum-likelihood estimate of θ obtained from Experiment II, then the equation for $\hat{\theta}_2$ is

$$340 + 313\hat{\theta}_2 - 803\hat{\theta}_2^2 = 0.$$

The relevant root of this equation gives $\hat{\theta}_2 = 0 \cdot 874\ 155$. Hence

$$\text{estimate var}(\hat{\theta}_2) = 10^{-3} \times 0 \cdot 286\ 539.$$

Hence

$$\text{s.e.}(\hat{\theta}_1 - \hat{\theta}_2) = [\text{Estimate var}(\hat{\theta}_1) + \text{Estimate var}(\hat{\theta}_2)]^{\frac{1}{2}} = 10^{-1} \times 0.337\ 860.$$

To test the null hypothesis that the true value of θ is the same in the two experimental populations, we have

$$z = \frac{\hat{\theta}_1 - \hat{\theta}_2}{\text{s.e.}(\hat{\theta}_1 - \hat{\theta}_2)} = -2.623 \text{ is } \sim N(0, 1).$$

The observed value of z is significant at the one per cent level, and we therefore reject the null hypothesis that the two experimental populations have the same value of θ.

(iii) By using the estimates $\hat{\theta}_1$ and $\hat{\theta}_2$, the estimated expected frequencies in Experiments I and II are as follows:

Class	Experiment I	Experiment II
(1)	297·3543	576·9866
(2) + (3)	45·7915	50·5268
(4)	83·8543	175·4866
Total	427·0001	803·0000

Hence the goodness-of-fit χ^2's for the two experiments are 0·0227 and 0·2397 respectively, and each χ^2 has 1 d.f. These χ^2's are not significant and so the fit is good.

(iv) If $\hat{\theta}$ denotes the joint maximum-likelihood estimate of θ, then the equation for $\hat{\theta}$ is $510 + 432\hat{\theta} - 1230\hat{\theta}^2 = 0$, whence $\hat{\theta} = 0.843\ 047$.

By using this estimate, the χ^2 for testing the equality of the true values of θ in the two experimental populations is

$$\left[\frac{X^2}{\text{var}(X)}\right]_{\text{I}} + \left[\frac{X^2}{\text{var}(X)}\right]_{\text{II}} = \frac{2198·9546}{431·4739} + \frac{2199·0294}{811·4134} = 7·8065 \text{ with 1 d.f.}$$

The one per cent table value of χ^2 with 1 d.f. is 6·635, and so the observed result is highly significant. We conclude that almost certainly the two experimental populations do not have the same value of θ. This confirms our conclusion in (ii), based on an approximate comparison of the estimates $\hat{\theta}_1$ and $\hat{\theta}_2$ obtained separately from the two experiments.

(v) By using the estimate $\hat{\theta}$, the estimated expected frequencies in Experiments I and II are as follows:

Class	Experiment I	Experiment II
(1)	303·4953	570·7417
(2) + (3)	33·5095	63·0166
(4)	89·9953	169·2417
Total	427·0001	803·0000

The overall goodness-of-fit χ^2 is 8·0735 with 3 d.f., since the two experimental totals are fixed and one parameter is estimated from the data. The 5 per cent table value of χ^2 with 3 d.f. is 7·815, and so the observed χ^2 is significant.

This χ^2 includes two superfluous d.f., and these may be identified with the 2 d.f. in which the two sets of data differ from expectation when each is given its own estimated value of θ. These χ^2's were found to be 0·0227 and 0·2397 in (iii) above. Therefore the residual χ^2 is 7·8111 with 1 d.f., and this tests the equality of the values of θ in the two experimental populations. This value is an approximation to the correct χ^2 value obtained in (iv) for testing the equality of the parameter values in the two experimental populations. The difference, though trifling here, is not due to arithmetical approximation but to the slight inaccuracy of using values of χ^2 based on different series of expected frequencies.

98 (i) To test the null hypothesis that the R and S seedlings in Class I are in the ratio 3 : 1, we have

$$\chi^2 = \frac{(3 \times 94 - 331)^2}{3 \times 425} = 1·883 \text{ with 1 d.f.}$$

Similarly, the test for the 3 : 1 ratio in Class II gives

$$\chi^2 = \frac{(3 \times 88 - 291)^2}{3 \times 379} = 0·641 \text{ with 1 d.f.}$$

The totals of R and S seedlings in Classes I and II taken together are 622 and 182 respectively. To test that these frequencies are in agreement with the 3 : 1 ratio, we have

$$\chi^2 = \frac{(3 \times 182 - 622)^2}{3 \times 804} = 2·395 \text{ with 1 d.f.}$$

Finally, to test the heterogeneity between the results for Classes I and II, we have the contingency

$$\chi^2 = \frac{804 (331 \times 88 - 291 \times 94)^2}{622 \times 182 \times 425 \times 379} = 0·014 \text{ with 1 d.f.}$$

All the above four χ^2's are not significant, and so the observed results in Classes I and II conform with the 3 : 1 expectation.

(ii) To test the null hypothesis that the R and S seedlings in Classes III, IV and V separately are in the 1 : 1 ratio, we have

$$\chi^2 = \frac{(36 - 41)^2}{77} = 0·325 \text{ with 1 d.f.}$$

$$\chi^2 = \frac{(146 - 148)^2}{294} = 0·014 \text{ with 1 d.f.}$$

$$\chi^2 = \frac{(130 - 133)^2}{263} = 0·034 \text{ with 1 d.f.}$$

The totals of the R and S seedlings in Classes III, IV and V taken together are 312 and 322 respectively. To test that these also conform with the $1:1$ expectation, we have

$$\chi^2 = \frac{(312 - 322)^2}{634} = 0 \cdot 158 \text{ with 1 d.f.}$$

Finally, the heterogeneity $\chi^2 = 0 \cdot 325 + 0 \cdot 014 + 0 \cdot 034 - 0 \cdot 158$

$$= 0 \cdot 215 \text{ with 2 d.f.}$$

All the above four χ^2's are not significant, and so the observed results in Classes III, IV and V conform with the $1:1$ expectation.

To test that the totals in Classes I and II, and Classes III, IV and V jointly agree with the $3:1$ and $1:1$ expectations, we have

$$\chi^2 = \frac{[(322 - 312) + 2(3 \times 182 - 622)/3]^{2(*)}}{634 + 4 \times 804/3} = 0 \cdot 969 \text{ with 1 d.f.}$$

This observed value of χ^2 is obviously not significant.

99 Since $\sum\limits_{i=1}^{k} m_i = N$, a constant, therefore $\sum\limits_{i=1}^{k} \frac{dm_i}{dp} = 0.$ (1)

The sample likelihood is

$$L = \frac{N!}{\prod\limits_{i=1}^{k} a_i!} \times \prod\limits_{i=1}^{k} (m_i/N)^{a_i},$$

so that

$$\log L = \text{constant} + \sum\limits_{i=1}^{k} a_i \log m_i$$

and

$$\frac{d \log L}{dp} = \sum\limits_{i=1}^{k} \frac{a_i}{m_i} \times \frac{dm_i}{dp},$$

whence the required equation for $\hat{p}$. Also,

$$E(X) = \sum\limits_{i=1}^{k} \frac{1}{m_i} \times \frac{dm_i}{dp} \times E(a_i) = \sum\limits_{i=1}^{k} \frac{dm_i}{dp} = 0 \quad \text{by (1)}.$$

To obtain var(X), we observe that

var$(a_i) = m_i(1 - m_i/N)$;   cov$(a_i, a_j) = - m_i m_j/N$, for $i \neq j$.

Therefore

$$\text{var}(X) = \sum\limits_{i=1}^{k} \left(\frac{1}{m_i} \times \frac{dm_i}{dp}\right)^2 \text{var}(a_i) + \sum\limits_{i \neq j} \left(\frac{1}{m_i} \times \frac{dm_i}{dp}\right)\left(\frac{1}{m_j} \times \frac{dm_j}{dp}\right) \text{cov}(a_i, a_j)$$

$$= \sum\limits_{i=1}^{k} \frac{1}{m_i}\left(\frac{dm_i}{dp}\right)^2 - \frac{1}{N}\left(\sum\limits_{i=1}^{k} \frac{dm_i}{dp}\right)^2 = \sum\limits_{i=1}^{k} \frac{1}{m_i}\left(\frac{dm_i}{dp}\right)^2.$$

(*)This χ^2 is obtained from the formula for χ^2_m derived in Exercise 95 above.

The joint likelihood of the six experiments is

$L =$

constant $\times (1 - p)^{394} p^{73} \times (1 - p)^{408} p^{81} \times (3 - 2p + p^2)^{298} \ (2p - p^2)^{88} \ (1 - p)^{272}$

$\times (2 + p^2)^{51} (1 - p^2)^{48} \ p^2 \times (2 - p)^9 (1 + p)^8 p^2 (1 - p)^{10}$

$\times (1 + p)^{48} (1 - p)^{20} (2 - p)^{63} p^5,$

so that

$\log L = $ constant $+ \ 1152 \log (1 - p) + 251 \log p + 298 \log (3 - 2p + p^2) +$

$+ \ 160 \log p + 51 \log (2 + p^2) + 104 \log (1 + p).$

Therefore

$$\frac{d \log L}{dp} = -\frac{1152}{1 - p} + \frac{251}{p} - \frac{596 (1 - p)}{3 - 2p + p^2} - \frac{160}{2 - p} + \frac{102p}{2 + p^2} + \frac{104}{1 + p},$$

whence the required equation for $\hat{p}$. For computation, this equation may be written as

$$\hat{p} = 0 \cdot 154 \ 986 + 0 \cdot 253 \ 679 \hat{p}^2 (1 + 0 \cdot 732 \ 860 \hat{p} - 2 \cdot 631 \ 643 \hat{p}^2 +$$

$$+ \ 2 \cdot 466 \ 531 \hat{p}^3 - 1 \cdot 520 \ 487 \hat{p}^4 + 0 \cdot 479 \ 716 \hat{p}^5),$$

whence, on iteration, $\hat{p} = 0 \cdot 162 \ 04.$

Let $n_1, n_2, \dots, n_6$ be the sample sizes in the six experiments and $X_1, X_2, \dots, X_6$ the corresponding values of X. Then

$$\mathrm{var}(X_1) = \frac{n_1}{p(1 - p)}; \quad \mathrm{var}(X_2) = \frac{n_2}{p(1 - p)}; \quad \mathrm{var}(X_3) = \frac{2n_3[1 + 2(1 - p)^2]}{p(2 - p)(3 - 2p + p^2)};$$

$$\mathrm{var}(X_4) = \frac{2n_4(1 + 2p^2)}{(1 - p^2)(2 + p^2)}; \quad \mathrm{var}(X_5) = \frac{n_5[1 + 2p(1 - p)]}{2p(2 - p)(1 - p^2)};$$

$$\mathrm{var}(X_6) = \frac{n_6[1 + 2p(1 - p)]}{2p(2 - p)(1 - p^2)}.$$

Using the estimate $\hat{p} = 0 \cdot 162 \ 04$, the corresponding estimates of the variances are:

$3439 \cdot 311; \ 3601 \cdot 334; \ 3119 \cdot 084; \ 106 \cdot 689; \ 63 \cdot 578; \ 298 \cdot 157.$

Again, $\quad X_1 = -\dfrac{a_1 + a_4}{1 - p} + \dfrac{a_2 + a_3}{p}; \quad X_2 = \dfrac{a_1 + a_4}{p} - \dfrac{a_2 + a_3}{1 - p};$

$$X_3 = 2(1 - p) \left[-\frac{a_1}{3 - 2p + p^2} + \frac{a_2 + a_3}{2p - p^2} - \frac{a_4}{(1 - p)^2} \right];$$

$$X_4 = 2p \left[\frac{a_1}{2 + p^2} - \frac{a_2 + a_3}{1 - p^2} + \frac{a_4}{p^2} \right]; \quad X_5 = -\frac{a_1}{2 - p} + \frac{a_2}{1 + p} + \frac{a_3}{p} - \frac{a_4}{1 - p};$$

$$X_6 = \frac{a_1}{1 + p} - \frac{a_2}{1 - p} - \frac{a_3}{2 - p} + \frac{a_4}{p}.$$

Hence, for $\hat{p} = 0.162\ 04$, the estimated values are:

$$\hat{X}_1 = -19.6835; \quad \hat{X}_2 = 12.9799; \quad \hat{X}_3 = -14.2242;$$
$$\hat{X}_4 = 4.5243; \quad \hat{X}_5 = 2.3966; \quad \hat{X}_6 = 14.0186.$$

Hence the goodness-of-fit $\chi^2 = \sum_{i=1}^{6} \hat{X}_i^2 / \text{Estimate var}(X_i)$

$$= 0.112\ 65 + 0.046\ 78 + 0.064\ 87 + 0.191\ 86 + 0.090\ 34 + 0.659\ 12$$

$$= 1.165\ 62 \text{ with 5 d.f.}$$

This χ^2 is obviously not significant and so the fit is good.

100 (i) To test the null hypothesis that the yellow and ivory totals in Experiment I are in the ratio $1:1$, we have

$$\chi^2 = (98 - 49)^2 / 147 = 16.3333 \text{ with 1 d.f.}$$

The 5 per cent table value of χ^2 with 1 d.f. is 3.841. Therefore the observed χ^2 is highly significant and so the observed totals do not agree with the Mendelian expectation.

(ii) To test the null hypothesis that the yellow and ivory totals in Experiments II and III taken together are in the ratio $3:1$, we have

$$\chi^2 = (201 - 183)^2 / 262 = 1.2366 \text{ with 1 d.f.}$$

This χ^2 is not significant, and so the observed totals agree with the Mendelian expectation.

(iii) To test the null hypothesis that the yellow and ivory totals considered in (i) and (ii) are jointly in agreement with their respective Mendelian expectations, we have

$$\chi^2 = \frac{[(98 - 49) + 2(201 - 183)/3]^{2(*)}}{147 + 262 \times 4/3} = 7.4970 \text{ with 1 d.f.}$$

This χ^2 is again significant, and so the null hypothesis of joint agreement with Mendelian expectation is rejected.

(iv) The joint likelihood of the three experiments is

$$L = \binom{147}{98}(1 - \theta)^{98}\theta^{49} \times \binom{116}{91}(1 - \theta^2)^{91}\theta^{50} \times \binom{146}{110}(1 - \theta^2)^{110}\theta^{72},$$

so that

$$\log L = \text{constant} + 98 \log(1 - \theta) + 171 \log \theta + 201 \log(1 - \theta^2)$$

and

$$\frac{d \log L}{d\theta} = -\frac{98}{1 - \theta} + \frac{171}{\theta} - \frac{402\theta}{1 - \theta^2}.$$

Hence the equation for the maximum-likelihood estimate $\hat{\theta}$ is

$$171 - 98\hat{\theta} - 671\hat{\theta}^2 = 0.$$

The appropriate root of this equation gives $\hat{\theta} = 0.437\ 049$.

(*) This χ^2 is obtained from the formula for χ_m^2 derived in Exercise 95 above.

Hence the heterogeneity χ^2 is

$$\frac{[98 - 147(1 - \hat{\theta})]^2}{147\hat{\theta}(1 - \hat{\theta})} + \frac{[201 - 262(1 - \hat{\theta}^2)]^2}{262\hat{\theta}^2(1 - \hat{\theta}^2)}$$

$$= 6\cdot4270 + 2\cdot9642 = 9\cdot3912 \text{ with 1 d.f.}$$

This χ^2 is highly significant, and so the data do not agree with the postulated model.

Experiment I

(v) As seen in (i), the observed proportions of yellow and ivory plants do not agree with the Mendelian expectation of $1:1$.

(vi) To test the individual families for heterogeneity, suppose p is the true proportion of yellow plants in the families of Experiment I. Then the maximum-likelihood estimate of p is $\hat{p} = 98/147 = 2/3$.

If n_i is the total and r_i the number of yellow plants in the ith family of Experiment I, then the heterogeneity χ^2 is

$$\sum_{i=1}^{3} (r_i - n_i\hat{p})^2/n_i\hat{p}(1 - \hat{p}) = \sum_{i=1}^{3} (3r_i - 2n_i)^2/2n_i = 1\cdot3071 \text{ with 2 d.f.}$$

This observed χ^2 value is obviously not significant. Therefore the three families of Experiment I are consistent in showing disagreement with the Mendelian expectation of $1:1$.

Experiment II

(v) To test the null hypothesis that the yellow and ivory totals are in agreement with the Mendelian expectation of $3:1$, we have

$$\chi^2 = (91 - 75)^2/348 = 0\cdot7356 \text{ with 1 d.f.}$$

The observed χ^2 value is not significant and so, on the whole, the yellow and ivory plants in Experiment II are in agreement with the Mendelian expectation of $3:1$.

(vi) To test whether each of the four families also conforms with the $3:1$ expectation, the appropriate χ^2's are

$$\chi_1^2 = (6 - 27)^2/87 = 5\cdot0690 \text{ with 1 d.f.}$$
$$\chi_2^2 = (36 - 15)^2/81 = 5\cdot4444 \text{ with 1 d.f.}$$
$$\chi_3^2 = (18 - 24)^2/90 = 0\cdot4000 \text{ with 1 d.f.}$$
$$\chi_4^2 = (15 - 25)^2/90 = 1\cdot1111 \text{ with 1 d.f.}$$

Thus the total χ^2 is $12\cdot0245$ with 4 d.f., and so, using (v), the heterogeneity χ^2 is $11\cdot2889$ with 3 d.f. The 5 per cent table value of χ^2 with 3 d.f. is $7\cdot815$, and so the heterogeneity χ^2 is significant. The individual families of Experiment II are not consistent in showing the same Mendelian expectation.

Experiment III

(v) To test the null hypothesis that the yellow and ivory totals are in agreement with the Mendelian expectation of $3:1$, we have

$$\chi^2 = (110 - 108)^2/438 = 0 \cdot 0091 \text{ with 1 d.f.}$$

The observed χ^2 value is not significant and so, on the whole, the yellow and ivory plants in Experiment III are in agreement with the Mendelian expectation of $3:1$.

(vi) To test whether each of the four families also conforms with the $3:1$ expectation, the appropriate χ^2's are

$$\chi_1^2 = (48 - 44)^2/180 = 0 \cdot 0889 \text{ with 1 d.f.}$$

$$\chi_2^2 = (21 - 21)^2/84 = 0 \text{ with 1 d.f.}$$

$$\chi_3^2 = (21 - 23)^2/90 = 0 \cdot 0444 \text{ with 1 d.f.}$$

$$\chi_4^2 = (18 - 22)^2/84 = 0 \cdot 1905 \text{ with 1 d.f.}$$

Thus the total χ^2 is $0 \cdot 3238$ with 4 d.f., and so, using (v), the heterogeneity χ^2 is $0 \cdot 3147$ with 3 d.f. The latter χ^2 is obviously not significant, and so the individual families of Experiment III are consistent in showing agreement with the Mendelian expectation of $3:1$.